THE WORKS OF
CHARLES DARWIN

Volume 25. *The effects of cross and self fertilization in the vegetable kingdom*

THE WORKS OF
CHARLES DARWIN

EDITED BY

PAUL H. BARRETT & R. B. FREEMAN

ADVISOR: PETER GAUTREY

VOLUME

25

THE EFFECTS OF
CROSS AND SELF FERTILIZATION
IN THE VEGETABLE KINGDOM

NEW YORK UNIVERSITY PRESS
WASHINGTON SQUARE, NEW YORK

Published in 1989 in the U.S.A. by New York University Press

Washington Square, New York, NY 10003

© Pickering & Chatto (Publishers) Limited, 1989

Library of Congress Cataloging-in-Publication Data

Darwin, Charles, 1809–1882.
  The effects of cross and self fertilization in the vegetable
kingdom.
    p. cm. — (The works of Charles Darwin; v. 25)
    ISBN 0–8147–1823–X
    1. Pollination. 2. Botany—Variation. I. Title. II. Series:
Darwin, Charles, 1809–1882. Works. 1987 ; v. 25.
QH365.A1   1987   vol. 25
[QK926]
575 s—dc20
[581.1′5]   89–12917
                                                                CIP

Printed and bound in Great Britain by
Redwood Burn Limited
Trowbridge

INTRODUCTION TO VOLUME TWENTY-FIVE

Effects of Cross and Self Fertilization in the Vegetable Kingdom. [Second edition] 1878. Freeman 1251.

Self fertilization is the purest form of incest. Close in-breeding had long been thought deleterious to progeny. Darwin conducted hundreds of plant experiments designed to test the hypothesis that in-breeding is in fact more harmful that 'out-breeding'. This book is a report of his experimental procedures and results.

From the extensive tests which he performed, and the vast quantity of data collected, the question arises, how did he fail to discover the Mendellian laws of inheritance? Especially in view of his proposal to keep records of numbers and types of offspring of successive generations produced by in-breeding. Most likely Darwin failed for the simple reason that he was attempting in the main to investigate vitality of offspring from different crosses, i.e., he was testing comparative survival and possible adaptive values of the various forms produced by self fertilization vis-à-vis cross fertilization. He kept detailed records of kinds of offspring, but not having Mendel's experience in viewing nature as composed of discreet particles, and not having Mendel's mathematical talents, was at a distinct philosophical disadvantage.

Darwin nevertheless made some remarkable discoveries, one of which in modern times is still a subject of continual ecological fascination involving complex population statistics.

By close observation he saw bees boring holes through the side of a flower and sucking the nectar through the hole, rather than obtaining the nectar in the usual manner, viz, by crawling down into the inside of the flower from the opening containing the pistil and stamens. When done normally, i.e., the latter way, the bees collected pollen from the anthers and when visiting other flowers later transported pollen from one flower to another – in effect assuring cross fertilization. But when 'feloniously' obtaining nectar by the hole-boring method, the mathematical probability of such a flower passing on to its progeny its peculiar set of genes were considerably reduced, thus making survival of such kinds of flowers less likely in the long run.

But Darwin noticed a curious fact correlated with the hole-boring habits. Only flowers which grew in abundance in close proximity were subject to hole-boring. Isolated plants of the same species were pollinated by bees in the normal fashion. Darwin explained the odd circumstance by suggesting that bees would visit many more flowers much more quickly, with the least expenditure of time and energy if the flowers growing near each other had holes through their corollas.

Thus a peculiar balance arose whereby survival advantage was against plants growing in thick clusters, but on the other hand plants growing in isolated places had reproductive advantage – leading to such increased survival that in a few generations their progeny in turn grew in thick clusters. Thus an ecological advantage led to an inevitable disadvantage, which in turn was followed by the previous advantage. But what mathematical sort of model would best represent the population dynamics of such nicely balanced relations?

*The effects of
cross and self-fertilization
in the
vegetable kingdom*

THE EFFECTS

OF

CROSS AND SELF FERTILISATION

IN THE

VEGETABLE KINGDOM.

By CHARLES DARWIN, LL.D., F.R.S.,

ETC.

SECOND EDITION.

LONDON:

JOHN MURRAY, ALBEMARLE STREET.

1878.

CONTENTS

CHAPTER I

Introductory remarks

Various means which favour or determine the cross-fertilization of plants – Benefits derived from cross-fertilization – Self fertilization favourable to the propagation of the species – Brief history of the subject – Object of the experiments, and the manner in which they were tried – Statistical value of the measurements – The experiments carried on during several successive generations – Nature of the relationship of the plants in the later generations – Uniformity of the conditions to which the plants were subjected – Some apparent and some real causes of error – Amount of pollen employed – Arrangement of the work – Importance of the conclusions [1] 1

CHAPTER II

Convolvulaceae

Ipomoea purpurea, comparison of the height and fertility of the crossed and self-fertilized plants during ten successive generations – Greater constitutional vigour of the crossed plants – The effects on the offspring of crossing different flowers on the same plant, instead of crossing distinct individuals – The effects of a cross with a fresh stock – The descendants of the self-fertilzed plant name 'Hero' – Summary on the growth, vigour, and fertility of the successive crossed and self-fertilized generations – Small amount of pollen in the anthers of the self-fertilized / plants of the later generations, and the sterility of their first-produced flowers – Uniform colour of the flowers produced by the self-fertilized plants – The advantage from a cross between two distinct plants depends on their differing in constitution [28] 20

CHAPTER III

Scrophulariaceae, Gesneriaceae, Labiatae, etc.

Mimulus luteus; height, vigour, and fertility of the crossed and self-fertilized plants of the first four generations – Appearance of a new, tall, and highly self-fertile variety – Offspring from a cross between self-fertilized plants – Effects of a cross with a fresh stock – Effects of crossing flowers on the same plant – Summary on *Mimulus luteus* – *Digitalis purpurea*, superiority of the crossed plants – Effects of crossing flowers on the same plant – Calceolaria – *Linaria vulgaris* – *Verbascum thapsus* – *Vandellia nummularifolia* – Cleistogamic flowers – *Gesneria pendulina* – *Salvia coccinea* – *Origanum vulgare*, a great increase of the crossed plants by stolons – *Thunbergia alata*   [63]   51

CHAPTER IV

Cruciferae, Papaveraceae, Resedaceae, etc.

Brassica oleracea, crossed and self-fertilized plants – Great effect of a cross with a fresh stock on the weight of the offspring – *Iberis umbellata* – *Papaver vagum* – *Eschscholtzia californica*, seedlings from a cross with a fresh stock not more vigorous, but more fertile than the self-fertilized seedlings – *Reseda lutea* and *odorata*, many individuals sterile with their own pollen – *Viola tricolor*, wonderful effects of a cross – *Adonis aestivalis* – *Delphinium consolida* – *Viscaria oculata*, crossed plants hardly taller, but more fertile than the self-fertilized – *Dianthus caryophyllus*, crossed and self-fertilized plants compared for four generations – Great effects of a cross with a fresh stock – Uniform colour of the flowers on the self-fertilized plants – *Hibiscus africanus*   [98] / 84

CHAPTER V

Geraniaceae, Leguminosae, Onagraceae, etc.

Pelargonium zonale, a cross between plants propagated by cuttings does no good – *Tropaeolum minus* – *Limnanthes douglasii* – *Lupinus luteus* and *pilosus* – *Phaseolus multiflorus* and *vulgaris* – *Lathyrus odoratus*, varieties of, never naturally intercross in England – *Pisum sativum*, varieties of, rarely intercross, but a cross between them highly beneficial – *Sarothamnus scoparius*, wonderful effects of a cross – *Ononis minutissima*, cleistogamic flowers of – Summary on the Leguminosae – *Clarkia elegans* – *Bartonia aurea* – *Passiflora gracilis* – *Apium petroselinum* – *Scabiosa atropurpurea* – *Lactuca sativa* – *Specularia speculum* – *Lobelia ramosa*, advantages of a cross during two generations – *Lobelia fulgens* – *Nemophila insignis*, great advantages of a cross – *Borago officinalis* – *Nolana prostrata*   [142]   126

CHAPTER VI

Solanaceae, Primulaceae, Polygoneae, etc.

Petunia violacea, crossed and self-fertilized plants compared for four generations – Effects of a cross with a fresh stock – Uniform colour of the flowers on the self-fertilized plants of the fourth generation – *Nicotiana tabacum*, crossed and self-fertilized plants of equal height – Great effects of a cross with a distinct sub-variety on the height, but not on the fertility, of the offspring – *Cyclamen persicum*, crossed seedlings greatly superior to the self-fertilized – *Anagallis collina* – *Primula veris* – Equal-styled variety of *Primula veris*, fertility of, greatly increased by a cross with a fresh stock – *Fagopyrum esculentum* – *Beta vulgaris* – *Canna warscewiczi*, crossed and self-fertilized plants of equal height – *Zea mays* – *Phalaris canariensis*   [188] / 170

CHAPTER VII

Summary of the heights and weights of the crossed and self-fertilized plants

Number of species and plants measured – Tables given – Preliminary remarks on the offspring of plants crossed by a fresh stock – Thirteen cases specially considered – The effects of crossing a self-fertilized plant either by another self-fertilized plant or by an intercrossed plant of the old stock – Summary of the results – Preliminary remarks on the crossed and self-fertilized plants of the same stock – The twenty-six exceptional cases considered, in which the crossed plants did not exceed greatly in height the self-fertilized – Most of these cases shown not to be real exceptions to the rule that cross-fertilization is beneficial – Summary of results – Relative weights of the crossed and self-fertilized plants   [238]   217

CHAPTER VIII

Difference between crossed and self-fertilized plants in constitutional vigour and in other respects

Greater constitutional vigour of crossed plants – The effects of great crowding – Competition with other kinds of plants – Self-fertilized plants more liable to premature death – Crossed plants generally flower before the self-sterilized – Negative effects of intercrossing flowers on the same plant – Cases described – Transmission of the good effects of a cross to later generations – Effects of crossing plants of closely related parentage – Uniform colour of the flowers on plants self-fertilized during several generations and cultivated under similar conditions   [285]   259

CHAPTER IX

The effects of cross-fertilization and self-fertilization on the production of seeds

Fertility of plants of crossed and self-fertilized parentage, both lots being fertilized in the same manner – Fertility of the parent-plants when first crossed and self-fertilized, and of their crossed / and self-fertilized offspring when again crossed and self-fertilized – Comparison of the fertility of flowers fertilized with their own pollen and with that from other flowers on the same plant – Self-sterile plants – Causes of self-sterility – The appearance of highly self-fertile varieties – Self-fertilization apparently in some respects beneficial, independently of the assured production of seeds – Relative weights and rates of germination of seeds from crossed and self-fertilized flowers   [312]   278

CHAPTER X

Means of fertilization

Sterility and fertility of plants when insects are excluded – The means by which flowers are cross-fertilized – Structures favourable to self-fertilization – Relation between the structure and conspicuousness of

flowers, the visits of insects, and the advantages of cross-fertilization –
The means by which flowers are fertilized with pollen from a distinct
plant – Greater fertilizing power of such pollen – Anemophilous
species – Conversion of anemophilous species into entomophilous –
Origin of nectar – Anemophilus plants generally have their sexes
separated – Conversion of diclinous into hermaphrodite flowers –
Trees often have their sexes separated   [356]   310

CHAPTER XI

The habits of insects in relation to the fertilization of flowers

Insects visit the flowers of the same species as long as they can – Cause
of this habit – Means by which bees recognize the flowers of the same
species – Sudden secretion of nectar – Nectar of certain flowers
unattractive to certain insects – Industry of bees, and the number of
flowers visited within a short time – Perforation of the corolla by bees
– Skill shown in the operation – Hive-bees profit by the holes made
by humble-bees – Effects of habit – The motive for perforating
flowers to save time – Flowers growing in crowded masses chiefly
perforated   [418] / 354

CHAPTER XII

General results

Cross-fertilization proved to be beneficial, and self-fertilization injurious – Allied species differ greatly in the means by which cross-
fertilization is favoured and self-fertilization avoided – The benefits
and evils of the two processes depend on the degree of differentiation
in the sexual elements – The evil effects not due to the combination of
morbid tendencies in the parents – Nature of the conditions to which
plants are subjected when growing near together in a state of nature or
under culture, and the effects of such conditions – Theoretical
considerations with respect to the interaction of differentiated sexual
elements – Practical lessons – Genesis of the two sexes – Close

correspondence between the effects of cross-fertilization and self-fertilization, and of the legitimate and illegitimate unions of hetero-styled plants, in comparison with hybrid unions   [439]   369

INDEX   [475] / 394

CHAPTER I

INTRODUCTORY REMARKS

Various means which favour or determine the cross-fertilization of plants – Benefits derived from cross-fertilization – Self-fertilization favourable to the propagation of the species – Brief history of the subject – Object of the experiments, and the manner in which they were tried – Statistical value of the measurements – The experiments carried on during several successive generations – Nature of the relationship of the plants in the later generations – Uniformity of the conditions to which the plants were subjected – Some apparent and some real causes of error – Amount of pollen employed – Arrangement of the work – Importance of the conclusions.

There is weighty and abundant evidence that the flowers of most kinds of plants are constructed so as to be occasionally or habitually cross-fertilized by pollen from another flower, produced either by the same plant, or generally, as we shall hereafter see reason to believe, by a distinct plant. Cross-fertilization is sometimes ensured by the sexes being separated, and in a large number of cases by the pollen and stigma of the same flower being matured at different times. Such plants are called dichogamous, and have been divided into two subclasses: proterandrous species, / in which the pollen is mature before the stigma, and proterogynous species, in which the reverse occurs; this latter form of dichogamy not being nearly so common as the other. Cross-fertilization is also ensured, in many cases, by mechanical contrivances of wonderful beauty, preventing the impregnation of the flowers by their own pollen. There is a small class of plants, which I have called dimorphic and trimorphic, but to which Hildebrand has given the more appropriate name of heterostyled; this class consists of plants presenting two or three different forms, adapted for reciprocal fertilization, so that, like plants with separate sexes, they can hardly fail to be intercrossed in each generation. The male and female organs of some flowers are irritable, and the insects which touch them get dusted with pollen, which is thus transported to

1

other flowers. Again, there is a class, in which the ovules absolutely refuse to be fertilized by pollen from the same plant, but can be fertilized by pollen from any other individual of the same species. There are also very many species which are partially sterile with their own pollen. Lastly, there is a large class in which the flowers present no apparent obstacle of any kind to self-fertilization, nevertheless these plants are frequently intercrossed, owing to the prepotency of pollen from another individual or variety over the plant's own pollen.

As plants are adapted by such diversified and effective means for cross-fertilization, it might have been inferred from this fact alone that they derived some great advantage from the process; and it is the object of the present work to show the nature and importance of the benefits thus derived. There are, however, some exceptions to the rule of plants being constructed so as to allow of or to favour cross-fertilization, for some / few plants seem to be invariably self-fertilized; yet even these retain traces of having been formerly adapted for cross-fertilization. These exceptions need not make us doubt the truth of the above rule, any more than the existence of some few plants which produce flowers, and yet never set seed, should make us doubt that flowers are adapted for the production of seed and the propagation of the species.

We should always keep in mind the obvious fact that the production of seed is the chief end of the act of fertilization; and that this end can be gained by hermaphrodite plants with incomparably greater certainty by self-fertilization, than by the union of the sexual elements belonging to two distinct flowers or plants. Yet it is as unmistakably plain that innumerable flowers are adapted for cross-fertilization, as that the teeth and talons of a carnivorous animal are adapted for catching prey; or that the plumes, wings, and hooks of a seed are adapted for its dissemination. Flowers, therefore, are constructed so as to gain two objects which are, to a certain extent, antagonistic, and this explains many apparent anomalies in their structure. The close proximity of the anthers to the stigma in a multitude of species favours, and often leads, to self-fertilization; but this end could have been gained far more safely if the flowers had been completely closed, for then the pollen would not have been injured by the rain or devoured by insects, as often happens. Moreover, in this case, a very small quantity of pollen would have been sufficient for fertilization, instead of millions of grains being produced. But the openness of the flower and the production of a great and apparently wasteful amount of pollen are necessary for cross-fertilization. These remarks are well illustrated by the plants called cleistogamic, which bear on the / same stock two kinds of flowers. The flowers of the one kind are minute and completely closed, so that they cannot possibly be crossed; but they are abundantly fertile, although producing an extremely small quantity of pollen. The flowers of the other kind produce much pollen and are open; and these can be, and often are, cross-fertilized. Hermann Müller has also made the remarkable discovery that there are some plants which exist under two forms; that is, produce on distinct stocks two kinds of hermaphrodite flowers. The one form bears small flowers constructed for self-fertilization; whilst the other bears larger and much more conspicuous flowers plainly constructed for cross-fertilization by the aid of insects; and without their aid these produce no seed.

The adaptation of flowers for cross-fertilization is a subject which has interested me for the last thirty-seven years, and I have collected a large mass of observations, but these are now rendered superfluous by the many excellent works which have been lately published. In the year 1857 I wrote[1] a short paper on the fertilization of the kidney bean; andin 1862 my work *On the Contrivances by which British and Foreign Orchids are Fertilised by Insects* appeared. It seemed to me a better plan to work out one group of plants as carefully as I could, rather than to publish many miscellaneous and imperfect observations. My present work is the complement of that one orchids, in which it was shown how admirably these plants are constructed so as to permit of, or to favour, or to necessitate cross-fertilization. The adaptations / for cross-fertilization are perhaps more obvious in the Orchideae than in any other group of plants, but it is an error to speak of them, as some authors have done, as an exceptional case. The lever-like action of the stamens of salvia (described by Hildebrand, Dr W. Ogle, and others), by which the anthers are depressed and rubbed on the backs of bees, shows as perfect a structure as can be found in any orchid. Papilionaceous flowers as described by various authors – for instance, Mr T. H. Farrer – offer innumerable curious adaptations for cross-fertilization. The case of *Posoqueria fragrans* (one of the Rubiaceae), is as wonderful as that of the most wonderful orchid. The stamens, according to Fritz

---

[1] *Gardeners' Chronicle*, 1857, p. 725, and 1858, p. 828. Also *Annals and Mag. of Nat. Hist.*, 3rd series, vol. ii, 1858, p. 462.

Müller,[2] are irritable, so that as soon as a moth visits a flower, the anthers explode and cover the insect with pollen; one of the filaments which is broader than the others then moves and closes the flower for about twelve hours, after which time it resumes its original position. Thus the stigma cannot be fertilized by pollen from the same flower, but only by that brought by a moth from some other flower. Endless other beautiful contrivances for this same purpose could be specified.

Long before I had attended to the fertilization of flowers, a remarkable book appeared in 1793 in Germany, *Das Entdeckte Geheimniss der Natur*, by C. K. Sprengel, in which he clearly proved by innumerable observations, how essential a part insects play in the fertilization of many plants. But he was in advance of his age, and his discoveries were for a long time neglected. Since the appearance of my book on orchids, many excellent works on the fertilization of flowers, such as those by Hildebrand, Delpino, Axell, / and Hermann Müller,[3] and numerous shorter papers, have been published. A list would occupy several pages, and this is not the proper place to give their titles, as we are not here concerned with the means, but with the results of cross-fertilization. No one who feels interest in the mechanism by which nature effects her ends, can read these books and memoirs without the most lively interest.

From my own observations on plants, guided to a certain extent by the experience of the breeders of animals, I became convinced many years ago that it is a general law of nature that flowers are adapted to be crossed, at least occasionally, by pollen from a distinct plant. Sprengel at times foresaw this law, but only partially, for it does not appear that he was aware that there was any difference in power between pollen from the same plant and from a distinct plant. In the introduction to his book (p. 4) he says, as the sexes are separated in

---

[2] *Botanische Zeitung*, 1866, p. 129.

[3] Sir John Lubbock has given an interesting summary of the whole subject in his *British Wild Flowers considered in relation to Insects*, 1875. Hermann Müller's work *Die Befruchtung der Blumen durch Insekten*, 1873, contains an immense number of original observations and generalizations. It is, moreover, invaluable as a repertory with references to almost everything which has been published on the subject. His work differs from that of all others in specifying what kinds of insects, as far as known, visit the flowers of each species. He likewise enters on new ground, by showing not only that flowers are adapted for their own good to the visits of certain insects; but that the insects themselves are excellently adapted for procuring nectar or pollen from certain flowers. The value of H. Müller's work can hardly be over-estimated, and it is much to be desired that it should be translated into English. Severin Axell's work is written in Swedish, so that I have not been able to read it.

so many flowers, and as so many other flowers are dichogamous, 'it appears that nature has not willed that any one flower should be fertilized by its own pollen'. Nevertheless, he was far from keeping this conclusion always before his mind, or he did not / see its full importance, as may be perceived by anyone who will read his observations carefully; and he consequently mistook the meaning of various structures. But his discoveries are so numerous and his work so excellent, that he can well afford to bear a small amount of blame. A most capable judge, H. Müller, likewise says:[4] 'It is remarkable in how very many cases Sprengel rightly perceived that pollen is necessarily transported to the stigmas of other flowers of the same species by the insects which visit them, and yet did not imagine that this transportation was of any service to the plants themselves.'

Andrew Knight saw the truth much more clearly, for he remarks,[5] 'Nature intended that a sexual intercourse should take place between neighbouring plants of the same species.' After alluding to the various means by which pollen is transported from flower to flower, as far as was then imperfectly known, he adds, 'Nature has something more in view than that its own proper males should fecundate each blossom.' In 1811 Kölreuter plainly hinted at the same law, as did afterwards another famour hybridiser of plants, Herbert.[6] But none of these distinguished observers appear to have been sufficiently impressed with the / truth and generality of the law, so as to insist on it and impress their belief on others.

In 1862 I summed up my observations on orchids by saying that nature 'abhors perpetual self-fertilization'. If the word perpetual had been omitted, the aphorism would have been false. As it stands, I believe that it is true, though perhaps rather too strongly expressed; and I should have added the self-evident proposition that the propagation of

[4] *Die Befruchtung der Blumen*, 1873, p. 4. His words are: 'Es ist merkwürdig, in wie zahlreichen Fäller Sprengel richtig erkannte, dass durch die Besuchenden Insekten der Blüthenstaub mit Nothwendigkeit auf die Narben anderer Blüthen derselben Art übertragen wird, ohne auf die Vermuthung zu kommen, dass in dieser Wirkung der Nutzen des Insektenbesuches für die Pflanzen selbst gesucht werden müsse.'

[5] *Philosophical Transactions*, 1799, p. 202.

[6] Kölreuter, *Mém. de l'Acad. de St. Pétersbourg*, vol. iii, 1809 (published 1811), p. 197. After showing how well the Malvaceae are adapted for cross-fertilization, he asks, 'An id aliquid in recessu habeat, quod hujuscemodi flores nunquam proprio suo pulvere, sed semper eo aliarum suae speciei impregnentur, merito quaeritur? Certe natura nil facit frustra.' Herbert, *Amaryllidaceae, with a Treatise on Cross-bred Vegetables*, 1837.

the species, whether by self-fertilization or by cross-fertilization, or asexually by buds, stolons, etc. is of paramount importance. Hermann Müller has done excellent service by insisting repeatedly on this latter point.

It often occurred to me that it would be advisable to try whether seedlings from cross-fertilized flowers were in any way superior to those from self-fertilized flowers. But as no instance was known with animals of any evil appearing in a single generation from the closest possible interbreeding, that is between brothers and sisters, I thought that the same rule would hold good with plants; and that it would be necessary at the sacrifice of too much time to self-fertilize and intercross plants during several successive generations, in order to arrive at any result. I ought to have reflected that such elaborate provisions favouring cross-fertilization, as we see in innumerable plants, would not have been acquired for the sake of gaining a distant and slight advantage, or of avoiding a distant and slight evil. Moreover, the fertilization of a flower by its own pollen corresponds to a closer form of interbreeding than is possible with ordinary bi-sexual animals; so that an earlier result might have been expected.

I was at last led to make the experiments recorded in the present volume from the following circumstance. / For the sake of determining certain points with respect to inheritance, and without any thought of the effects of close interbreeding, I raised close together two large beds of self-fertilized and crossed seedlings from the same plant of *Linaria vulgaris*. To my surprise, the crossed plants when fully grown were plainly taller and more vigorous than the self-fertilized ones. Bees incessantly visit the flowers of this linaria and carry pollen from one to the other; and if insects are excluded, the flowers produce extremely few seeds; so that the wild plants from which my seedlings were raised must have been intercrossed during all previous generations. It seemed therefore quite incredible that the difference between the two beds of seedlings could have been due to a single act of self-fertilization; and I attributed the result to the self-fertilized seeds not having been well ripened, improbably as it was that all should have been in this state, or to some other accidental and inexplicable cause. During the next year, I raised for the same purpose as before two large beds close together of self-fertilized and crossed seedlings from the carnation, *Dianthus caryophyllus*. This plant, like the linaria, is almost sterile if insects are excluded; and we may draw the same inference as before, namely, that the parent-plants must have been intercrossed

6

during every or almost every previous generation. Nevertheless, the self-fertilized seedlings were plainly inferior in height and vigour to the crossed.

My attention was now thoroughly aroused, for I could hardly doubt that the difference between the two beds was due to the one set being the offspring of crossed, and the other of self-fertilized flowers. Accordingly I selected almost by hazard to other plants, which happened to be in flower in the greenhouse, namely, / *Mimulus luteus* and *Ipomoea purpurea*, both of which, unlike the linaria and dianthus, are highly self-fertile if insects are excluded. Some flowers on a single plant of both species were fertilized with their own pollen, and others were crossed with pollen from a distinct individual; both plants being protected by a net from insects. The crossed and self-fertilized seeds thus produced were sown on opposite sides of the same pots, and treated in all respects alike; and the plants when fully grown were measured and compared. With both species, as in the cases of the linaria and dianthus, the crossed seedlings were conspicuously superior in height and in other ways to the self-fertilized. I therefore determined to begin a long series of experiments with various plants, and these were continued for the following eleven years; and we shall see that in a large majority of cases the crossed beat the self-fertilized plants. Several of the exceptional cases, moreover, in which the crossed plants were not victorious, can be explained.

It should be observed that I have spoken for the sake of brevity, and shall continue to do so, of crossed and self-fertilized seeds, seedlings, or plants; these terms implying that they are the product of crossed or self-fertilized flowers. cross-fertilization always means a cross between distinct plants which were raised from seeds and not from cuttings or buds. Self-fertilization always implies that the flowers in question were impregnated with their own pollen.

My experiments were tried in the following manner. A single plant, if it produced a sufficiency of flowers, or two or three plants were placed under a net stretched on a frame, and large enough to cover the plant (together with the pot, when one was used) without touching it. This latter point is important, for if / the flowers touch the net they may be cross-fertilized by bees, as I have known to happen; and when the net is wet the pollen may be injured. I used at first 'white cotton net', with very fine meshes, but afterwards a kind of net with meshes one-tenth of an inch in diameter; and this I found by experience effectually excluded all insects excepting thrips, which no net will

7

exclude. On the plants thus protected several flowers were marked, and were fertilized with their own pollen; and an equal number on the same plants, marked in a different manner, were at the same time crossed with pollen from a distinct plant. The crossed flowers were never castrated, in order to make the experiments as like as possible to what occurs under nature with plants fertilized by the aid of insects. Therefore, some of the flowers which were crossed may have failed to be thus fertilized, and afterwards have been self-fertilized. But this and some other sources of error will presently be discussed. In some few cases of spontaneously self-fertile species, the flowers were allowed to fertilize themselves under the net; and in still fewer cases uncovered plants were allowed to be freely crossed by the insects which incessantly visited them. There are some great advantages and some disadvantages in my having occasionally varied my method of proceeding but when there was any difference in the treatment, it is always so stated under the head of each species.

Care was taken that the seeds were thoroughly ripened before being gathered. Afterwards the crossed and self-fertilized seeds were in most cases placed on damp sand on opposite sides of a glass tumbler covered by a glass plate, with a partition between the two lots; and the glass was placed on the chimney-piece in a warm room. I could thus observe the germination of / the seeds. Sometimes a few would germinate on one side before any on the other, and these were thrown away. But as often as a pair germinated at the same time, they were planted on opposite sides of a pot, with a superficial partition between the two; and I thus proceeded until from half-a-dozen to a score or more seedlings of exactly the same age were planted on the opposite sides of several pots. If one of the young seedlings became sickly or was in any way injured, it was pulled up and thrown away, as well as its antagonist on the opposite side of the same pot.

As a large number of seeds were placed on the sand to germinate, many remained after the pairs had been selected, some of which were in a state of germination and others not so; and these were sown crowded together on the opposite sides of one or two rather larger pots, or sometimes in two long rows out of doors. In these cases there was the most severe struggle for life among the crossed seedlings on one side of the pot, and the self-fertilized seedlings on the other side, and between the two lots which grew in competition in the same pot. A vast number soon perished, and the tallest of the survivors on both sides when fully grown were measured. Plants treated in this manner, were subjected to nearly the same conditions as those growing in a state of nature, which have to struggle to maturity in the midst of a host of competitors.

On other occasions, from the want of time, the seeds, instead of being allowed to germinate on damp sand, were sown on the opposite sides of pots, and the fully grown plants measured. But this plan is less accurate, as the seeds sometimes germinated more quickly on one side than on the other. It was however necessary to act in this manner with some few species, as certain / kinds of seeds would not germinate well when exposed to the light; though the glasses containing them were kept on the chimney-piece on one side of a room, and some way from the two windows which faced the N.E.[7]

The soil in the pots in which the seedlings were planted, or the seeds sown, was well mixed, so as to be uniform in composition. The plants on the two sides were always watered at the same time and as equally as possible; and even if this had not been done, the water would have spread almost equally to both sides, as the pots were not large. The crossed and self-fertilized plants were separated by a superficial partition, which was always kept directed towards the chief source of the light, so that the plants on both sides were equally illuminated. I do not believe it possible that two sets of plants could have been subjected to more closely similar conditions, than were my crossed and self-fertilized seedlings, as grown in the above described manner.

In comparing the two sets, the eye alone was never trusted. Generally the height of every plant on both sides was carefully measured, often more than once, viz., whilst young, sometimes again when older, and finally when fully or almost fully grown. But in some cases, which are always specified, owing to the want of time, only one or two of the tallest plants on each side was measured. This plan, which is not a good one, was never followed (except with the crowded / plants raised from the seeds remaining after the pairs had been planted) unless the tallest plants on each side seemed fairly to represent the average difference between those on both sides. It has, however, some great advantages, as sickly or accidentally injured plants, or the offspring

---

[7] This occurred in the plainest manner with the seeds of *Papaver vagum* and *Delphinium consolida*, and less plainly with those of *Adonis aestivalis* and *Ononis minutissima*. Rarely more than one or two of the seeds of these four species germinated on the bare sand, though left there for some weeks; but when these same seeds were placed on earth in pots, and covered with a thin layer of sand, they germinated immediately in large numbers.

9

of ill-ripened seeds, are thus eliminated. When the tallest plants alone on each side were measured, their average height of course exceeds that of all the plants on the same side taken together. But in the case of the much crowded plants raised from the remaining seeds, the average height of the tallest plants was less than that of the plants in pairs, owing to the unfavourable conditions to which they were subjected from being greatly crowded. For our purpose, however, of the comparison of the crossed and self-fertilized plants, their absolute height signifies little.

As the plants were measured by an ordinary English standard divided into inches and eighths of an inch, I have not thought it worth while to change the fractions into decimals. The average or mean heights were calculated in the ordinary rough method by adding up the measurements of all, and dividing the product by the number of plants measured; the result being here given in inches and decimals. As the different species grow to various heights, I have always for the sake of easy comparison given in addition the average height of the crossed plants of each species taken as 100, and have calculated the average height of the self-fertilized plant in relation to this standard. With respect to the crowded plants raised from the seeds remaining after the pairs had been planted, and of which only some of the tallest on each side were measured, I have not thought it worth while to complicate the results by giving separate averages / for them and for the pairs, but have added up all their heights, and thus obtained a single average.

I long doubted whether it was worth while to give the measurements of each separate plant, but have decided to do so, in order that it may be seen that the superiority of the crossed plants over the self-fertilized, does not commonly depend on the presence of two or three extra fine plants on the one side, or of a few very poor plants on the other side. Although several observers have insisted in general terms on the offspring from intercrossed varieties being superior to either parent-form, no precise measurements have been given;[8] and I have met with no observations on the effects of crossing and self-fertilizing the individuals of the same variety. Moreover, experiments of this kind require so much time – mine having been continued during eleven years – that they are not likely soon to be repeated.

[8] A summary of these statements, with references, may be found in my *Variations of Animals and Plants under Domestication*, chap. xvii, 2nd edit., 1875, vol. ii, p. 109.

As only a moderate number of crossed and self-fertilized plants were measured, it was of great importance to me to learn how far the averages were trustworthy. I therefore asked Mr Galton, who has had much experience in statistical researches, to examine some of my tables of measurements, seven in number, namely, those of ipomoea, digitalis, *reseda lutea*, viola, limnanthes, petunia, and zea. I may premise that if we took by chance a dozen or score of men belonging to two nations and measured them, it would I presume be very rash to form any judgement from such small numbers on their average heights. But the case is somewhat different with my crossed and self-fertilized plants, as they were of exactly the same / age, were subjected from first to last to the same conditions, and were descenced from the same parents. When only two to six pairs of plants were measured, the results are manifestly of little or no value, except in so far as they confirm and are confirmed by experiments made on a larger scale with other species. I will now give the report on the seven tables of measurements, which Mr Galton has had the great kindness to draw up for me.

Zea mays (young plants)

			ARRANGED IN ORDER OF MAGNITUDE				
AS RECORDED BY MR DARWIN			*In separate pots*		*In a single series*		
Column I	II	III	IV	V	VI	VII	VIII
	Crossed	Self-fert.	Crossed	Self-fert.	Crossed	Self-fert.	Difference
	Inches	*Inches*	*Inches*	*Inches*	*Inches*	*Inches*	*Inches*
	$23\frac{4}{8}$	$17\frac{3}{8}$	$23\frac{4}{8}$	$20\frac{3}{8}$	$23\frac{4}{8}$	$20\frac{3}{8}$	$-3\frac{1}{8}$
Pot I	12	$20\frac{3}{8}$	21	20	$23\frac{2}{8}$	20	$-3\frac{2}{8}$
	21	20	12	$17\frac{3}{8}$	23	20	-3
					$22\frac{1}{8}$	$18\frac{5}{8}$	$-3\frac{4}{8}$
	22	20	22	20	$22\frac{1}{8}$	$18\frac{5}{8}$	$-3\frac{4}{8}$
Pot II	$19\frac{1}{8}$	$18\frac{3}{8}$	$21\frac{4}{8}$	$18\frac{5}{8}$	22	$18\frac{3}{8}$	$-3\frac{5}{8}$
	$21\frac{4}{8}$	$18\frac{5}{8}$	$19\frac{1}{8}$	$18\frac{3}{8}$	$21\frac{5}{8}$	18	$-3\frac{5}{8}$
					$21\frac{4}{8}$	18	$-3\frac{4}{8}$
	$22\frac{1}{8}$	$18\frac{5}{8}$	$23\frac{2}{8}$	$18\frac{5}{8}$	21	18	-3
	$20\frac{3}{8}$	$15\frac{2}{8}$	$22\frac{1}{8}$	18	21	$17\frac{3}{8}$	$-3\frac{5}{8}$
Pot III	$18\frac{2}{8}$	$16\frac{4}{8}$	$21\frac{5}{8}$	$16\frac{4}{8}$	$20\frac{3}{8}$	$16\frac{4}{8}$	$-3\frac{7}{8}$
	$21\frac{5}{8}$	18	$20\frac{3}{8}$	$16\frac{2}{8}$	$19\frac{1}{8}$	$16\frac{2}{8}$	$-2\frac{7}{8}$
	$23\frac{2}{8}$	$16\frac{2}{8}$	$18\frac{2}{8}$	$15\frac{2}{8}$	$18\frac{2}{8}$	$15\frac{4}{8}$	$-2\frac{6}{8}$
					12	$15\frac{4}{8}$	$+3\frac{2}{8}$
	21	18	23	18	12	$12\frac{6}{8}$	$+0\frac{6}{8}$
Pot IV	$22\frac{1}{8}$	$12\frac{6}{8}$	$22\frac{1}{8}$	18			
	23	$15\frac{4}{8}$	21	$15\frac{4}{8}$			
	12	18	12	$12\frac{6}{8}$			

11

I have examined the measurements of the plants with care, and by many statistical methods, to find out how far the means of the several sets represent constant realities, such as would come out the same so long as the general conditions of growth remained unaltered. The principal methods that were adopted are easily explained by selecting one of the shorter series of plants, say of *Zea mays*, for an example. /

The observations as I received them are shown in columns II and III, where they certainly have no *prima facie* appearance of regularity. But as soon as we arrange them in the order of their magnitudes, as in columns IV and V, the case is materially altered. We now see, with few exceptions, that the largest plant on the crossed side in each pot exceeds the largest plant on the self-fertilized side, that the second exceeds the second, the third the third, and so on. Out of the fifteen cases in the table, there are only two exceptions to this rule. We may therefore confidently affirm that a crossed series will always be found to exceed a self-fertilized series, within the range of the conditions under which the present experiment has been made.

Pot	*Crossed*	*Self-fert*	*Difference*
I	17⅞	19 2/8	+0 3/8
II	20⅞	19	−1⅞
III	21⅛	16⅞	−4 2/8
IV	19 6/8	16	−3 6/8

Next as regards the numerical estimate of this excess. The mean values of the several groups are so discordant, as is shown in the table just given, that a fairly precise numerical estimate seems impossible. But the consideration arises, whether the difference between pot and pot may not be of much the same order of importance as that of the other conditions upon which the growth of the plants has been modified. If so, and only on that condition, it would follow that when all the measurements, either of the crossed or the self-fertilized plants, were combined into a single series, that series would be statistically regular. The experiment is tried in columns VII and VIII, where the regularity is abundantly clear, and justifies us in considering its mean as perfectly reliable. I have protracted these measurements, and revised them in the usual way, but drawing a curve through them with a free hand, but the revision barely modifies the means derived from the original observations. In the present, and in nearly all the other cases, the difference between the original and revised means is under 2 per cent of their value. It is a very remarkable coincidence / that in the seven kinds of plants, whose measurements I have examined, the ratio between the heights of the crossed and of the self-fertilized ranges in five cases within very narrow limits. In *Zea mays* it is as 100 to 84, and in the others it ranges between 100 to 76 and 100 to 86.

The determination of the variability (measured by what is technically called the 'probable error') is a problem of more delicacy than that of determining the means, and I doubt, after making many trials, whether it is possible to derive useful conclusions from these few observations. We ought to have measurements of at least fifty plants in each case, in order to be in a position to deduce fair results. One fact, however, bearing on variability, is very evident in most cases, though not in *Zea mays*, viz., that the self-fertilized plants include the larger number of exceptionally small specimens, while the crossed are more generally full grown.

Those groups of cases in which measurements have been made of a few of the tallest plants that grew in rows, each of which contained a multitude of plants, show very clearly that the crossed plants exceed the self-fertilized in height, but they do not tell by inference anything about their respective mean values. If it should happen that a series is known to follow the law of error or any other law, and if the number of individuals in the series is known, it would be always possible to reconstruct the whole series when a fragment of it has been given. But I find no such method to be applicable in the present case. The doubt as to the number of plants in each row is of minor importance; the real difficulty lies in our ignorance of the precise law followed by the series. The experience of the plants in pots does not help us to determine that law, because the observations of such plants are too few to enable us to lay down more than the middle terms of the series to which they belong with any sort of accuracy, whereas the cases we are now considering refer to one of its extremities. There are other special difficulties which need not be gone into, as the one already mentioned is a complete bar.

Mr Galton sent me at the same time graphical representations which he had made of the measurements, and they evidently form fairly regular curves. He appends the words 'very good' to those of zea and / limnanthes. He also calculated the average height of the crossed and self-fertilized plants in the seven tables by a more correct method than that followed by me, namely, by including the heights, as estimated in accordance with statistical rules, of a few plants which died before they were measured; whereas I merely added up the heights of the survivors, and divided the sum by their number. The difference in our results is in one way highly satisfactory, for the average heights of the self-fertilized plants, as deduced by Mr Galton, is less than mine in all the cases excepting one, in which our averages are the same; and this shows that I have by no means exaggerated the superiority of the crossed over the self-fertilized plants.

After the heights of the crossed and self-fertilized plants had been taken, they were sometimes cut down close to the ground, and an equal number of both weighed. This method of comparison gives very striking results, and I wish that it had been oftener followed. Finally a record was often kept of any marked difference in the rate of germination of the crossed and self-fertilized seeds – of the relative periods of flowering of the plants raised from them – and of their productiveness, that is, of the number of seed-capsules which they produced and of the average number of seeds which each capsule contained.

When I began my experiments I did not intend to raise crossed and self-fertilized plants for more than a single generation; but as soon as the plants of the first generation were in flower I thought that I would raise one more generation, and acted in the following manner. Several flowers on one or more of the self-fertilized plants were again self-fertilized; and several / flowers on one or more of the crossed plants were fertilized with pollen from another crossed plant of the same lot. Having thus once begun, the same method was followed for as many as ten successive generations with some of the species. The seeds and seedlings were always treated in exactly the same manner as already described. The self-fertilized plants, whether originally descended from one or two mother-plants, were thus in each generation as closely interbred as was possible; and I could not have improved on my plan. But instead of crossing one of the crossed plants with another crossed plant, I ought to have crossed the self-fertilized plants of each generation with pollen taken from a non-related plant – that is, one belonging to a distinct family or stock of the same species and variety. This was done in several cases as an additional experiment, and gave very striking results. But the plan usually followed was to put into competition and compare intercrossed plants, which were almost always the offspring of more or less closely related plants, with the self-fertilized plants of each succeeding generation; all having been grown under closely similar conditions. I have, however, learnt more by this method of proceeding, which was begun by an oversight and then necessarily followed, than if I had always crossed the self-fertilized plants of each succeeding generation with pollen from a fresh stock.

I have said that the crossed plants of the successive generations were almost always inter-related. When the flowers on an hermaphrodite plant are crossed with pollen taken from a distinct plant, the seedlings thus raised may be considered as hermaphrodite brothers or sisters; those raised from the same capsule being as close as twins or animals of the same litter. But in one sense the flowers on the same plant are distinct / individuals, and as several flowers on the mother-plant were crossed by pollen taken from several flowers on the father-plant, such seedlings would be in one sense half-brothers or sisters, but more closely related than are the half-brothers and sisters of ordinary animals. The flowers on the mother-plant were, however, commonly crossed by pollen taken from two or more distinct plants; and in these cases the seedlings might be called with more truth half-brothers or sisters. When two or three mother-plants were crossed, as often happened, by pollen taken from two or three father-plants (the seeds being all intermingled), some of the seedlings of the first generation would be in no way related, whilst many others would be whole or half-brothers and sisters. In the second generation a large number of the seedlings would be what may be called whole or half first-cousins, mingled with whole and half-brothers and sisters, and with some plants not at all related. So it would be in the succeeding generations, but there would also be many cousins of the second and more remote degrees. The relationship will thus have become more and more inextricably complex in the later generations; with most of the plants in some degree and many of them closely related.

I have only one other point to notice, but this is one of the highest importance; namely, that the crossed and self-fertilized plants were subjected in the same generation to as nearly similar and uniform conditions as was possible. In the successive generations they were exposed to slightly different conditions as the seasons varied, and they were raised at different periods. But in other respects all were treated alike, being grown in pots in the same artificially prepared soil, being watered at the same time, and kept close together in the same greenhouse or hothouse. They were / therefore not exposed during successive years to such great vicissitudes of climate as are plants growing out of doors.

On some apparent and real causes of error in my experiments

It has been objected to such experiments as mine, that covering plants with a net, although only for a short time whilst in flower, may affect their health and fertility. I have seen no such effect except in one instance with a myosotis, and the covering may not then have been the real cause of injury. But even if the net were slightly injurious, and certainly it was not so in any high degree, as I could judge by the appearance of the plants and by comparing their fertility with that of neighbouring uncovered plants, it would not have vitiated my experiments; for in all the more important cases the flowers were crossed as well as self-fertilized under a net, so that they were treated in this respect exactly alike.

As it is impossible to exclude such minute pollen-carrying insects as thrips, flowers which it was intended to fertilize with their own pollen may sometimes have been afterwards crossed with pollen brought by these insects from another flower on the same plant; but as we shall hereafter see, a cross of this kind does not produce any effect, or at most only a slight one. When two or more plants were placed near one another under the same net, as was often done, there is some real though not great danger of the flowers which were believed to be self-fertilized being afterwards crossed with pollen brought by thrips from a distinct plant. I have said that the danger is not great, because I have often found that plants which are self-sterile, unless aided by insects, remained sterile when several plants of the same species were placed / under the same net. If, however, the flowers which had been presumably self-fertilized by me were in any case afterwards crossed by thrips with pollen brought from a distinct plant, crossed seedlings would have been included among the self-fertilized; but it should be especially observed that this occurrence would tend to diminish and not to increase any superiority in average height, fertility, etc., of the crossed over the self-fertilized plants.

As the flowers which were crossed were never castrated, it is probable or even almost certain that I sometimes failed to cross-fertilize them effectually, and that they were afterwards spontaneously self-fertilized. This would have been most likely to occur with dichogamous species, for without much care it is not easy to perceive whether their stigmas are ready to be fertilized when the anthers open. But in all cases, as the flowers were protected from wind, rain, and the access of insects, any pollen placed by me on the stigmatic surface whilst it was immature, would generally have remained there until the stigma was mature; and the flowers would then have been crossed as was intended. Nevertheless, it is highly probable that self-fertilized seedlings have sometimes by this means got included among the crossed seedlings. The effect would be, as in the former case, not to exaggerate but to diminish any average superiority of the crossed over the self-fertilized plants.

Errors arising rom the two causes just named, and from others – such as some of the seeds not having been thoroughly ripened, though care was taken to avoid this error – the sickness or unperceived injury of any of the plants – will have been to a large extent eliminated, in those cases in which many crossed and self-fertilized plants were measured and an average / struck. Some of these causes of error will also have been eliminated by the seeds having been allowed to germinate on bare damp sand, and being planted in pairs; for it is not likely that ill-matured and well-matured, or diseased and healthy seeds, would germinate at exactly the same time. The same result will have been gained in the several cases in which only a few of the tallest, finest, and healthiest plants on each side of the pots were measured.

Kölreuter and Gärtner[9] have proved that with some plants several, even as many as from fifty to sixty, pollen-grains are necessary for the fertilization of all the ovules in the ovarium. Naudin also found in the case of mirabilis that if only one or two of its very large pollen-grains were placed on the stigma, the plants raised from such seeds were dwarfed. I was therefore careful to give an imply sufficient supply of pollen, and generally covered the stigma with it; but I did not take any special pains to place exactly the same amount on the stigmas of the self-fertilized and crossed flowers. After having acted in this manner during two seasons, I remembered that Gärtner thought, though without any direct evidence, that an excess of pollen was perhaps injurious; and it has been proved by Spallanzani, Quatrefages, and Newport,[10] that with various animals an excess of the seminal fluid entirely prevents fertilization. It was therefore necessary to ascertain whether the fertility of the flowers was affected by applying a rather small and an extremely large quantity of pollen to the stigma. Accordingly a very small mass of pollen-grains was / placed on one side of the large stigma in sixty-four flowers of *Ipomoea purpurea*, and a great mass of pollen over the whole surface of the stigma in sixty-four other flowers. In order to vary the experiment, half the flowers of both lots were on plants produced from self-fertilized seeds, and the other half on plants from crossed seeds. The sixty-four flowers with an excess of pollen yielded sixty-one capsules; and excluding four capsules, each of which contained only a single poor seed, the remainder contained on an average 5·07 seeds per capsule. The sixty-four flowers with only a little pollen placed on one side of the stigma yielded sixty-three capsules, and excluding one from the same cause as before, the remainder contained on an average 5·129 seeds. So that the flowers fertilized with little pollen yielded rather more capsules and seeds than did those fertilized with an excess; but the difference is too slight to be of any significance. On the other hand, the seeds

[9] *Kenntniss der Befruchtung*, 1844, p. 345. Naudin, *Nouvelles Archives du Muséum*, vol. i, p. 27.
[10] *Transactions Philosophical Soc.*, 1853, pp. 253–8.

produced by the flowers with an excess of pollen were a little heavier of the two; for 170 of them weighed 79·67 grains, whilst 170 seeds from the flowers with very little pollen weighed 79·20 grains. Both lots of seeds having been placed on damp sand presented no difference in their rate of germination. We may therefore conclude that my experiments were not affected by any slight difference in the amount of pollen used; a sufficiency having been employed in all cases.

The order in which our subject will be treated in the present volume is as follows. A long series of experiments will first be given in chapters II to VI. Tables will afterwards be appended, showing in a condensed form the relative heights, weights, and fertility of the offspring of the various crossed and self-fertilized / species. Another table exhibits the striking results from fertilizing plants, which during several generations had either been self-fertilized or had been crossed with plants kept all the time under closely similar conditions, with pollen taken from plants of a distinct stock and which had been exposed to different conditions. In the concluding chapters various related points and questions of general interest will be discussed.

Anyone not specially interested in the subject need not attempt to read all the details; though they possess, I think, some value, and cannot be all summarized. But I would suggest to the reader to take as an example the experiments on ipomoea in chapter II; to which may be added those on digitalis, origanum, viola, or the common cabbage, as in all these cases the crossed plants are superior to the self-fertilized in a marked degree, but not in quite the same manner. As instances of self-fertilized plants being equal or superior to the crossed, the experiments on bartonia, canna, and the common pea ought to be read; but in the last case, and probably in that of canna, the want of any superiority in the crossed plants can be explained.

Species were selected for experiment belonging to widely distinct families, inhabiting various countries. In some few cases several genera belonging to the same family were tried, and these are grouped together; but the families themselves have been arranged not in any natural order, but in that which was the most convenient for my purpose. The experiments have been fully given, as the results appear to me of sufficient value to justify the details. Plants bearing hermaphrodite flowers can be interbred more closely than is possible with the higher animals, and are therefore / well-fitted to throw light on the nature and extent of the good effects of crossing, and on the evil effects of close interbreeding or self-fertilization. The most important conclusion at which I have arrived is that the mere act of crossing by itself does no good. The good depends on the individuals which are crossed differing slightly in constitution, owing to their progenitors having been subjected during several generations to slightly different conditions, or to what we call in our ignorance spontaneous variation. This conclusion, as we shall hereafter see, is closely connected with various important physiological problems, such as the benefit derived from slight changes in the conditions of life, and this stands in the closest connection with life itself. It throws light on the origin of the two sexes and on their separation or union in the same individual, and lastly on the whole subject of hybridism, which is one of the greatest obstacles to the general acceptance and progress of the great principle of evolution.

In order to avoid misapprehension, I beg leave to repeat that throughout this volume a crossed plant, seedling, or seed, means one of crossed *parentage*, that is, one derived from a flower fertilized with pollen from a distinct plant of the same species. And that a self-fertilized plant, seedling, or seed, means one of self-fertilized *parentage*, that is, one derived from a flower fertilized with pollen from the same flower, or sometimes, when thus stated, from another flower on the same plant. /

CHAPTER II

CONVOLVULCEAE

Ipomoea purpurea, comparison of the height and fertility of the crossed and self-fertilized plants during ten successive generations – Greater constitutional vigour of the crossed plants – The effects on the offspring of crossing different flowers on the same plant, instead of crossing distinct individuals – The effects of a cross with a fresh stock – The descendants of the self-fertilized plant named 'Hero' – Summary on the growth, vigour, and fertility of the successive crossed and self-fertilized generations – Small amount of pollen in the anthers of the self-fertilized plants of the later generations, and the sterility of their first-produced flowers – Uniform colour of the flowers produced by the self-fertilized plants – The advantage from a cross between two distinct plants depends on their differing in constitution.

A plant of *Ipomoea purpurea*, or as it is often called in England the convolvulus major, a native of South America, grew in my greenhouse. Ten flowers on this plant were fertilized with pollen from the same flower; and ten other flowers on the same plant were crossed with pollen from a distinct plant. The fertilization of the flowers with their own pollen was superfluous, as this convolvulus is highly self-fertile; but I acted in this manner to make the experiments correspond in all respects. Whilst the flowers are young the stigma projects beyond the anthers; and it might have been thought that it could not be fertilized without the aid of humble-bees, which often visit the flowers; but as the flower grows older the stamens increase in length, and their anthers brush against the stigma, which thus / received some pollen. The number of seeds produced by the crossed and self-fertilized flowers differed very little.

Crossed and self-fertilized seeds obtained in the above manner were allowed to germinate on damp sand, and as often as pairs germinated at the same time they were planted in the manner described in the Introduction, on the opposite sides of two pots. Five pairs were thus planted; and all the remaining seeds, whether or not in a state of

20

germination, were planted on the opposite sides of a third pot, so that the young plants on both sides were here greatly crowded and exposed to very severe competition. Rods of iron or wood of equal diameter were given to all the plants to twine up; and as soon as one of each pair reached the summit both were measured. A single rod was placed on each side of the crowded pot, No. III, and only the tallest plant on each side was measured.

TABLE I
(First generation)

No. of pot	Seedlings from crossed plants	Seedlings from self-fertilized plants
	Inches	Inches
I	87⅜	69
	87⅛	66
	89	73
II	88	68⅜
	87	60⅜
III Plants crowded; the tallest one measured on each side	77	57
Total in inches	516	394

The average height of the six crossed plants is here 86 inches, whilst that of the six self-fertilized plants in only 65·66 inches, so that the crossed plants are to the self-fertilized in height as 100 to 76. It should be oberved that this difference is not due to a few of the crossed plants being extremely tall, or to a few of the self-fertilized being extremely short, but to all the crossed plants attaining a greater height than their antagonists. The three pairs in Pot I were measured at two earlier periods, and the difference was sometimes greater and sometimes less than that / at the final measuring. But it is an interesting fact, of which I have seen several other instances, that one of the self-fertilized plants, when nearly a foot in height, was half an inch taller than the crossed plant; and again, when two feet high, it was 1⅜ inch taller, but during the ten subsequent days the crossed plant began to gain on its antagonist, and ever afterward asserted its supremacy, until it exceeded its self-fertilized opponent by 16 inches.

The five crossed plants in Pots I and II were covered with a net, and produced 121 capsules; the five self-fertilized plants produced eighty-four capsules, so that the numbers of capsules were as 100 to 69. Of the 121 capsules on the crossed plants sixty-five were the product of flowers crossed with pollen from a distinct plant, and these contained on an average 5·23 seeds per capsule; the remaining fifty-six capsules were spontaneously self-fertilized. Of the eighty-four capsules on the self-fertilized plants, all the product of renewed self-fertilization, fifty-five (which were alone examined) contained on an average 4·85 seeds per capsule. Therefore the cross-fertilized capsules, compared with the self-fertilized capsules, yielded seeds in the proportion of 100 to 93. The crossed seeds were relatively heavier than the self-fertilized seeds. Combining the above data (i.e., number of capsules and average number of contained seeds), the crossed plants, compared with the self-fertilized, yielded seeds in the ratio of 100 to 64.

These crossed plants produced, as already stated, fifty-six spontaneously self-fertilized capsules, and the self-fertilized plants produced twenty-nine such capsules. The former contained on an average, in comparison with the latter, seeds in the proportion of 100 to 99.

In Pot III, on the opposite sides of which a large number of crossed and self-fertilized seeds had been sown and the seedlings allowed to struggle together, the crossed plans had at first no great advantage. At one time the tallest crossed was 25⅛ inches high, and the tallest self-fertilized plants 21⅜. But the difference afterwards became much greater. The plants on both sides, from being so crowded, were poor specimens. The flowers were allowed to fertilize themselves spontaneously under a net; the crossed plants produced thirty-seven capsules, the self-fertilized plants only eighteen, or as 100 to 47. The former contained on an average 3·62 seeds per capsule; and the latter 3·38 seeds, or as 100 to 93. Combining these data (i.e., number / of capsules and average number of seeds), the crowded crossed plants produced seeds compared with the self-fertilized as 100 to 45. These latter seeds, however, were decidedly heavier, a hundred weighing 41·64 grains, than those from the capsules on the crossed plants, of which a hundred weighed 36·79 grains; and this probably was due to the fewer capsules borne by the self-fertilized plants having been better nourished. We thus see that the crossed plants in this the first generation, when grown under favourable conditions, and when grown under unfavourable conditions from being much crowded, greatly exceeded in height, and in the number of capsules produced, and slightly in the number of seeds per capsule, the self-fertilized plants.

Crossed and self-fertilized plants of the second generation. Flowers on the crossed plants of the last generation (Table I) were crossed by pollen from distinct plants of the same generation; and flowers on the self-fertilized plants were fertilized by pollen from the same flower. The seeds thus produced were treated in every respect as before, and we have in Table II the result.

TABLE II
(Second generation)

No. of pot	Crossed plants	Self-fertilized plants
	Inches	Inches
I	87	67⅛
	83	68⅜
	83	80⅜
II	85⅜	61⅜
	89	79
	77⅜	41
Total inches	505	398

Here again every single crossed plant is taller than its antagonist. The self-fertilized plant in Pot I, which ultimately reached the unusual height of 80⅜ inches, was for a long time taller than the opposed crossed plant, though at last beaten by it. The average height of the six crossed plants is 84·16 inches, whilst that of the six self-fertilized plants is 66·33 inches, or as 100 to 79.

Crossed and self-fertilized plants of the third generation. Seeds from the crossed plants of the last generation (Table II) again / crossed, and from the self-fertilized plants again self-fertilized, were treated in all respects exactly as before, with the following result:

23

TABLE III
(Third generation)

No. of pot	Crossed plants	Self-fertilized plants
	Inches	Inches
I	74	56⅛
	72	51⅛
	73⅜	54
II	82	59
	81	30
	82	66
Total inches	464·5	317·0

Again all the crossed plants are higher than their antagonists: their average height is 77·41 inches, whereas that of the self-fertilized is 52·83 inches, or 100 to 68.

I attended closely to the fertility of the plants of this third generation. Thirty flowers on the crossed plants were crossed with pollen from other crossed plants of the same generation, and the twenty-six capsules thus produced contained, on an average, 4·73 seeds; whilst thirty flowers on the self-fertilized plants, fertilized with the pollen from the same flower, produced twenty-three capsules, each containing 4·43 seeds. Thus the average number of seeds in the crossed capsules was to that in the self-fertilized capsules as 100 to 94. A hundred of the crossed seeds weighed 43·27 grains, whilst a hundred of the self-fertilized seeds weighed only 37·63 grains. Many of these lighter self-fertilized seeds placed on damp sand germinated before the crossed; thus thirty-six of the former germinated whilst only thirteen of the latter or crossed seeds germinated. In Pot I the three crossed plants produced spontaneously under the net (besides the twenty-six artificially cross-fertilized capsules) seventy-seven self-fertilized capsules containing on an average 4·41 seeds; whilst the three self-fertilized plants produced spontaneously (besides the twenty-three artificially self-fertilized capsules) only twenty-nine self-fertilized capsules, containing on an average 4·14 seeds. Therefore the average number of seeds in the two lots of spontaneously self-fertilized capsules was as / 100 to 94. Taking into consideration the number of capsules together with the average number of seeds, the crossed plants (spontaneously self-fertilized) produced seeds in comparison with the self-fertilized plants (spontaneously self-fertilized) in the proportion of 100 to 35. By whatever method the fertility of these plants is compared, the crossed are more fertile than the self-fertilized plants.

I tried in several ways the comparative vigour and powers of growth of the crossed and self-fertilized plants of this third generation. Thus, four self-fertilized seeds which had just germinated were planted on one side of a pot, and after an interval of forty-eight hours, four crossed seeds in the same state of germination were planted on the opposite side; and the pot was kept in the hothouse. I thought that the advantage thus given to the self-fertilized seedlings would have been so great that they would never have been beaten by the crossed ones. They were not beaten until all had grown to a height of 18 inches; and the degree to which they were finally beaten is shown in the following table (No. IV). We here see that the average height of the four crossed plants is 76·62, and of the four self-fertilized plants 65·87 inches, or as 100 to 86; therefore less than when both sides started fair.

TABLE IV

(Third generation, the self-fertilized plants having had a start of forty-eight hours)

No. of pot	Crossed plants	Self-fertilized plants
	Inches	*Inches*
III	78⁴⁄₈	73⁴⁄₈
	77⁴⁄₈	53
	73	61⁴⁄₈
	77⁴⁄₈	75⁴⁄₈
Total inches	306·5	263·5

Crossed and self-fertilized seeds of the third generation were also sown out of doors late in the summer, and therefore under unfavourable conditions, and a single stick was given to each lot of plants to twine up. The two lots were sufficiently separate so as not to interfere with each other's growth, and the ground was clear of weeds. As soon as they were killed by the first frost (and there was no difference in their hardiness), the two tallest crossed plants were found to be 24·5 and 22·5 inches, / whilst the two tallest self-fertilized plants were only 15 and 12·5 inches in height, or as 100 to 59.

I likewise sowed at the same time two lots of the same seeds in a part of the garden which was shady and covered with weeds. The crossed seedlings from the first looked the most healthy, but they twined up a stick only to a height of 7¼ inches; whilst the self-fertilized were not able to twine at all; and the tallest of them was only 3½ inches in height.

Lastly, two lots of the same weeds were sown in the midst of a bed of candy-tuft (iberis) growing vigorously. The seedlings came up, but all the self-fertilized ones soon died excepting one, which never twined and grew to a height of only 4 inches. Many of the crossed seedlings, on the other hand, survived; and some twined up the stems of the iberis to the height of 11 inches. These cases prove that the crossed seedlings have an immense advantage over the self-fertilized, both when growing isolated under very unfavourable conditions, and when put into competition with each other or with other plants, as would happen in a state of nature.

Crossed and self-fertilized plants of the fourth generation. Seedlings raised as before from the crossed and self-fertilized plants of the third generation in Table III, gave results as follows:

TABLE V
(Fourth generation)

No. of pot	Crossed plants	Self-fertilized plants
	Inches	*Inches*
I	84	80
	47	44½
II	83	73½
	59	51½
III	82	56½
	65½	63
	68	52
Total inches	488·5	421·0

Here the average height of the seven crossed plants is 69·78 inches, and that of the seven self-fertilized plants 60·14; or as 100 to 86. This smaller difference relatively to that in the former generations, may be attributed to the plants having been raised during the depth of winter, and consequently to their not / having grown vigorously, as was shown by their general appearance and from several of them never reaching the summits of the rods. In Pot II, one of the self-fertilized plants was for a long time taller by two inches than its opponent, but was ultimately beaten by it, so that all the crossed plants exceeded their opponents in height. Of twenty-eight capsules produced by the crossed plants fertilized by pollen from a distinct plant, each contained on an average 4·75 seeds; of twenty-seven self-fertilized capsules on the self-fertilized plants, each contained on an average 4·47 seeds; so that the proportion of seeds in the crossed and self-fertilized capsules was as 100 to 94.

Some of the same seeds, from which the plants in the last Table I had been raised, were planted, after they had germinated on damp sand, in a square tub, in which a large Brugmansia had long been growing. The soil was extremely poor and full of roots; six crossed seeds were planted in one corner, and six self-fertilized seeds in the opposite corner. All the seedlings from the latter soon died excepting one, and this grew to the height of only 1½ inch. Of the crossed plants three survived, and they grew to the height of 2½ inches, but were not able to twine round a stick; nevertheless, to my surprise, they produced some small miserable flowers. The crossed plants thus had a decided advantage over the self-fertilized plants under this extremity of bad conditions.

Crossed and self-fertilized plants of the fifth generation. These were raised in the same manner as before, and when measured gave the following results:

TABLE VI
(Fifth generation)

No. of pot	Crossed plants	Self-fertilized plants
	Inches	Inches
I	96	73
	86	78
	69	29
II	84	51
	84	84
	76¼	59
Total inches	495·25	374·00

The average height of the six crossed plants is 82·54 inches, / and that of the six self-fertilized plants 62·33 inches, or as 100 to 75. Every crossed plant exceeded its antagonist in height. In Pot I the middle plant on the crossed side was slightly injured whilst young by a blow, and was for a time beaten by its opponent, but ultimately recovered the usual superiority. The crossed plants produced spontaneously a vast number more capsules than did the self-fertilized plants; and the

27

capsules of the former contained on an average 3·37 seeds, whilst those of the latter contained only 3·0 per capsule, or as 100 to 89. But looking only to the artificially fertilized capsules, those on the crossed plants again crossed contained on an average 4·46 seeds, whilst those on the self-fertilized plants again self-fertilized contained 4·77 seeds; so that the self-fertilized capsules were the more fertile of the two, and of this unusual fact I can offer no explanation.

Crossed and self-fertilized plants of the sixth generation. These were raised in the usual manner, with the following result. I should state that there were originally eight plants on each side; but as two of the self-fertilized became extremely unhealthy and never grew to near their full height, these as well as their opponents have been struck out of the list. If they had been retained, they would have made the average height of the crossed plants unfairly greater than that of the self-fertilized. I have acted in the same manner in a few other instances, when one of a pair plainly became very unhealthy.

TABLE VII
(Sixth generation)

No. of pot	Crossed plants	Self-fertilized plants
	Inches	Inches
I	93	50½
	91	65
II	79	50
	86½	87
	88	62
III	87½	64½
Total inches	525	379

The average height of the six crossed plants is here 87·5, and of the six self-fertilized plants 63·16, or as 100 to 72. This large difference was chiefly due to most of the plants, especially the / self-fertilized ones, having become unhealthy towards the close of their growth, and they were severely attacked by aphides. From this cause nothing can be inferred with respect to their relative fertility. In this generation we have the first instance of a self-fertilized plant in Pot II exceeding (though only by half an inch) its crossed opponent. This victory was fairly won after a long struggle. At first the self-fertilized plant was several inches taller than its opponent, but when the latter was 4½ feet high it had grown equal; it then grew a little taller than the self-fertilized plant, but was ultimately beaten by it to the extent of half an inch, as shown in the table. I was so much surprised at this case that I saved the self-fertilized seeds of this plant, which I will call the 'Hero', and experimented on its descendants, as will hereafter be described.

Besides the plants included in Table VII, nine crossed and nine self-fertilized plants of the same lot were raised in two other pots, IV and V. These pots had been kept in the hothouse, but from want of room were, whilst the plants were young, suddenly moved during very cold weather into the coldest part of the greenhouse. They all suffered greatly, and never quite recovered. After a fortnight only two of the nine self-fertilized seedlings were alive, whilst seven of the crossed survived. The tallest of these latter plants when measured was 47 inches in height, whilst the tallest of the two surviving self-fertilized plants was only 32 inches. Here again we see how much more vigorous the crossed plants are than the self-fertilized.

Crossed and self-fertilized plants of the seventh generation. These were raised as heretofore with the following result:

TABLE VIII
(Seventh generation)

No. of pot	Crossed plants	Self-fertilized plants
	Inches	Inches
I	84⁴⁄₈	74⁶⁄₈
	84⁶⁄₈	84
	76²⁄₈	55⁴⁄₈
II	84⁴⁄₈	65
	90	51²⁄₈
	82²⁄₈	80⁴⁄₈
III	83	67⁶⁄₈
	86	60²⁄₈
IV	84²⁄₈	75²⁄₈
Total inches	755·50	614·25 /

Each of these nine crossed plants is higher than its opponent, though in one case only by three-quarters of an inch. Their average height is 83·94 inches, and that of the self-fertilized plants 68·25, or as 100 to 81. These plants, after growing to their full height, became very unhealthy and infested with aphides, just when the seeds were setting, so that many of the capsules failed, and nothing can be said on their relative fertility.

Crossed and self-fertilized plants of the eighth generation. As just stated, the plants of the last generation, from which the present ones were raised, were very unhealthy and their seeds of unusually small size; and this probably accounts, through abnormal premature growth, for the two lots behaving differently to what they did in any of the previous or succeeding generations. Many of the self-fertilized seeds germinated before the crossed ones, and these were of course rejected. When the crossed seedlings in Table IX had grown to a height of between 1 and 2 feet, they were all, or almost all, shorter than their self-fertilized opponents, but were not then measured. When they had acquired an average height of 32·28 inches, that of the self-fertilized plants was 40·68, or as 100 to 122. Moreover, every one of the self-fertilized plants, with a single exception, exceeded its crossed opponent. When, however, the crossed plants had grown to an average height of 77·56 inches, they just exceeded (viz., by 0·7 of an inch) the average height of the self-fertilized plants; but two of the latter were still taller than their crossed opponents. I was so much astonished at this whole case, that I tied string to the summits of the rods; the plants being thus allowed to continue climbing upwards. When their growth was complete they were untwined, stretched straight, and measured. The crossed plants had now almost regained their accustomed superiority, as may be seen in Table IX.

The average height of the eight crossed plants is here 113·25 inches, and that of the self-fertilized plants 96·65, or as 100 to 85. Nevertheless two of the self-fertilized plants, as may be seen in the table, were still higher than their crossed opponents. The latter manifestly had much thicker stems and many more lateral branches, and looked altogether more vigorous than the self-fertilized plants, and generally flowered before them. The earlier flowers produced by these self-fertilized plants did not set any capsules, and their anthers contained only a small amount of pollen; but to this subject I shall return. Nevertheless / capsules produced by two other self-fertilized plants of the same lot, not included in Table IX, which had been highly favoured by being grown in separate pots, contained the large average number of 5·1 seeds per capsule.

30

TABLE IX
(Eighth generation)

No. of pot	*Crossed plants*	*Self-fertilized plants*
	Inches	*Inches*
I	111 6/8	96
	127	54
	130 6/8	93 4/8
II	97 2/8	94
	89 4/8	125 6/8
III	103 6/8	115 4/8
	100 6/8	84 6/8
	147 4/8	109 6/8
Total inches	908·25	773·25

Crossed and self-fertilized plants of the ninth generation. The plants of this generation were raised in the same manner as before, with the result shown in Table X.

The fourteen crossed plants average in height 81·39 inches and the fourteen self-fertilized plants 64·07, or as 100 to 79. One self-fertilized plant in Pot III exceeded, and one in Pot IV equalled in height, its opponent. The self-fertilized plants showed no sign of inheriting the precocious growth of their parents; this having been due, as it would appear, to the abnormal state of the seeds from the unhealthiness of their parents. The fourteen self-fertilized plants yielded only forty spontaneously self-fertilized capsules, to which must be added seven, the product of ten flowers artificially self-fertilized. On the other hand, the fourteen crossed plants yielded 152 spontaneously self-fertilized capsules; but thirty-six flowers on these plants were crossed (yielding thirty-three capsules), and these flowers would probably have produced about thirty spontaneously self-fertilized capsules. Therefore an equal number of the crossed and self-fertilized plants would have produced capsules in the proportion of about 182 to 47, or as 100 to 26. Another phenomenon was well pronounced in this generation, but I believe had occurred previously to a slight extent; / namely, that most of the flowers on the self-fertilized plants were somewhat monstrous. The monstrosity consisted in the corolla being irregularly split so that it did not open properly, with one or two of the stamens slightly foliaceous, coloured, and firmly coherent to the corolla. I observed this monstrosity in only one flower on the crossed plants. The self-fertilized plants, if well nourished, would almost certainly, in a few more generations, have produced double flowers, for they had already become in some degree sterile.[11]

TABLE X
(Ninth generation)

No. of pot	Crossed plants	Self-fertilized plants
	Inches	Inches
I	83⁴⁄₈	57
	85⁴⁄₈	71
	83⁴⁄₈	48³⁄₈
II	83²⁄₈	45
	64²⁄₈	43⁶⁄₈
	64³⁄₈	38⁴⁄₈
III	79	63
	88⅛	71
	61	89⁴⁄₈
IV	82⁴⁄₈	82⁴⁄₈
	90	76⅛
V	89⁴⁄₈	67
Crowded plants	92⁴⁄₈	74²⁄₈
	92⁴⁄₈	70
Total inches	1139·5	897·0

Crossed and self-fertilized plants of the tenth generation. Six plants were raised in the usual manner from the crossed plants of the last generation (Table I) again intercrossed, and from the self-fertilized again self-fertilized. As one of the crossed plants in Pot I in the following table became much diseased, having crumpled leaves, and producing hardly any capsules, it and its opponent have been struck out of the table. /

TABLE XI
(Tenth generation)

No. of pot	Crossed plants	Self-fertilized plants
	Inches	Inches
I	92³⁄₈	47²⁄₈
	94⁴⁄₈	34⁶⁄₈
II	87	54⁴⁄₈
	89⁵⁄₈	49²⁄₈
	105	66²⁄₈
Total inches	468·5	252·0

[11] See on this subject *Variation of Animals and Plants under Domestication*, chap. xviii, 2nd edit., vol. ii, p. 152.

The five crossed plants average 93·7 inches, and the five self-fertilized only 50·4, or as 100 to 54. This difference, however, is to great that it must be looked at as in part accidental. The six crossed plants (the diseased one here included) yielded spontaneously 101 capsules, and the six self-fertilized plants 88, the latter being chiefly produced by one of the plants. But as the diseased plant, which yielded hardly any seed, is here included, the ratio of 101 to 88 does not fairly give the relative fertility of the two lots. The stems of the six crossed plants looked so much finer than those of the six self-fertilized plants, that after the capsules had been gathered and most of the leaves had fallen off, they were weighed. Those of the crossed plants weighed 2,693 grains, whilst those of the self-fertilized plants weighed only 1,173 grains, or as 100 to 44; but as the diseased and dwarfed crossed plant is here included, the superiority of the former in weight was really greater.

The effects on the offspring of crossing different flowers on the same plant, instead of crossing distinct individuals

In all the foregoing experiments, seedlings from flowers crossed by pollen from a distinct plant (though in the later generations more or less closely related) were put into competition with, and almost invariably proved markedly superior in height to the offspring from self-fertilized flowers. I wished, therefore, to ascertain whether a cross between two flowers on the same plant would give to the offspring any superiority / over the offspring from flowers fertilized with their own pollen. I procured some fresh seed and raised two plants, which were covered with a net; and several of their flowers were crossed with pollen from a distinct flower on the same plant. Twenty-nine capsules thus produced contained on an average 4·86 seeds per capsule· and 100 of these seeds weighed 36·77 grains. Several other flowers were fertilized with their own pollen, and twenty-six capsules thus produced contained on an average 4·42 seeds per capsule; 100 of which weighed 42·61 grains. So that a cross of this kind appears to have increased slightly the number of seeds per capsule, in the ratio of 100 to 91; but these crossed seeds were lighter than the self-fertilized in the ratio of 86 to 100. I doubt, however, from other observations, whether these results are fully trustworthy. The two lots of seeds, after germinating on sand, were planted in pairs on the opposite sides of nine pots, and were treated in every respect like the plants in the previous experiments. The remaining seeds, some in a state of germination and some not so, were sown on the opposite sides of a large pot (No. X); and the four tallest plants on each side of this pot were measured. The result is shown in Table XII.

The average height of the thirty-one crossed plants is 73·23 inches,

TABLE XII

No. of pot	Crossed plants	Self-fertilized plants
	Inches	*Inches*
I	82	77^{4}/$_{8}$
	75	87
	65	64
	76	87^{2}/$_{8}$
II	78^{4}/$_{8}$	84
	43	86^{4}/$_{8}$
	65^{4}/$_{8}$	90^{4}/$_{8}$
III	61^{2}/$_{8}$	86
	85	69^{4}/$_{8}$
	89	87^{4}/$_{8}$
IV	83	80^{4}/$_{8}$
	73^{4}/$_{8}$	88^{4}/$_{8}$
	67	84^{4}/$_{8}$
V	78	66^{4}/$_{8}$
	76^{6}/$_{8}$	77^{4}/$_{8}$
	57	81^{4}/$_{8}$
VI	70^{4}/$_{8}$	80
	79	82^{4}/$_{8}$
	79^{6}/$_{8}$	55^{4}/$_{8}$
VII	76	77
	84^{4}/$_{8}$	83^{4}/$_{8}$
	79	73^{4}/$_{8}$
VIII	73	76^{4}/$_{8}$
	67	82
	83	80^{4}/$_{8}$
IX	73^{2}/$_{8}$	78^{4}/$_{8}$
	78	67^{4}/$_{8}$
X	34	82^{4}/$_{8}$
Crowded plants	82	36^{6}/$_{8}$
	84^{6}/$_{8}$	69^{4}/$_{8}$
	71	75^{2}/$_{8}$
Total inches	2270·25	2399·75

and that of the thirty-one self-fertilized plants 77·41 inches; or as 100 to 106. Looking to each pair, it may be seen that only thirteen of the crossed plants, whilst eighteen of the self-fertilized plants exceed their opponents. A record was kept with respect to the plant which flowered first in each pot; and only two of the crossed flowered before one of the self-fertilized in the same pot; whilst eight of the self-fertilized flowered first. It thus appears that the / crossed plants are slightly inferior in height and in earliness of flowering to the self-fertilized. But the inferiority in height is so small, namely as 100 to 106, that I should have felt very doubtful on this head, had I not cut down all the plants (except those in the crowded pot No. X) close to the ground and weighed them. The twenty-seven crossed plants weighed 16½ ounces, and the twenty-seven self-fertilized plants 20½ ounces; and this gives a ratio of 100 to 124.

A self-fertilized plant of the same parentage as those in Table XII had been raised in a separate pot for a distinct purpose; and it proved partially sterile, the anthers containing very little pollen. Several flowers on this plant were crossed with the little pollen which could be obtained from the other flowers on the same plant; and other flowers were self-fertilized. From the seeds thus produced four crossed and four self-fertilized plants were raised, which were planted in the usual manner on the opposite sides of two pots. All these four crossed plants were inferior in height to their opponents; they averaged 78·18 inches, whilst the four self-fertilized plants averaged 84·8 inches; or as 100 to 108.[12] This case, therefore, confirms the last. Taking all the evidence together, we must conclude that these strictly self-fertilized plants grew a little taller, were heavier, and generally flowered before those derived from a cross between two flowers on the same plant. These latter plants thus present a wonderful contrast with those derived from a cross between two distinct individuals. /

The effects on the offspring of a cross with a distinct or fresh stock belonging to the same variety

From the two foregoing series of experiments we see, first, the good effects during several successive generations of a cross between distinct plants, although these were in some degree inter-related and had been

---

[12] From one of these self-fertilized plants, spontaneously self-fertilized, I gathered twenty-four capsules, and they contained on an average only 3·2 seeds per capsule; so that this plant had apparently inherited some of the sterility of its parent.

grown under nearly the same conditions; and, secondly, the absence of all such good effects from a cross between flowers on the same plant; the comparison in both cases being made with the offspring of flowers fertilized with their own pollen. The experiments now to be given show how powerfully and beneficially plants, which have been inter-crossed during many successive generations, having been kept all the time under nearly uniform conditions, are affected by a cross with another plant belonging to the same variety, but to a distinct family or stock, which had grown under different conditions.

Several flowers on the crossed plants of the ninth generation in Table X, were crossed with pollen from another crossed plant of the same lot. The seedlings thus raised formed the tenth intercrossed generation, and I will call them the '*intercrossed plants*'. Several other flowers on the same crossed plants of the ninth generation were fertilized (not having been castrated) with pollen taken from plants of the same variety, but belonging to a distinct family, which had been grown in a distant garden at Colchester, and therefore under somewhat different conditions. The capsules produced by this cross contained, to my surprise, fewer and lighter seeds than did the capsules of the intercrossed plants; but this, I think must have been accidental. The seedlings raised from them I will call the '*Colchester-crossed*'. The two lots of seeds, after germinating on sand, were planted in the usual manner on the opposite sides of five pots, and the remaining seeds, whether or not in a state of germination, were thickly sown on the opposite sides of a very large pot, No. VI, in Table XIII. In three of the six pots, after the young plants had twined a short way up their sticks, one of the / Colchester-crossed plants was much taller than any one of the intercrossed plants on the opposite side of the same pot; and in the three other pots somewhat taller. I should state that two of the Colchester-crossed plants in Pot IV, when about two-thirds grown, became much diseased, and were, together with their intercrossed opponents, rejected. The remaining nineteen plants, when almost fully grown, were measured, with the following result in Table XIII.

In sixteen out of these nineteen pairs, the Colchester-crossed plant exceeded in height its intercrossed opponent. The average height of the Colchester-crossed is 84·03 inches, and that of the intercrossed 65·78 inches; or as 100 to 78. With respect / to the fertility of the two lots, it was too troublesome to collect and count the capsules on all the plants; so I selected two of the best pots, V and VI, and in these the

TABLE XIII

No. of pot	Colchester-crossed plants	Intercrossed plants of the tenth generation
	Inches	Inches
I	87	78
	87 4/8	68 4/8
	85 1/8	94 4/8
II	93 6/8	60
	85 4/8	87 2/8
	90 5/8	45 4/8
III	84 2/8	70 1/8
	92 4/8	81 6/8
	85	86 2/8
IV	95 6/8	65 1/8
V	90 4/8	85 6/8
	86 6/8	63
	84	62 6/8
VI	90 4/8	43 4/8
Crowded plants in a very large pot	75	39 6/8
	71	30 2/8
	83 6/8	86
	63	53
	65	48 6/8
Total inches	1596·50	1249·75

Colchester-crossed produced 269 mature and half-mature capsules, whilst an equal number of the intercrossed plants produced only 154 capsules; or as 100 to 57. By weight the capsules from the Colchester-crossed plants were to those from the intercrossed plants as 100 to 51; so that the former probably contained a somewhat larger average number of seeds.

We learn from this important experiment that plants in some degree related, which had been intercrossed during the nine previous generations, when they were fertilized with pollen from a fresh stock, yielded seedlings as superior to the seedlings of the tenth intercrossed generation, as these latter were to the self-fertilized plants of the corresponding generation. For if we look to the plants of the ninth generation in Table X (and these offer in most respects the fairest standard of comparison) we find that the intercrossed plants were in height to the self-fertilized as 100 to 79, and in fertility as 100 to 26; whilst the Colchester-crossed plants are in height to the intercrossed as 100 to 78, and in fertility as 100 to 51.

The descendants of the self-fertilized plant, named Hero, which appeared in the sixth self-fertilized generation

In the five generations before the sixth, the crossed plant of each pair was taller than its self-fertilized opponent; but in the sixth generation (Table VII, Pot II) the Hero appeared, which after a long and dubious struggle conquered its crossed opponent, though by only half an inch. I was so much surprised at this fact, that I resolved to ascertain whether this plant would transmit its powers of growth to its seedlings. Several flowers on Hero were therefore fertilized with their own pollen, and the seedlings thus raised were put into competition with self-fertilized and intercrossed plants of the corresponding generation. The three lots of seedlings thus all belong to the seventh generation. / Their relative heights are shown in the two following tables:

TABLE XIV

No. of pot	Self-fertilized plants of the seventh generation, children of Hero	Self-fertilized plants of the seventh generation
	Inches	Inches
I	74	89⅜
	60	61
	55⅜	49
II	92	82
	91⅜	56
	74⅜	38
Total inches	447·25	375·50

The average height of the six self-fertilized children of Hero is 74·54 inches, whilst that of the ordinary self-fertilized plants of the corresponding generation is only 62·58 inches, or as 100 to 84.

TABLE XV

No. of pot	Self-fertilized plants of the seventh generation, children of Hero	Intercrossed plants of the seventh generation
	Inches	Inches
III	92	76 6/8
IV	87	89
	87 6/8	86 6/8
Total inches	266·75	252·50

Here the average height of the three self-fertilized children of Hero is 88·91 inches, whilst that of the intercrossed plants is 84·16; or as 100 to 95. We thus see that the self-fertilized children of Hero certainly inherit the powers of growth of their parents; for they greatly exceed in height the self-fertilized offspring of the other self-fertilized plants, and even exceed by a trifle the intercrossed plants – all of the corresponding generation. /

Several flowers on the self-fertilized children of Hero in Table XIV were fertilized with pollen from the same flower; and from the seeds thus produced, self-fertilized plants of the eighth generation (grandchildren of Hero) were raised. Several other flowers on the same plants were crossed with pollen from the other children of Hero. The seedlings raised from this cross may be considered as the offspring of the union of brothers and sisters. The result of the competition between these two sets of seedlings (namely self-fertilized and the offspring of brothers and sisters) is given in the following table:

TABLE XVI

No. of pot	Self-fertilized grandchildren of Hero, from the self-fertilized children. Eighth generation	Grandchildren from a cross between the self-fertilized children of Hero. Eighth generation
	Inches	Inches
I	86 6/8	95 6/8
	90 3/8	95 3/8
II	96	85
	77 2/8	93
III	73	86 2/8
	66	82 2/8
	84 4/8	70 6/8

TABLE XVI – *continued*

No. of pot	*Self-fertilized grand-children of Hero, from the self-fertilized children. Eighth generation*	*Grandchildren from a cross between the self-fertilized children of Hero. Eighth generation*
	Inches	*Inches*
IV	88⅛	66⅜
	84	15⁴⁄₈
	36²⁄₈	38
	74	78⅜
V	90⅛	82⁶⁄₈
	90⅝	83⁶⁄₈
Total inches	1037·00	973·13

The average height of the thirteen self-fertilized grandchildren of Hero is 79·76 inches, and that of the grandchildren from a cross between the self-fertilized children is 74·85; or as 100 to 94. But in Pot IV one of the crossed plants grew only to a height of 15½ inches; and if this plant and its opponent are struck out, as would be the fairest plan, the average height of the crossed plants exceeds, but only by a fraction of an inch, that of the self-fertilized / plants. It is therefore clear that a cross between the self-fertilized children of Hero did not produce any beneficial effect worth notice; and it is very doubtful whether this negative result can be attributed merely to the fact of brothers and sisters having been united, for the ordinary intercrossed plants of the several successive generations must often have been derived from the union of brothers and sisters (as shown in chap. I), and yet all of them were greatly superior to the self-fertilized plants. We are therefore driven to the suspicion, which we shall soon see strengthened, that Hero transmitted to its offspring a peculiar constituton adapted for self-fertilization.

It would appear that the self-fertilized descendants of Hero have not only inherited from Hero a power of growth equal to that of the ordinary intercrossed plants, but have become more fertile when self-fertilized than is usual with the plants of the present species. The flowers on the self-fertilized grandchildren of Hero in Table XVI (the eighth generation of self-fertilized plants) were fertilized with their own pollen and produced plenty of capsules, ten of which (though this is too few a number for a safe average) contained 5·2 seeds per capsule – a higher average than was observed in any other case with the self-fertilized plants. The anthers produced by these self-fertilized grand-children were also as well developed and contained as much pollen as those on the intercrossed plants of the corresponding generation; whereas this was not the case with the ordinary self-fertilized plants of the later generations. Nevertheless some few of the flowers produced by the grandchildren of Hero were slightly monstrous, like those of the ordinary self-fertilized plants of the later generations. In order not to recur to the subject of fertility, I may add that twenty-one self-fertilized capsules, spontaneously produced by the great-grandchildren of Hero (forming the ninth generation of self-fertilized plants), contained on an average 4·47 seeds; and this is as high an average as the self-fertilized flowers of any generation usually yielded.

Several flowers on the self-fertilized grandchildren of Hero in Table XVI were fertilized with pollen from the same flower; and the seedlings raised from them (great-grandchildren of Hero) formed the ninth self-fertilized generation. Several other flowers were crossed with pollen from another grandchild, so that they may be considered as the offspring of brothers and sisters, and the seedlings thus raised may be called the *intercrossed* great-grandchildren. And lastly, other flowers were fertilized with pollen / from a distinct stock, and the seedlings thus raised may be called the *Colchester-crossed* great-grandchildren. In my anxiety to see that the result would be, I unfortunately planted the three lots of seeds (after they had germinated on sand) in the hothouse in the middle of winter, and in consequence of this the seedlings (twenty in number of each kind) became very unhealthy, some growing only a few inches in height, and very few to their proper height. The result, therefore, cannot be fully trusted; and it would be useless to give the measurements in detail. In order to strike as fair an average as possible, I first excluded all the plants under 50 inches in height, thus rejecting all the most unhealthy plants. The six self-fertilized thus left were on an average 66·86 inches high; the eight intercrossed plants 63·2 high; and the seven Colchester-crossed 65·37 high; so that there was not much difference between the three sets, the self-fertilized plants having a slight advantage. Nor was there any great difference when only the plants under 36 inches in height were excluded. Nor again when all the plants, however much dwarfed and unhealthy, were included. In this latter case the Colchester-crossed gave the lowest average of all; and if these plants had been in any marked manner superior to the other two lots, as from my former experience I fully expected they would have been, I cannot but think that some vestige of such superiority would have been evident, notwithstanding the very unhealthy condition of most of the plants. No advantage, as far as we can judge, was derived from intercrossing two of the grandchildren of Hero, any more than when two of the children were crossed. It appears therefore that Hero and its descendants have varied from the common type, not only in acquiring great power of growth, and increased fertility when subjected to self-fertilization, but in not profiting from a cross with a distinct stock; and this latter fact, if trustworthy, is a unique case, as far as I have observed in all my experiments.

Summary on the growth, vigour, and fertility of the successive generations of the crossed and self-fertilized plants of Ipomoea purpurea, together with some miscellaneous observations

In the following table, No. XVII, we see the average or mean heights of the ten successive generations of the intercrossed and self-fertilized plants, grown in / competition with each other; and in the right-hand column we have the ratios of the one to the other, the height of the intercrossed plants being taken at 100. In the bottom line the mean height of the seventy-three intercrossed plants is shown to be 85·84 inches, and that of the seventy-three self-fertilized plants 66·02 inches, or as 100 to 77.

TABLE XVII

Ipomoea purpurea. Summary of measurements (in inches) of the ten generations

Number of the generation	Number of crossed plants	Average height of crossed plants	Number of self-fertilized plants	Average height of self-fertilized plants	Ratio between average heights of crossed and self-fertilized plants
First generation Table I	6	86·00	6	65·66	as 100 to 76
Second generation Table II	6	84·16	6	66·33	as 100 to 79
Third generation Table III	6	77·41	6	52·83	as 100 to 68
Fourth generation Table V	7	69·78	7	60·14	as 100 to 86
Fifth generation Table VI	6	82·54	6	62·33	as 100 to 75

TABLE XVII – *continued*

Ipomoea purpurea. Summary of measurements (in inches) of the ten generations

Number of the generation	Number of crossed plants	Average height of crossed plants	Number of self-fertilized plants	Average height of self-fertilized plants	Ratio between average heights of crossed and self-fertilized plants
Sixth generation Table VII	6	87·50	6	63·16	as 100 to 72
Seventh generation Table VIII	9	83·94	9	68·25	as 100 to 81
Eighth generation Table IX	8	113·25	8	96·65	as 100 to 85
Ninth generation Table X	14	81·39	14	64·07	as 100 to 79
Tenth generation Table XI	5	93·70	5	50·40	as 100 to 54
All the ten generations taken together	73	85·84	73	66·02	as 100 to 77 /

The mean height of the self-fertilized plants in each of the ten generations is also shown in the accompanying diagram, that of the intercrossed plants being taken at 100; and on the right side we see the relative heights of the seventy-three intercrossed plants, and of the seventy-three self-fertilized plants. The difference in height between the crossed and self-fertilized plants will perhaps be best appreciated by an illustration: If all the men in a country were on an average 6 feet high, and there were some families which had been long and closely interbred, these would be almost dwarfs, their average height during ten generations being only 4 feet 8¼ inches. /

It should be especially observed that the average difference between the crossed and self-fertilized plants is not due to a few of the former having grown to an extraordinary height, or to a few of the self-fertilized being extremely short, but to all the crossed plants having surpassed their self-fertilized opponents, with the few following exceptions. The first occurred in the sixth generation, in which the plant named 'Hero' appeared; two in the eighth generation, but the self-fertilized plants in this generation were in an anomalous condition, as they grew at first at an unusual rate and conquered for a time

43

the opposed crossed plants; and two exceptions in the ninth generation, though one of these plants only equalled its crossed opponent.
Therefore, of the seventy-three crossed plants, sixty-eight grew to a
greater height than the self-fertilized plants, to which they were
opposed.

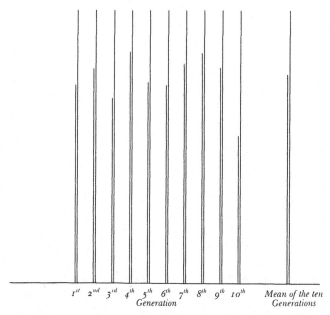

1st 2nd 3rd 4th 5th 6th 7th 8th 9th 10th *Mean of the ten*
Generation   *Generations*

Diagram showing the mean heights of the crossed and self-fertilized plants
of *Ipomoea purpurea* in the ten generations; the mean height of the crossed
plants being taken as 100. On the right hand, the mean heights of the crossed
and self-fertilized plants of all the generations taken together are shown.

In the right-hand column of figures, the difference in height
between the crossed and self-fertilized plants in the successive generations is seen to fluctuate much, as might indeed have been expected
from the small number of plants measured in each generation being
insufficient to give a fair average. It should be remembered that the
absolute height of the plants goes for nothing, as each pair was
measured as soon as one of them had twined up to the summit of its
rod. The great difference in the tenth generation, viz., 100 to 154, no
doubt was partly accidental, though, when these plants were weighed,
the difference was even greater, viz., 100 to 44. The smallest amount
of difference occurred in the fourth and the eighth generations, and

this was apparently due to both the crossed and self-fertilized plants having become unhealthy, which prevented the former attaining their usual degree of superiority. This was an / unfortunate circumstance, but my experiments were not thus vitiated, as both lots of plants were exposed to the same conditions, whether favourable or unfavourable.

There is reason to believe that the flowers of this Ipomoea, when growing out of doors, are habitually crossed by insects, so that the first seedlings which I raised from purchased seeds were probably the offspring of a cross. I infer that this is the case, firstly from humble-bees often visiting the flowers, and from the quantity of pollen left by them on the stigmas of such flowers; and, secondly, from the plants raised from the same lot of seed varying greatly in the colour of their flowers, for as we shall hereafter see, this indicates much intercrossing.[13] It is, therefore, remarkable that the plants raised by me from flowers which were, in all probability, self-fertilized for the first time after many generations of crossing, should have been so markedly inferior in height to the intercrossed plants as they were, namely, as 76 to 100. As the plants which were self-fertilized in each succeeding generation necessarily became much more closely interbred in the later than in the earlier generations, it might have been expected that the difference in height between them and the crossed plants would have gone on increasing; but, so far is this from being the case, that the difference between the two sets of plants in the seventh, eighth, and ninth generations taken together is less than in the first and second generations together. When, however, we remember that the self-fertilized and crossed plants are all descended from the same / mother-plant, that many of the crossed plants in each generation were related, often closely related, and that all were exposed to the same conditions, which, as we shall hereafter find, is a very important circumstance, it is not at all surprising that the difference between them should have somewhat decreased in the later generations. It is, on the contrary, an astonishing fact, that the crossed plants should have been victorious, even to a slight degree, over the self-fertilized plants of the later generations.

The much greater constitutional vigour of the crossed than of the self-fertilized plants, was proved on five occasions in various ways; namely, by exposing them, while young, to a low temperature or to a sudden change of temperature, or by growing them, under very

[13] Verlot says (*Sur la Production des Variétés*, 1865, p. 66) that certain varieties of a closely allied plant, the *Convolvulus tricolor*, cannot be kept pure unless grown at a distance from all other varieties.

unfavourable conditions, in competition with full-grown plants of other kinds.

With respect to the productiveness of the crossed and self-fertilized plants of the successive generations, my observations unfortunately were not made on any uniform plan, partly from the want of time, and partly from not having at first intended to observe more than a single generation. A summary of the results is here given in a tabulated form, the fertility of the crossed plants being taken as 100.

First generation of crossed and self-fertilized plants growing in competition with one another

Sixty-five capsules produced from flowers on five crossed plants fertilized by pollen from a distinct plant, and fifty-five capsules produced from flowers on five self-fertilized plants fertilized by their own pollen, contained seeds in the proportion of .... 100 to 93

Fifty-six spontaneously self-fertilized capsules on the above five crossed plants, and twenty-five spontaneously self-fertilized capsules on the above five self-fertilized plants, yielded seeds in the proportion of .... 100 to 99 /

Combining the total number of capsules produced by these plants, and the average number of seeds in each, the above crossed and self-fertilized plants yielded seeds in the proportion of .... 100 to 64

Other plants of this first generation grown under unfavourable conditions and spontaneously self-fertilized, yielded seeds in the proportion of .... 100 to 45

Third generation of crossed and self-fertilized plants

Crossed capsules compared with self-fertilized capsules contained seeds in the ratio of .... 100 to 94

An equal number of crossed and self-fertilized plants, both spontaneously self-fertilized, produced capsules in the ratio of .... 100 to 38

And these capsules contained seeds in the ratio of .... 100 to 94

Combining these data, the productiveness of the crossed and self-fertilized plants, both spontaneously self-fertilized, was as .... 100 to 35

Fourth generation of crossed and self-fertilized plants

Capsules from flowers on the crossed plants fertilized by pollen from another plant, and capsules from flowers on the self-fertilized plants fertilized with their own pollen, contained seeds in the proportion of .... 100 to 94

Fifth generation of crossed and self-fertilized plants

The crossed plants produced spontaneously a vast number more pods (not actually counted) than the self-fertilized, and these contained seeds in the proportion of .... 100 to 89

Ninth generation of crossed and self-fertilized plants

Fourteen crossed plants, spontaneously self-fertilized, and fourteen self-fertilized plants spontaneously self-fertilized, yielded capsules (the average number of seeds per capsule not having been ascertained) in the proportion of   100 to 26

Plants derived from a cross with a fresh stock compared with intercrossed plants

The offspring of intercrossed plants of the ninth generation, crossed by a fresh stock, compared with plants of the same stock intercrossed during ten generations, both sets of plants left uncovered and naturally fertilized, produced capsules by weight as   100 to 51

We see in this table that the crossed plants are / always in some degree more productive than the self-fertilized plants, by whatever standard they are compared. The degree differs greatly; but this depends chiefly on whether an average was taken of the seeds alone, or of the capsules alone, or of both combined. The relative superiority of the crossed plants is chiefly due to their producing a much greater number of capsules, and not to each capsule containing a larger average number of seeds. For instance, in the third generation the crossed and self-fertilized plants produced capsules in the ratio of 100 to 38, whilst the seeds in the capsules on the crossed plants were to those on the self-fertilized plants only as 100 to 94. In the eighth generation the capsules on two self-fertilized plants (not included in the above table), grown in separate pots and thus not subjected to any competition, yielded the large average of 5·1 seeds. The smaller number of capsules produced by the self-fertilized plants may be in part, but not altogether, attributed to their lessened size or height; this being chiefly due to their lessened constitutional vigour, so that they were not able to compete with the crossed plants growing in the same pots. The seeds produced by the crossed flowers on the crossed plants were not always heavier than the self-fertilized seeds on the self-fertilized plants. The lighter seeds, whether produced from crossed or self-fertilized flowers, generally germinated before the heavier seeds. I may add that the crossed plants, with very few exceptions, flowered before their self-fertilized opponents, as might have been expected from their greater height and vigour.

The impaired fertility of the self-fertilized plants was shown in another way, namely, by their anthers being smaller than those in the flowers on the crossed plants. This was first observed in the seventh generation, but / may have occurred earlier. Several anthers from

flowers on the crossed and self-fertilized plants of the eighth generation were compared under the microscope; and those from the former
were generally longer and plainly broader than the anthers of the self-
fertilized plants. The quantity of pollen contained in one of the latter
was, as far as could be judged by the eye, about half of that contained
in one from a crossed plant. The impaired fertility of the self-fertilized
plants of the eighth generation was also shown in another manner,
which may often be observed in hybrids – namely, by the first-formed
flowers being sterile. For instance, the fifteen first flowers on a self-
fertilized plant of one of the later generations were carefully fertilized
with their own pollen, and eight of them dropped off; at the same time
fifteen flowers on a crossed plant growing in the same pot were self-
fertilized, and only one dropped off. On two other crossed plants of
the same generation, several of the earliest flowers were observed to
fertilize themselves and to produce capsules. In the plants of the ninth,
and I believe of some previous generations, very many of the flowers,
as already stated, were slightly monstrous; and this probably was
connected with their lessened fertility.

All the self-fertilized plants of the seventh generation, and I believe
of one or two previous generations, produced flowers of exactly the
same tint, namely, of a rich dark purple. So did all the plants, without
any exception, in the three succeeding generations of self-fertilized
plants; and very many were raised on account of other experiments in
progress not here recorded. My attention was first called to this fact by
my gardener remarking that there was no occasion to label the self-
fertilized plants, as they could always be known by their colour. The
flowers were as uniform in tint / as those of a wild species growing in a
state of nature; whether the same tint occurred, as is probable, in the
earlier generations, neither my gardener nor self could recollect. The
flowers on the plants which were first raised from purchased seed, as
well as during the first few generations, varied much in the depth of
the purple tint; many were more or less pink, and occasionally a white
variety appeared. The crossed plants continued to the tenth generation to vary in the same manner as before, but to a much less degree,
owing, probably, to their having become more or less closely inter-
related. We must therefore attribute the extraordinary uniformity of
colour in the flowers on the plants of the seventh and succeeding self-
fertilized generations, to inheritance not having been interfered with
by crosses during several preceding generations, in combination with
the conditions of life having been very uniform.

A plant appeared in the sixth self-fertilized generation, named the Hero, which exceeded by a little in height its crossed antagonist, and which transmitted its powers of growth and increased self-fertility to its children and grandchildren. A cross between the children of Hero did not give to the grandchildren any advantage over the self-fertilized grandchildren raised from the self-fertilized children. And as far as my observations can be trusted, which were made on very unhealthy plants, the great-grandchildren raised from intercrossing the grandchildren had not advantage over the seedlings from the grandchildren the product of continued self-fertilization; and what is far more remarkable, the great-grandchildren raised by crossing the grandchildren with a fresh stock, had no advantage over either the intercrossed or self-fertilized great-grandchildren. It thus appears that Hero and its / descendants differed in constitution in an extraordinary manner from ordinary plants of the present species.

Although the plants raised during ten successive generations from crosses between distinct yet inter-related plants almost invariably exceeded in height, constitutional vigour, and fertility their self-fertilized opponents, it has been proved that seedlings raised by intercrossing flowers on the same plant are by no means superior, on the contrary are somewhat inferior in height and weight, to seedlings raised from flowers fertilized with their own pollen. This is a remarkable fact, which seems to indicate that self-fertilization is in some manner more advantageous than crossing, unless the cross brings with it, as is generally the case, some decided and preponderant advantage; but to this subject I shall recur in a future chapter.

The benefits which so generally follow from a cross between two plants apparently depend on the two differing somewhat in constitution or character. This is shown by the seedlings from the intercrossed plants of the ninth generation, when crossed with pollen from a fresh stock, being as superior in height and almost as superior in fertility to the again intercrossed plants, as these latter were to seedlings from self-fertilized plants of the corresponding generation. We thus learn the important fact that the mere act of crossing two distinct plants, which are in some degree inter-related and which have been long subjected to nearly the same conditions, does little good as compared with that from a cross between plants belonging to different stocks or families, and which have been subjected to somewhat different conditions. We may attribute the good derived from the crossing of the intercrossed plants during the ten successive generations to their still differing somewhat / in constitution or character, as was indeed proved by their flowers still differing somewhat in colour. But the several conclusions which may be deduced from the experiments on Ipomoea will be more fully considered in the final chapters, after all my other observations have been given. /

CHAPTER III

SCROPHULARIACEAE, GESNERIACEAE, LABIATAE, ETC.

Mimulus luteus; height, vigour, and fertility of the crossed and self-fertilized plants of the first four generations – Appearance of a new, tall, and highly self-fertile variety – Offspring from a cross between self-fertilized plants – Effects of a cross with a fresh stock – Effects of crossing flowers on the same plant – Summary on *Mimulus luteus – Digitalis purpurea*, superiority of the crossed plants – Effects of crossing flowers on the same plant – Calceolaria – *Linaria vulgaris* – *Verbascum thapsus* – *Vandellia nummularifolia* – Cleistogamic flowers – *Gesneria pendulina* – *Salvia coccinea* – *Origanum vulgare*, great increase of the crossed plants by stolons – *Thunbergia alata*.

In the family of the Scrophulariaceae I experimented on species in the six following genera: Mimulus, Digitalis, Calceolaria, Linaria, Verbascum, and Vandellia.

II. SCROPHULARIACEAE

MIMULUS LUTEUS

The plants which I raised from purchased seed varied greatly in the colour of their flowers, so that hardly two individuals were quite alike; the corolla being of all shades of yellow, with the most diversified blotches of purple, crimson, orange, and coppery brown. But these plants differed in no other respect.[1] The flowers are evidently well adapted for fertilization by the agency of insects; and in the case of a closely allied species, *M. roseus*, I have watched bees entering the flowers, thus getting their backs well dusted with pollen; and when they entered another flower the pollen was licked off their backs by /

[1] I sent several specimens with variously coloured flowers to Kew, and Dr Hooker informs me that they all consisted of *M. luteus*. The flowers with much red have been named by horticulturists as var. *Youngiana*.

the two-lipped stigma, the lips of which are irritable and close like a forceps on the pollen-grains. If no pollen is enclosed between the lips, these open again after a time. Mr Kitchener has ingeniously explained[2] the use of these movements, namely, to prevent the self-fertilization of the flower. If a bee with no pollen on its back enters a flower it touches the stigma, which quickly closes, and when the bee retires dusted with pollen, it can leave none on the stigma of the same flower. But as soon as it enters any other flower, plenty of pollen is left on the stigma, which will be thus cross-fertilized. Nevertheless, if insects are excluded, the flowers fertilize themselves perfectly and produce plenty of seed; but I did not ascertain whether this is effected by the stamens increasing in length with advancing age, or by the bending down of the pistil. The chief interest in my experiments on the present species, lies in the appearance in the fourth self-fertilized generation of a variety which bore large peculiarly-coloured flowers, and grew to a greater height than the other varieties; it likewise became more highly self-fertile, so that this variety resembles the plant named Hero, which appeared in the sixth self-fertilized generation of Ipomoea.

Some flowers on one of the plants raised from the purchased seeds were fertilized with their own pollen; and others on the same plant were crossed with pollen from a distinct plant. The seeds from twelve capsules thus produced were placed in separate watch-glasses for comparison; and those from the six crossed capsules appeared to the eye hardly more numerous than those from the six self-fertilized capsules. But when the seeds were weighed, those from the crossed capsules amounted to 1·02 grain, whilst those from the self-fertilized capsules were only 0·81 grain; so that the former were either heavier or more numerous than the latter, in the ratio of 100 to 79.

Crossed and self-fertilized plants of the first generation. Having ascertained, by leaving crossed and self-fertilized seed on a damp sand, that they germinated simultaneously, both kinds were thickly sown on opposite sides of a broad and rather shallow pan; so that the two sets of seedlings, which came up at the same time, were subjected to the same unfavourable conditions. This was a bad method of treatment, but this species was one of the first on which I experimented. When the crossed seedlings / were on an average half an inch high, the self-

[2] *A Year's Botany,* 1874, p. 118.

fertilized ones were only a quarter of an inch high. When grown to their full height under the above unfavourable conditions, the four tallest crossed plants averaged 7·62, and the four tallest self-fertilized 5·87 inches in height; or as 100 to 77. Ten flowers on the crossed plants were fully expanded before one on the self-fertilized plants. A few of these plants of both lots were transplanted into a large pot with plenty of good earth, and the self-fertilized plants, not now being subjected to severe competition, grew during the following year as tall as the crossed plants; but from a case which follows it is doubtful whether they would have long continued equal. Some flowers on the crossed plants were crossed with pollen from another plant, and the capsules thus produced contained a rather greater weight of seed than those on the self-fertilized plants again self-fertilized.

Crossed and self-fertilized plants of the second generation. Seeds from the foregoing plants, fertilized in the manner just stated, were sown on the opposite sides of a small pot (I) and came up crowded. The four tallest crossed seedlings, at the time of flowering, averaged 8 inches in height, whilst the four tallest self-fertilized plants averaged only 4 inches. Crossed seeds were sown by themselves in a second small pot, and self-fertilized seeds were sown by themselves in a third small pot; so that there was no competition whatever between these two lots. Nevertheless the crossed plants grew from 1 to 2 inches higher on an average than the self-fertilized. Both lots looked equally vigorous, but the crossed plants flowered earlier and more profusely than the self-fertilized. In Pot I, in which the two lots competed with each other, the crossed plants flowered first and produced a large number of capsules, whilst the self-fertilized produced only nineteen. The contents of twelve capsules from the crossed flowers on the crossed plants, and of twelve capsules from self-fertilized flowers on the self-fertilized plants, were placed in separate watch-glasses for comparison; and the crossed seeds seemed more numerous by half than the self-fertilized.

The plants on both sides of Pot I, after they had seeded, were cut down and transplanted into a large pot with plenty of good earth, and in the following spring, when they had grown to a height of between 5 and 6 inches, the two lots were equal, as occurred in a similar experiment in the last generation. But after some weeks the crossed plants exceeded the self-fertilized / ones on the opposite side of the same pot, though not nearly to so great a degree as before, when they were subjected to very severe competition.

53

Crossed and self-fertilized plants of the third generation. Crossed seeds from the crossed plants, and self-fertilized seeds from the self-fertilized plants of the last generation, were sown thickly on opposite sides of a small pot, No. I. The two tallest plants on each side were measured after they had flowered, and the two crossed ones were 12 and 7½ inches, and the two self-fertilized ones 8 and 5½ inches in height; that is, in the ratio of 100 to 69. Twenty flowers on the crossed plants were again crossed and produced twenty capsules; ten of which contained 1·33 grain weight of seeds. Thirty flowers on the self-fertilized plants were again self-fertilized and produced twenty-six capsules; ten of the best of which (many being very poor) contained only 0·87 grain weight of seeds; that is, in the ratio of 100 to 65 by weight.

The superiority of the crossed over the self-fertilized plants was proved in various ways. Self-fertilized seeds were sown on one side of a pot, and two days afterwards crossed seeds on the opposite side. The two lots of seedlings were equal until they were above half an inch high; but when fully grown the two tallest crossed plants attained a height of 12½ and 8¾ inches, whilst the two tallest self-fertilized plants were only 8 and 5½ inches high.

In a third pot, crossed seeds were sown four days after the self-fertilized, and the seedlings from the latter had at first, as might have been expected, an advantage; but when the two lots were between 5 and 6 inches in height, they were equal, and ultimately the three tallest crossed plants were 11, 10, and 8 inches, whilst the three tallest self-fertilized were 12, 8½, and 7½ inches in height. So that there was not much difference between them, the crossed plants having an average advantage of only the third of an inch. The plants were cut down, and without being disturbed were transplanted into a larger pot. Thus the two lots started fair in the following spring, and now the crossed plants showed their inherent superiority, for the two tallest were 13 inches, whilst the two tallest self-fertilized plants were only 11 and 8½ inches in height; or as 100 to 75. The two lots were allowed to fertilize themselves spontaneously: the crossed plants produced a large number of capsules, whilst the self-fertilized produced very few and poor ones. The seeds / from eight of the capsules on the crossed plants weighed 0·65 grain, whilst those from eight of the capsules on the self-fertilized plants weighed only 0·22 grain; or as 100 to 34.

The crossed plants in the above three pots, as in almost all the previous experiments, flowered before the self-fertilized. This occurred even in the third pot in which the crossed seeds were sown four days after the self-fertilized seeds.

Lastly, seeds of both lots were sown on opposite sides of a large pot in which a Fuchsia had long been growing, so that the earth was full of roots. Both lots grew miserably; but the crossed seedlings had an advantage at all times, and ultimately attained to a height of 3½ inches, whilst the self-fertilized seedlings never exceeded 1 inch. The several foregoing experiments prove in a decisive manner the superiority in constitutional vigour of the crossed over the self-fertilized plants.

In the three generations now described and taken together, the average height of the ten tallest crossed plants was 8·19 inches, and that of the ten tallest self-fertilized plants 5·29 inches (the plants having been grown in small pots), or as 100 to 65.

In the next or fourth self-fertilized generation, several plants of a new and tall variety appeared, which increased in the later self-fertilized generations, owing to its great self-fertility, to the complete exclusion of the original kinds. The same variety also appeared among the crossed plants, but as it was not at first regarded with any particular attention, I know now how far it was used for raising the intercrossed plants; and in the later crossed generations it was rarely present. Owing to the appearance of this tall variety, the comparison of the crossed and self-fertilized plants of the fifth and succeeding generations was rendered unfair, as all the self-fertilized and only a few or none of the crossed plants consisted of it. Nevertheless, the results of the later experiments are in some respects well worth giving.

Crossed and self-fertilized plants of the fourth generation. Seeds of the two kinds, produced in the usual way from the two sets of plants of the third generation, were sown on opposite sides of two pots (I and II); but the seedlings were not thinned enough and did not grow well. Many of the self-fertilized plants, especially in one of the pots, consisted of the new and tall variety above referred to, which bore large and almost white flowers marked with crimson blotches. I will call it the *White variety.* I believe that it first appeared among both the / crossed and self-fertilized plants of the last generation; but neither my gardener nor myself could remember any such variety in the seedlings raised from the purchased seed. It must therefore have arisen either through ordinary variation, or, judging from its appearance among both the crossed and self-fertilized plants, more probably through reversion to a formerly existing variety.

55

In Pot I the tallest crossed plant was 8½ inches, and the tallest self-fertilized 5 inches in height. In Pot II the tallest crossed plant was 6½ inches, and the tallest self-fertilized plant, which consisted of the white variety, 7 inches in height; and this was the first instance in my experiments on Mimulus in which the tallest self-fertilized plant exceeded the tallest crossed. Nevertheless, the two tallest crossed plants taken together were to the two tallest self-fertilized plants in height as 100 to 80. As yet the crossed plants were superior to the self-fertilized in fertility; for twelve flowers on the crossed plants were crossed and yielded ten capsules, the seeds of which weighed 1·71 grain. Twenty flowers on the self-fertilized plants were self-fertilized, and produced fifteen capsules, all appearing poor; and the seeds from ten of them weighed only 0·68 grain, so that from an equal number of capsules the crossed seeds were to the self-fertilized in weight as 100 to 40.

Crossed and self-fertilized plants of the fifth generation. Seeds from both lots of the fourth generation, fertilized in the usual manner, were sown on opposite sides of three pots. When the seedlings flowered, most of the self-fertilized plants were found to consist of the tall white variety. Several of the crossed plants in Pot I likewise belonged to this variety, as did a very few in Pots II and III. The tallest crossed plant in Pot I was 7 inches, and the tallest self-fertilized plant on the opposite side 8 inches; in Pots II and III the tallest crossed were 4½ and 5½, and the tallest self-fertilized 7 and 6½ inches in height; so that the average height of the tallest plants in the two lots was as 100 for the crossed to 126 for the self-fertilized; and thus we have a complete reversal of what occurred in the four previous generations. Nevertheless, in all three pots the crossed plants retained their habit of flowering before the self-fertilized. The plants were unhealthy from being crowded and from the extreme heat of the season, and were in consequence more or less sterile; but the crossed plants were somewhat less sterile than the self-fertilized plants. /

Crossed and self-fertilized plants of the sixth generation. Seeds from plants of the fifth generation crossed and self-fertilized in the usual manner were sown on opposite sides of several pots. On the self-fertilized side every single plant belonged to the tall white variety. On the crossed side. some plants belonged to this variety, but the greater number approached in character to the old and shorter kinds with smaller yellowish flowers blotched with coppery brown. When the plants on both sides were from 2 to 3 inches in height they were equal, but when fully grown the self-fertilized were decidedly the tallest and finest plants, but, from want of time, they were not actually measured. In half the pots the first plant which flowered was a self-fertilized one, and in the other half a crossed one. And now anther remarkable change was clearly perceived, namely, that the self-fertilized plants had become more self-fertile than the crossed. The pots were all put under a net to exclude insects, and the crossed plants produced spontaneously only fifty-five capsules, whilst the self-fertilized plants produced eighty-one capsules, or as 100 to 147. The seeds from nine capsules of both lots were placed in separate watch-glasses for comparison, and the self-fertilized appeared rather more numerous. Besides these spontaneously self-fertilized capsules, twenty flowers on the crossed plants again crossed yielded sixteen capsules; twenty-five flowers on the self-fertilized plants again self-fertilized yielded seven-teen capsules, and this is a larger proportional number of capsules than was produced by the self-fertilized flowers on the self-fertilized plants in the previous generations. The contents of ten capsules of both these lots were compared in separate watch-glasses, and the seeds from the self-fertilized appeared decidedly more numerous than those from the crossed plants.

Crossed and self-fertilized plants of the seventh generation. Crossed and self-fertilized seeds from the crossed and self-fertilized plants of the sixth generation were sown in the usual manner on opposite sides of three pots, and the seedlings were well and equally thinned. Every one of the self-fertilized plants (and many were raised) in this, as well as in the eighth and ninth generations, belonged to the tall white variety. Their uniformity of character, in comparison with the seedlings first raised from the purchased seed, was quite remarkable. On the other hand, the crossed plants differed much in the tints of their flowers, but not, I think to so great a degree as those first raised. I determined this time to measure the plants on both sides / carefully. The self-fertilized seedlings came up rather before the crossed, but both lots were for a time of equal height. When first measured, the average height of the six tallest crossed plants in the three pots was 7·02, and that of the six tallest self-fertilized plants 8·97 inches, or as 100 to 128. When fully grown the same plants were again measured, with the result shown in the following table:

TABLE XVIII
(Seventh generation)

No. of pot	Crossed plants	Self-fertilized plants
	Inches	*Inches*
I	11²⁄₈	19⅛
	11⅞	18
II	12⁶⁄₈	18²⁄₈
	11²⁄₈	14⁶⁄₈
III	9⁶⁄₈	12⁶⁄₈
	11⁶⁄₈	11
Total inches	68·63	93·88

The average height of the six crossed is here 11·43, and that of the six self-fertilized 15·64, or as 100 to 137.

As it was now evident that the tall white variety transmitted its characters faithfully, and as the self-fertilized plants consisted exclusively of this variety, it was manifest that they would always exceed in height the crossed plants which belonged chiefly to the original shorter varieties. This line of experiment was therefore discontinued, and I tried whether intercrossing two self-fertilized plants of the sixth generation, growing in distinct pots, would give their offspring any advantage over the offspring of flowers on one of the same plants fertilized with their own pollen. These latter seedlings formed the seventh generation of self-fertilized plants, like those in the right-hand column in Table XVIII; the crossed plants were the product of six previous self-fertilized generations with an intercross in the last generation. The seeds were allowed to germinate on sand, and were planted in pairs on opposite sides of four pots, all the remaining seeds being sown crowded on opposite sides of Pot V in Table XIX; the three tallest on each side in this latter pot being alone measured. All the plants were twice measured – the first time whilst young, and the average height of the crossed plants / to that of the self-fertilized was then as 100 to 122. When fully grown they were again measured, as in the following table on the opposite page.

The average height of the sixteen intercrossed plants is here 9·96 inches, and that of the sixteen self-fertilized plants 10·96, or as 100 to 110; so that the intercrossed plants, the progenitors of which had been self-fertilized for the six previous generations, and had been exposed during the whole time to remarkably uniform conditions, were

TABLE XIX

No. of pot	Intercrossed plants from self-fertilized plants of the sixth generation	Self-fertilized plants of the seventh generation
	Inches	Inches
I	12$\frac{6}{8}$	15$\frac{2}{8}$
	10$\frac{4}{8}$	11$\frac{5}{8}$
	10	11
	14$\frac{5}{8}$	11
II	10$\frac{2}{8}$	11$\frac{3}{8}$
	7$\frac{6}{8}$	11$\frac{4}{8}$
	12$\frac{1}{8}$	8$\frac{5}{8}$
	7	14$\frac{3}{8}$
III	13$\frac{5}{8}$	10$\frac{3}{8}$
	12$\frac{2}{8}$	11$\frac{6}{8}$
IV	7$\frac{1}{8}$	14$\frac{6}{8}$
	8$\frac{2}{8}$	7
	7$\frac{2}{8}$	8
V	8$\frac{5}{8}$	10$\frac{2}{8}$
Crowded	9	9$\frac{3}{8}$
	8$\frac{2}{8}$	9$\frac{2}{8}$
Total inches	159·38	175·50

somewhat inferior in height to the plants of the seventh self-fertilized generation. But as we shall presently see that a similar experiment made after two additional generations of self-fertilized gave a different result, I know not how far to trust the present one. In three of the five pots in Table XIX a self-fertilized plant flowered first, and in the other two a crossed plant. These self-fertilized plants were remarkably fertile, for twenty flowers fertilized with their own pollen produced no less than nineteen very fine capsules! /

The effects of a cross with a distinct stock. Some flowers on the self-fertilized plants in Pot IV in Table XIX were fertilized with their own pollen, and the plants of the eighth self-fertilized generations were thus raised, merely to serve as parents in the following experiment. Several flowers on these plants were allowed to fertilize themselves spontaneously (insects being of course excluded), and the plants raised from these seeds formed the ninth self-fertilized generation; they consisted wholly of the tall white variety with crimson blotches. Other

59

flowers on the same plants of the eighth self-fertilized generation were crossed with pollen taken from another plant of the same lot; so that the seedlings thus raised were the offspring of eight previous generations of self-fertilization with an intercross in the last generation; these I will call the *intercrossed plants*. Lastly, other flowers on the same plants of the eighth self-fertilized generation were crossed with pollen taken from plants which had been raised from seed procured from a garden at Chelsea. The Chelsea plants bore yellow flowers blotched with red, but differed in no other respect. They had been grown out of doors, whilst mine had been cultivated in pots in the greenhouse for the last eight generations, and in a different kind of soil. The seedlings raised from this cross with a wholly different stock may be called the '*Chelsea-crossed*'. The three lots of seeds thus obtained were allowed to germinate on bare sand; and whenever a seed in all three lots, or in only two, germinated at the same time, they were planted in pots superficially divided into three or two compartments. The remaining seeds, whether or not in a state of germination, were thickly sown in three divisions in a large pot, X, in Table XX. When the plants had grown to their full height they were measured, as shown in the following table; but only the three tallest plants in each of the three divisions in Pot X were measured.

In this table the average height of the twenty-eight Chelsea-crossed plants in 21·62 inches; that of the twenty-seven intercrossed plants 12·2; and that of the nineteen self-fertilized 10·44. But with respect to the latter it will be the fairest plan to strike out two dwarfed ones (only 4 inches in height), so as not to exaggerate the inferiority of the self-fertilized plants; and this will raise the average height of the seventeen remaining self-fertilized plants to 11·2 inches. Therefore the Chelsea-crossed are to the intercrossed in height as 100 to 56; the Chelsea-crossed to the self-fertilized as 100 to 52; and the intercrossed to the self-fertilized / as 100 to 92. We thus see how immensely superior in height the Chelsea-crossed are to the intercrossed and to the self-fertilized plants. They began to show their superiority when only one inch high. They were also, when fully grown, much more branched with larger leaves and somewhat larger flowers than the plants of the other two lots, so that if they had been weighed, the ratio would certainly have been much higher than that of 100 to 56 and 52.

The intercrossed plants are here to the self-fertilized in height as 100 to 92; whereas in the analogous experiment given in Table XIX the intercrossed plants from the self-fertilized plants of the sixth

TABLE XX

No. of pot	*Plants from self-fertilized plants of the eighth generation crossed by Chelsea plants*	*Plants from an intercross between the plants of the eighth self-fertilized generation*	*Self-fertilized plants of the ninth generation from plants of the eighth self-fertilized generation*
	Inches	*Inches*	*Inches*
I	30 7/8	14	9 4/8
	28 3/8	13 6/8	10 5/8
	..	13 7/8	10
II	29 6/8	11 4/8	11 6/8
	22 2/8	12	12 3/8
	..	9 1/8	..
III	23 6/8	12 2/8	8 5/8
	24 1/8	..	11 4/8
	25 6/8	..	6 7/8
IV	22 5/8	9 2/8	4
	22	8 1/8	13 3/8
	17	..	11
V	22 3/8	9	4 4/8
	19 5/8	11	13
	23 4/8	..	13 4/8
VI	28 2/8	18 6/8	12
	22	7	16 1/8
	..	12 4/8	..
VII	12 4/8	15	..
	24 3/8	12 3/8	..
	20 4/8	11 2/8	..
	26 4/8	15 2/8	..
VIII	17 2/8	13 3/8	..
	22 6/8	14 5/8	..
	27	14 3/8	..
IX	22 6/8	11 6/8	..
	6	17	..
	20 2/8	14 7/8	..
X	18 1/8	9 2/8	10 3/8
Crowded plants	16 2/8	8 2/8	8 1/8
	17 4/8	10	11 2/8
Total inches	605·38	329·50	198·50 /

generation were inferior in height to the self-fertilized plants in the ratio of 100 to 110. I doubt whether this discordance in the results of the two experiments can be explained by the self-fertilized plants in the present case having been raised from spontaneously self-fertilized seeds, whereas in the former case they were raised from artificially self-fertilized seeds; nor by the present plants having been self-fertilized during two additional generations, though this is a more probable explanation.

With respect to fertility, the twenty-eight Chelsea-crossed plants produced 272 capsules; the twenty-seven intercrossed plants produced 24; and the seventeen self-fertilized plants 17 capsules. All the plants were left uncovered so as to be naturally fertilized, and empty capsules were rejected.

			Capsules
Therefore 20 Chelsea-crossed plants would have produced			194·29
„ 20 Intercrossed plants	„	„	17·77
„ 20 Self-fertilized plants	„	„	20·00
			Grains
The seeds contained in 8 capsules from the Chelsea-crossed plants weighed			1·1
The seeds contained in 8 capsules from the Intercrossed plants weighed			0·51
The seeds contained in 8 capsules from the Self-fertilized plants weighed			0·33

If we combine the number of capsules produced together with the average weight of contained seeds, we get the following extraordinary ratios:

Weight of seed produced by the same number of Chelsea-crossed and intercrossed plants	as 100 to 4 /
Weight of seed produced by the same number of Chelsea-crossed and self-fertilized plants	as 100 to 3
Weight of seeds produced by the same number of intercrossed and self-fertilized plants	as 100 to 73

It is also a remarkable fact that the Chelsea-crossed plants exceeded the two other lots in hardiness, as greatly as they did in height, luxuriance, and fertility. In the early autumn most of the pots were bedded out in the open ground; and this always injures plants which have been long kept in a warm greenhouse. All three lots consequently suffered greatly, but the Chelsea-crossed plants much less than the other two lots. On the 3rd of October the Chelsea-crossed plants began to flower again, and continued to do so for some time; whilst not a single flower was produced by the plants of the other two lots, the stems of which were cut almost down to the ground and seemed half dead. Early in December there was a sharp frost, and the stems of Chelsea-crossed were now cut down; but on the 23rd of December they began to shoot up again from the roots, whilst all the plants of the other two lots were quite dead.

Although several of the self-fertilized seeds, from which the plants in the right-hand column in Table XX were raised, germinated (and were of course rejected) before any of those of the other two lots, yet in only one of the ten pots did a self-fertilized plant flower before the Chelsea-crossed or the intercrossed plants growing in the same pots. The plants of these two latter lots flowered at the same time, though the Chelsea-crossed grew so much taller and more vigorously than the intercrossed.

As already stated, the flowers of the plants originally raised from the Chelsea seeds were yellow; and it deserves notice that every one of the twenty-eight seedlings raised from the tall white variety fertilized, without being castrated, with pollen from the Chelsea plants, produced yellow-flowers; and this shows how prepotent this colour, which is the natural one of the species, is over the white colour.

The effects on the offspring of intercrossing flowers on the same plant, instead of crossing distinct individuals. In all the foregoing experiments the crossed plants were the product of a cross between distinct plants. I now selected a very vigorous plant in Table XX, raised by fertilizing a plant of the eighth self-fertilized generation with pollen from the Chelsea stock. / Several flowers on this plant were crossed with pollen from other flowers on the same plant, and several other flowers were fertilized with their own pollen. The seed thus produced was allowed to germinate on bare sand; and the seedlings were planted in the usual manner on the opposite sides of six pots. All the remaining seeds, whether or not in a state of germination, were sown thickly in Pot VII; the three tallest plants on each side of this latter pot being alone measured. As I was in a hurry to learn the result, some of these seeds were sown late in the autumn, but the plants grew so irregularly during the winter, that one crossed plant was 28½ inches, and two others only 4, or less than 4 inches in height, as may be seen in Table XXI. Under such circumstances, as I have observed in many other cases, the result is not in the least trustworthy; nevertheless I feel bound to give the measurements.

TABLE XXI

No. of pot	*Plants raised from a cross between different flowers on the same plant*	*Plants raised from flowers fertilized with their own pollen*
	Inches	*Inches*
I	17	17
	9	3⅛
II	28⅜	19⅛
	16⁴⁄₈	6
	13⅝	2
III	4	15⁶⁄₈
	2²⁄₈	10
IV	23⁴⁄₈	6²⁄₈
	15⁴⁄₈	7⅛
V	7	13⁴⁄₈
VI	18⅜	1⁴⁄₈
	11	2
VII	21	15⅛
Crowded	11⁶⁄₈	11
	12⅛	11²⁄₈
Total inches	210·88	140·75 /

The fifteen crossed plants here average 14·05, and the fifteen self-fertilized plants 9·38 in height, or as 100 to 67. But if all the plants under ten inches in height are struck out, the ratio of the eleven crossed plants to the eight self-fertilized plants is as 100 to 82.

On the following spring, some remaining seeds of the two lots were treated in exactly the same manner; and the measurements of the seedlings are given in the table on the opposite page.

Here the average height of the twenty-two crossed plants is 16·85, and that of the twenty-two self-fertilized plants 16·07; or as 100 to 95. But if four of the plants in Pot VII, which are much shorter than any of the others, are struck out (and this would be the fairest plan), the twenty-one crossed are to the nineteen self-fertilized plants in height as 100 to 100·6 – that is, are equal. All the plants, except the crowded ones in Pot VIII, after being measured were cut down, and the eighteen crossed plants weighed 10 oz, whilst the same number of self-fertilized plants weighed 10¼ oz, or as 100 to 102·5; but if the dwarfed

TABLE XXII

No. of pot	Plants raised from a cross between different flowers on the same plant	Plants raised from flowers fertilized with their own pollen
	Inches	Inches
I	$15\frac{1}{8}$	$19\frac{1}{8}$
	12	$20\frac{5}{8}$
	$10\frac{1}{8}$	$12\frac{6}{8}$
II	$16\frac{2}{8}$	$11\frac{2}{8}$
	$13\frac{5}{8}$	$19\frac{3}{8}$
	$20\frac{1}{8}$	$17\frac{4}{8}$
III	$18\frac{7}{8}$	$12\frac{6}{8}$
	15	$15\frac{6}{8}$
	$13\frac{7}{8}$	17
IV	$19\frac{2}{8}$	$16\frac{2}{8}$
	$19\frac{6}{8}$	$21\frac{5}{8}$
V	$25\frac{3}{8}$	$22\frac{5}{8}$
VI	15	$19\frac{5}{8}$
	$20\frac{2}{8}$	$16\frac{2}{8}$
	$27\frac{2}{8}$	$19\frac{5}{8}$
VII	$7\frac{6}{8}$	$7\frac{6}{8}$
	14	8
	$13\frac{4}{8}$	7
VIII Crowded	$18\frac{2}{8}$	$20\frac{3}{8}$
	$18\frac{6}{8}$	$17\frac{6}{8}$
	$18\frac{3}{8}$	$15\frac{4}{8}$
	$18\frac{3}{8}$	$15\frac{1}{8}$
Total in inches	370·88	353·63 /

plants in Pot VII had been excluded, the self-fertilized would have exceeded the crossed in weight in a higher ratio. In all the previous experiments in which seedlings were raised from a cross between distinct plants, and were put into competition with self-fertilized plants, the former generally flowered first; but in the present case, in seven out of the eight pots a self-fertilized plant flowered before a crossed one on the opposite side. Considering all the evidence with respect to the plants in Table XXII, a cross between two flowers on the same plant seems to give no advantage to the offspring thus produced, the self-fertilized plants being in weight superior. But this conclusion cannot be absolutely trusted, owing to the measurements given in Table XXI, though these latter, from the cause already assigned, are very much less trustworthy than the present ones.

Summary of observations on Mimulus luteus

In the three first generations of crossed and self-fertilized plants, the tallest plants alone on each side of the several pots were measured; and the average height of the ten crossed to that of the ten self-fertilized plants was as 100 to 64. The crossed were also much more fertile than the self-fertilized, and so much more vigorous that they exceeded them in height, even when sown on the opposite side of the same pot after an interval of four days. The same superiority was likewise shown in a remarkable manner when both kinds of seeds were sown on the opposite sides of a pot with very poor earth full of the roots of another plant. / In one instance crossed and self-fertilized seedlings, grown in rich soil and not put into competition with each other, attained to an equal height. When we come to the fourth generation the two tallest crossed plants taken together exceeded by only a little the two tallest self-fertilized plants, and one of the latter beat its crossed opponent – a circumstance which had not occurred in the previous generations. This victorious self-fertilized plant consisted of a new white-flowered variety, which grew taller than the old yellowish varieties. From the first it seemed to be rather more fertile, when self-fertilized, than the old varieties, and in the succeeding self-fertilized generations became more and more self-fertile. In the sixth generation the self-fertilized plants of this variety compared with the crossed plants produced capsules in the proportion of 147 to 100, both lots being allowed to fertilize themselves spontaneously. In the seventh generation twenty flowers on one of these plants artificially self-fertilized yielded no less than nineteen very fine capsules!

This variety transmitted its characters so faithfully to all the succeeding self-fertilized generations, up to the last or ninth, that all the many plants which were raised presented a complete uniformity of character; thus offering a remarkable contrast with the seedlings raised from the purchased seeds. Yet this variety retained to the last a latent tendency to produce yellow flowers; for when a plant of the eighth self-fertilized generation was crossed with pollen from a yellow-flowered plant of the Chelsea stock, every single seedling bore yellow flowers. A similar variety, at least in the colour of its flowers, also appeared among the crossed plants of the third generation. No attention was at first paid to it, and I know not / how far it was at first used either for crossing or self-fertilization. In the fifth generation most of the self-fertilized plants, and in the sixth and all the succeeding generations every single plant consisted of this variety; and this no doubt was partly due to its great and increasing self-fertility. On the other hand, it disappeared from among the crossed plants in the later generations; and this was probably due to the continued intercrossing of the several plants. From the tallness of this variety, the self-fertilized plants exceeded the crossed plants in height in all the generations from the fifth to the seventh inclusive; and no doubt would have done so in the later generations, had they been grown in competition with one another. In the fifth generation the crossed plants were in height to the self-fertilized, as 100 to 126; in the sixth, as 100 to 147; and in the seventh generation, as 100 to 137. This excess of height may be attributed not only to this variety naturally growing taller than the other plants, but to its possessing a peculiar constitution, so that it did not suffer from continued self-fertilization.

This variety presents a strikingly analogous case to that of the plant called the Hero, which appeared in the sixth self-fertilized generation of Ipomoea. If the seeds produced by Hero had been as greatly in excess of those produced by the other plants, as was the case with Mimulus, and if all the seeds had been mingled together, the offspring of Hero would have increased to the entire exclusion of the ordinary plants in the later self-fertilized generations, and from naturally growing taller would have exceeded the crossed plants in height in each succeeding generation.

Some of the self-fertilized plants of the sixth generation were intercrossed, as were some in the eighth / generation; and the seedlings from these crosses were grown in competition with self-fertilized plants of the two corresponding generations. In the first trial the intercrossed plants were less fertile than the self-fertilized, and less tall in the ratio of 100 to 110. In the second trial, the intercrossed plants were more fertile than the self-fertilized in the ratio of 100 to 73, and taller in the ratio of 100 to 92. Notwithstanding that the self-fertilized plants in the second trial were the product of two additional generations of self-fertilization, I cannot understand this discordance in the results of the two analogous experiments.

The most important of all the experiments on Mimulus are those in which flowers on plants of the eighth self-fertilized generation were

again self-fertilized; other flowers on distinct plants of the same lot
were intercrossed; and others were crossed with a new stock of plants
from Chelsea. The Chelsea-crossed seedlings were to the intercrossed
in height as 100 to 56, and in fertility as 100 to 4; and they were to the
self-fertilized plants, in height as 100 to 52, and in fertility as 100 to 3.
These Chelsea-crossed plants were also much more hardy than the
plants of the other two lots; so that altogether the gain from the cross
with a fresh stock was wonderfully great.

Lastly, seedlings raised from a cross between flowers on the same
plant were not superior to those from flowers fertilized with their own
pollen; but this result cannot be absolutely trusted, owing to some
previous observations, which, however, were made under very unfavourable circumstances.

DIGITALIS PURPUREA

The flowers of the common Foxglove are proterandrous; that is, the
pollen is mature and mostly shed before the stigma of the same flower
is ready for fertilization. This is effected by / the larger humble-bees,
which, whilst in search of nectar, carry pollen from flower to flower.
The two upper and longer stamens shed their pollen before the two
lower and shorter ones. The meaning of this fact probably is, as Dr
Ogle remarks,[3] that the anthers of the longer stamens stand near to
the stigma, so that they would be the most likely to fertilize it; and as it
is an advantage to avoid self-fertilization, they shed their pollen first,
thus lessening the chance. There is, however, but little danger of self-
fertilization until the bifid stigma opens; for Hildebrand[4] found that
pollen placed on the stigma before it had opened produced no effect.
The anthers, which are large, stand at first transversely with respect to
the tubular corolla, and if they were to dehisce in this position they
would, as Dr Ogle also remarks, smear with pollen the whole back and
sides of an entering humble-bee in a useless manner; but the anthers
twist round and place themselves longitudinally before they dehisce.
The lower and inner side of the mouth of the corolla is thickly clothed
with hairs, and these collect so much of the fallen pollen that I have
seen the under surface of a humble-bee thickly dusted with it; but this

[3] *Popular Science Review*, January, 1870, p. 50.
[4] *Geschlechter – Vertheilung bei den Pflanzen*, 1867, p. 20.

can never be applied to the stigma, as the bees in retreating do not turn their under surfaces upwards. I was therefore puzzled whether these hairs were of any use; but Mr Belt has, I think, explained their use: the smaller kinds of bees are not fitted to fertilize the flowers, and if they were allowed to enter easily they would steal much nectar, and fewer large bees would haunt the flowers. Humble-bees can crawl into the dependent flowers with the greatest ease, using the 'hairs as footholds while sucking the honey; but the smaller bees are impeded by them, and when, having at length struggled through them, they reach the slipper precipice above, they are completely baffled'. Mr Belt says that he watched many flowers during a whole season in North Wales, and 'only once saw a small bee reach the nectary, though many were seen trying in vain to do so'.[5]

I covered a plant growing in its native soil in North Wales with a net, and fertilized six flowers each with its own pollen, / and six others with pollen from a distinct plant growing within the distance of a few feet. The covered plant was occasionally shaken with violence, so as to imitate the effects of a gale of wind, and thus to facilitate as far as possible self-fertilization. It bore ninety-two flowers (besides the dozen artificially fertilized), and of these only twenty-four produced capsules; whereas almost all the flowers on the surrounding uncovered plants were fruitful. Of the twenty-four spontaneously self-fertilized capsules, only two contained their full complement of seed; six contained a moderate supply; and the remaining sixteen extremely few seeds. A little pollen adhering to the anthers after they had dehisced, and accidentally falling on the stigma when mature, must have been the means by which the above twenty-four flowers were partially self-fertilized; for the margins of the corolla in withering do not curl inwards, nor do the flowers in dropping off turn round on their axes, so as to bring the pollen-covered hairs, with which the lower surface is clothed, into contact with the stigma – by either of which means self-fertilization might be effected.

Seeds from the above crossed and self-fertilized capsules, after germinating on bare sand, were planted in pairs on the opposite sides of five moderately-sized pots, which were kept in the greenhouse. The plants after a time appeared starved, and were therefore, without being disturbed, turned out of their pots, and planted in the open

[5] *The Naturalist in Nicaragua*, 1874, p. 132. But it appears from H. Müller (*Die Befruchtung der Blumen*, 1873, p. 285), that small insects sometimes succeed in entering the flowers.

ground in two close parallel rows. They were thus subjected to tolerably severe competition with one another but not nearly so severe as if they had been left in the pots. At the time when they were turned out, their leaves were between 5 and 8 inches in length, and the longest leaf on the finest plant on each side of each pot was measured, with the result that the leaves of the crossed plants exceeded, on an average, those of the self-fertilized plants by 0·4 of an inch.

In the following summer the tallest flower-stem on each plant, when fully grown, was measured. There were seventeen crossed plants; but one did not produce a flower-stem. There were also, originally, seventeen self-fertilized plants, but these had such poor constitutions that no less than nine died in the course of the winter and spring, leaving only eight to be measured, as in the following table: /

TABLE XXIII

The tallest flower-stem on each plant measured: 0 means that the plant died before a flower-stem was produced

No. of pot	Crossed plants	Self-fertilized plants
	Inches	Inches
I	53⁶/₈	27⁴/₈
	57⁴/₈	55⁶/₈
	57⁶/₈	0
	65	0
II	34⁴/₈	39
	52⁴/₈	32
	63⁶/₈	21
III	57⁴/₈	53⁴/₈
	53⁴/₈	0
	50⁶/₈	0
	37²/₈	0
IV	64⁴/₈	34⁴/₈
	37⁴/₈	23⁶/₈
	. .	0
V	53	0
	47⁶/₈	0
	34⁶/₈	0
Total in inches	821·25	287·00

The average height of the flower-stems of the sixteen crossed plants is here 51·33 inches; and that of the eight self-fertilized plants, 35·87;

or as 100 to 70. But this difference in height does not give at all a fair idea of the vast superiority of the crossed plants. These latter produced altogether sixty-four flower-stems, each plant producing, on an average, exactly four flower-stems; whereas the eight self-fertilized plants produced only fifteen flower-stems, each producing an average only of 1·87 stems, and these had a less luxuriant appearance. We may put the result in another way: the number of flower-stems on the crossed plants was to those on an equal number self-fertilized plants as 100 to 48.

Three crossed seeds in a state of germination were also planted in three separate pots; and three self-fertilized seeds in the same state in three other pots. These plants were therefore at first exposed to no competition with one another, and when / turned out of their pots into the open ground they were planted at a moderate distance apart, so that they were exposed to much less severe competition than in the last case. The longest leaves on the three crossed plants, when turned out, exceeded those on the self-fertilized plants by a mere trifle, viz., on an average by 0·17 of an inch. When fully grown the three crossed plants produced twenty-six flower-stems; the two tallest of which on each plant were on an average 54·04 inches in height. The three self-fertilized plants produced twenty-three flower-stems, the two tallest of which on each plant had an average height of 46·18 inches. So that the difference between these two lots, which hardly competed together, is much less than in the last case when there was moderately severe competition, namely, as 100 to 85, instead of as 100 to 70.

The effects on the offspring of intercrossing different flowers on the same plant, instead of crossing distinct individuals. A fine plant growing in my garden (one of the foregoing seedlings) was covered with a net, and six flowers were crossed with pollen from another flower on the same plant, and six others were fertilized with their own pollen. All produced good capsules. The seeds from each were placed in separate watch-glasses, and no difference could be perceived by the eye between the two lots of seeds; and when they were weighed there was no difference of any significance, as the seeds from the self-fertilized capsules weighed 7·65 grains, whilst those from the crossed capsules weighed 7·7 grains. Therefore the sterility of the present species, when insects are excluded, is not due to the impotence of pollen on the stigma of the same flower. Both lots of seeds and seedlings were treated in exactly the same manner as in the previous table (XXIII), excepting that after the pairs of germinating seeds had been planted on the opposite sides of eight pots, all the remaining seeds were thickly sown on the opposite sides of Pots IX and X in Table XXIV. The young plants during the following spring were turned out of their pots, without being disturbed, and planted in the open ground in two rows, not very close together, so that they were subjected to only moderately severe competition with one another. Very differently to what occurred in the first experiment, when the plants were subjected to somewhat severe mutual competition, an equal number on each side either died or did not produce flower-stems. The tallest flower-stems on the surviving plants were measured, as shown in the following table: /

TABLE XXIV

N.B. o signifies that the plant died, or did not produce a flower-stem

No. of pot	Plants raised from a cross between different flowers on the same plant	Plants raised from flowers fertilized with their own pollen
	Inches	*Inches*
I	49⁴/₈	45⁵/₈
	46⁷/₈	52
	43⁶/₈	0
II	38⁴/₈	54⁴/₈
	47⁴/₈	47⁴/₈
	0	32⁵/₈
III	54⁷/₈	46⁵/₈
IV	32¹/₈	41³/₈
	0	29⁷/₈
	43⁷/₈	37¹/₈
V	46⁶/₈	42¹/₈
	40⁴/₈	42¹/₈
	43	0
VI	48²/₈	47⁷/₈
	46²/₈	48³/₈
VII	48⁵/₈	25
	42	40⁵/₈
VIII	46⁷/₈	39¹/₈

TABLE XXIV – *continued*

N.B. 0 *signifies that the plant died, or did not produce a flower-stem*

No. of pot	*Plants raised from a cross between different flowers on the same plant*	*Plants raised from flowers fertilized with their own pollen*
	Inches	*Inches*
IX	49	30 3/8
Crowded plants	50 3/8	15
	46 3/8	36 7/8
	47 6/8	44 1/8
	0	31 6/8
X	46 4/8	47 7/8
Crowded plants	35 2/8	0
	24 5/8	34 7/8
	41 4/8	40 7/8
	17 3/8	41 1/8
Total inches	1078·00	995·38 /

The average height of the flower-stems on the twenty-five crossed plants in all the pots taken together is 43·12 inches, and that of the twenty-five self-fertilized plants 39·82, or as 100 to 92. In order to test this result, the plants planted in pairs in Pots I to VIII were considered by themselves, and the average height of the sixteen crossed plants is here 44·9, and that of the sixteen self-fertilized plants 42·03, or as 100 to 94. Again, the plants raised from the thickly sown seed in Pots XI and X, which were subjected to very severe mutual competition, were taken by themselves, and the average height of the nine crossed plants is 39·86, and that of the nine self-fertilized plants 35·88, or as 100 to 90. The plants in these two latter pots (IX and X), after being measured were cut down close to the ground and weighed: the nine crossed plants weighed 57·66 ounces, and the nine self-fertilized plants 45·25 ounces, or as 100 to 78. On the whole we may conclude, especially from the evidence of weight, that seedlings from a cross between flowers on the same plant have a decided, though not great, advantage over those from flowers fertilized with their own pollen, more especially in the case of the plants subjected to severe mutual competition. But the advantage is much less than that exhibited by the crossed offspring of distinct plants, for these exceeded the self-fertilized plants in height as 100 to 70, and in the number of flower-stems as 100 to 48. Digitalis thus differs from Ipomoea, and almost certainly from Mimulus, as with these two species a cross between flowers on the same plant did no good.

CALCEOLARIA

A bushy greenhouse variety, with yellow flowers blotched with purple

The flowers in this genus are constructed so as to favour or almost ensure cross-fertilization;[6] and Mr Anderson remarks[7] that extreme care is necessary to exclude insects in order to preserve any kind true. He adds the interesting statement, that when the corolla is cut quite away, insects, as far as he has seen, never discover or visit the flowers. This plant is, however, self-fertile if insects are excluded. So few experiments were made by me, that they are hardly worth giving. Crossed and self-fertilized seeds were sown on opposite sides of a pot, and / after a time the crossed seedlings slightly exceeded the self-fertilized in height. When a little further grown, the longest leaves on the former were very nearly 3 inches in length, whilst those on the self-fertilized plants were only 2 inches. Owing to an accident, and to the pot being too small, only one plant on each side grew up and flowered; the crossed plant was 19½ inches in height, and the self-fertilized one 15 inches; or as 100 to 77.

LINARIA VULGARIS

It has been mentioned in the introductory chapter that two large beds of this plant were raised by me many years ago from crossed and self-fertilized seeds, and that there was a conspicuous difference in height and general appearance between the two lots. The trial was afterwards repeated with more care; but as this was one of the first plants experimented on, my usual method was not followed. Seeds were taken from wild plants growing in this neighbourhood and sown in poor soil in my garden. Five plants were covered with a net, the others being left exposed to the bees, which incessantly visit the flowers of this species, and which, according to H. Müller, are the exclusive fertilizers. This excellent observer remarks[8] that, as the stigma lies between the anthers and is mature at the same time with them, self-

---

[6] Hildebrand, as quoted by H. Müller, *Die Befruchtung der Blumen*, 1873, p. 277.
[7] *Gardeners' Chronicle*, 1853, p. 534.
[8] *Die Befruchtung*, etc., p. 279.

fertilization is possible. But so few seeds are produced by protected plants, that the pollen and stigma of the same flower seem to have little power of mutual interaction. The exposed plants bore numerous capsules forming solid spikes. Five of these capsules were examined and appeared to contain an equal number of seeds; and these being counted in one capsule, were found to be 166. The five protected plants produced altogether only twenty-five capsules, of which five were much finer than all the others, and these contained an average of 23·6 seeds, with a maximum in one capsule of fifty-five. So that the number of seeds in the capsules on the exposed plants to the average number in the finest capsules on the protected plants was as 100 to 14.

Some of the spontaneously self-fertilized seeds from under the net, and some seeds from the uncovered plants naturally fertilized and almost certainly intercrossed by the bees, were sown separately in two large pots of the same size; so that the / two lots of seedlings were not subjected to any mutual competition. Three of the crossed plants when in full flower were measured, but no care was taken to select the tallest plants; their heights were 7⁴⁄₈, 7²⁄₈, and 6⁴⁄₈ inches; averaging 7·08 in height. The three tallest of all the self-fertilized plants were then carefully selected, and their heights were 6³⁄₈, 5⁵⁄₈, and 5²⁄₈, averaging 5·75 in height. So that the naturally crossed plants were to the spontaneously self-fertilized plants in height, at least as much as 100 to 81.

VERBASCUM THAPSUS

The flowers of this plant are frequented by various insects, chiefly by bees, for the sake of the pollen. H. Müller, however, has shown (*Die Befruchtung*, etc., p. 277) that *V. nigrum* secretes minute drops of nectar. The arrangement of the reproductive organs, though not at all complex, favours cross-fertilization; and even distinct species are often crossed, for a greater number of naturally produced hybrids have been observed in this genus than in almost any other.[9] Nevertheless the present species is perfectly self-fertile, if insects are excluded; for a plant protected by a net was as thickly loaded with fine capsules as the surrounding uncovered plants. *Verbascum lychnitis* is rather less self-

[9] I have given a striking case of a large number of such hybrids between *V. thapsus* a *lychnitis* found growing wild: *Journal of Linn. Soc. Bot.*, vol. x, p. 451.

fertile, for some protected plants did not yield quite so many capsules as the adjoining uncovered plants.

Plants of V. *thapsus* had been raised for a distinct purpose from self-fertilized seeds; and some flowers on these plants were again self-fertilized; yielding seed of the second self-fertilized generation; and other flowers were crossed with pollen from a distinct plant. The seeds thus produced were sown on the opposite sides of four large pots. They germinated, however, so irregularly (the crossed seedlings generally coming up first) that I was able to save only six pairs of equal age. These when in full flower were measured, as in the following table (XXV).

We here see that two of the self-fertilized plants exceed in height their crossed opponents. Nevertheless the average height of the six crossed plants is 65·34 inches, and that of the six self-fertilized plants 56·5 inches; or as 100 to 86. /

TABLE XXV

No. of pot	Crossed plants	Self-fertilized plants of the second generation
	Inches	Inches
I	76	53⅜
II	54	66
III	62	75
	60⅝	30⅛
IV	73	62
	66⅛	52
Total in inches	392·13	339·00

VANDELLIA NUMMULARIFOLIA

Seeds were sent to me by Mr J. Scott from Calcutta of this small Indian weed, which bears perfect and cleistogamic[10] flowers. The latter are extremely small, imperfectly developed, and never expand, yet yield plenty of seeds. The perfect and open flowers are also small, of a white colour with purple marks; they generally produce seed, although the

[10] The convenient term of *cleistogamic* was proposed by Kuhn in an article on the present genus in *Bot. Zeitung*, 1867, p. 65.

contrary has been asserted; and they do so even if protected from insects. They have a rather complicated structure, and appear to be adapted for cross-fertilization, but were not carefully examined by me. They are not easy to fertilize artificially, and it is possible that some of the flowers which I thought that I had succeeded in crossing were afterwards spontaneously self-fertilized under the net. Sixteen capsules from the crossed perfect flowers contained on an average ninety-three seeds (with a maximum in one capsule of 137), and thirteen capsules from the self-fertilized perfect flowers contained sixty-two seeds (with a maximum in one capsule of 135); or as 100 to 67. But I suspect that this considerable excess was accidental, as on one occasion nine crossed capsules were compared with seven self-fertilized capsules (both included in the above number), and they contained almost exactly the same average number of seed. I may add / that fifteen capsules from self-fertilized cleistogamic flowers contained on an average sixty-four seeds, with a maximum in one of eighty-seven.

Crossed and self-fertilized seeds from the perfect flowers, and other seeds from the self-fertilized cleistogamic flowers, were sown in five pots, each divided superficially into three compartments. The seedlings were thinned at an early age, so that twenty plants were left in each of the three divisions. The crossed plants when in full flower averaged 4·3 inches, and the self-fertilized plants from the perfect flowers 4·27 inches in height; or as 100 to 99. The self-fertilized plants from the cleistogamic flowers averaged 4·06 inches in height; so that the crossed were in height to these latter plants as 100 to 94.

I determined to compare again the growth of plants raised from crossed and self-fertilized perfect flowers, and obtained two fresh lots of seeds. These were sown on opposite sides of five pots, but they were not sufficiently thinned, so that they grew rather crowded. When fully grown, all those above 2 inches in height were selected, all below this standard being rejected; the former consisted of forty-seven crossed and forty-one self-fertilized plants; thus a greater number of the crossed than of the self-fertilized plants grew to a height of above 2 inches. Of the crossed plants, the twenty-four tallest were on an average 3·6 inches in height; whilst the twenty-four tallest self-fertilized plants were 3·38 inches in average height; or as 100 to 94. All these plants were then cut down close to the ground, and the forty-seven crossed plants weighed 1090·3 grains, and the forty-one self-fertilized plants weighed 887·4 grains. Therefore an equal number of crossed and self-fertilized would have been to each other in weight as

77

100 to 97. From these several facts we may conclude that the crossed plants had some real, though very slight, advantage in height and weight over the self-fertilized plants, when grown in competition with one another.

The crossed plants were, however, inferior in fertility to the self-fertilized. Six of the finest plants were selected out of the forty-seven crossed plants, and six out of the forty-one self-fertilized plants; and the former produced 598 capsules, whilst the latter or self-fertilized plants produced 752 capsules. All these capsules were the product of cleistogamic flowers, for the plants did not bear during the whole of this season any perfect flowers. The seeds were counted in ten cleistogamic capsules / produced by the crossed plants, and their average number was 46·4 per capsule; whilst the number in ten cleistogamic capsules produced by the self-fertilized plants was 49·4; or as 100 to 106.

III. GESNERIACEAE

GESNERIA PENDULINA

In Gesneria the several parts of the flower are arranged on nearly the same plan as in Digitalis,[11] and most or all of the species are dichogamous. Plants were raised from seed sent me by Fritz Müller from South Brazil. Seven flowers were crossed with pollen from a distinct

TABLE XXVI

No. of pot	*Crossed plants*	*Self-fertilized plants*
	Inches	*Inches*
I	42 2/8	39
	24 4/8	27 3/8
II	33	30 6/8
	27	19 2/8
III	33 4/8	31 7/8
	29 4/8	28 6/8
IV	30 6/8	29 6/8
	36	26 3/8
Total inches	256·50	233·13

[11] Dr Ogle, *Popular Science Review*, January, 1870, p. 51.

plant, and produced seven capsules containing by weight 3·01 grains of seeds. Seven flowers on the same plants were fertilized with their own pollen, and their seven capsules contained exactly the same weight of seeds. Germinating seeds were planted on opposite sides of four pots, and when fully grown measured to the tips of their leaves.

The average height of the eight crossed plants is 32·06 inches, and that of the eight self-fertilized plants 29·14; or as 100 to 90. /

IV. LABIATAE[12]

SALVIA COCCINEA

This species, unlike most of the others in the same genus, yields a good many seeds when insects are excluded. I gathered ninety-eight capsules produced by flowers spontaneously self-fertilized under a net, and they contained on an average 1·45 seeds, whilst flowers artificially fertilized with their own pollen, in which case the stigma will have received plenty of pollen, yielded on an average 3·3 seeds, or more than twice as many. Twenty flowers were crossed with pollen from a distinct plant, and twenty-six were self-fertilized. There was no great difference in the proportional number of flowers which produced capsules by these two processes, or in the number of the contained seeds or in the weight of an equal number of seeds.

Seeds of both kinds were sown rather thickly on opposite sides of three pots. When the seedlings were about 3 inches in height, the crossed showed a slight advantage over the self-fertilized. When two-thirds grown, the two tallest plants on each side of each pot were measured; the crossed averaged 16·37 inches, and the self-fertilized 11·75 in height; or as 100 to 71. When the plants were fully grown and had done flowering, the two tallest plants on each side were again measured, with the results shown in the Table XXVII overleaf. /

It may be here seen that each of the six tallest crossed plants exceeds in height its self-fertilized opponent; the former averaged 27·85 inches, whilst the six tallest self-fertilized plants averaged 21·16 inches; or as 100 to 76. In all three pots the first plant which flowered was a crossed one. All the crossed plants together produced 409 flowers, whilst all the self-fertilized together produced only 232 flowers; or as

[12] The admirable mechanical adaptations in this genus for favouring or ensuring cross-fertilization, have been fully described by Sprengel, Hildebrand, Delpino, H. Müller, Ogle, and others, in their several works.

TABLE XXVII

No. of pot	*Crossed plants*	*Self-fertilized plants*
	Inches	*Inches*
I	32 6/8	25
	20	18 6/8
II	32 3/8	20 6/8
	24 4/8	19 4/8
III	29 4/8	25
	28	18
Total inches	167·13	127·00

100 to 57. So that the crossed plants in this respect were far more productive than the self-fertilized.

ORIGANUM VULGARE

This plant exists, according to H. Müller, under two forms; one hermaphrodite and strong proterandrous, so that it is almost certain to be fertilized by pollen from another flower; the other form is exclusively female, has a small corolla, and must of course be fertilized by pollen from a distinct plant in order to yield any seeds. The plants on which I experimented were hermaphrodites; they had been cultivated for a long period as a pot-herb in my kitchen garden, and were, like so many long-cultivated plants, extremely sterile. As I felt doubtful about the specific name I sent specimens to Kew, and was assured that the species was *O. vulgare.* My plants formed one great clump, and had evidently spread from a single root by stolons. In a strict sense, therefore, they all belonged to the same individual. My object in experimenting on them was, firstly, to ascertain whether crossing flowers borne by plants having distinct roots, but all derived asexually from the same individual, would be in any respect more advantageous than self-fertilization; and, secondly, to raise for future trial seedlings which would constitute really distinct individuals. Several plants in the above clump were covered by a net, and about two dozen seeds (many of which, however, were small and withered) were obtained from the flowers thus spontaneously self-fertilized. The remainder of the plants were left uncovered and were incessantly visited by bees, so that they were doubtless crossed by them. These exposed plants yielded rather more and finer seed (but still very few) than did the covered plants. The two lots of seeds thus obtained were sown on opposite sides of two pots; the seedlings were carefully observed from their first growth to maturity, but they did not differ at any period in height or in vigour, the importance of which latter observation we shall presently see. When fully grown, the tallest crossed / plant in one pot was a very little taller than the tallest self-fertilized plant on the opposite side, and in the other pot exactly the reverse occurred. So that the two lots were in fact equal; and a cross of this kind did no more good than crossing two flowers on the same plant of Ipomoea or Mimulus.

The plants were turned out of the two pots without being disturbed and planted in the open ground, in order that they might grow more vigorously. In the following summer all the self-fertilized and some of the quasi-crossed plants were covered by a net. Many flowers on the latter were crossed by me with pollen from a distinct plant, and others were left to be crossed by the bees. These quasi-crossed plants produced rather more seed than did the original ones in the great clump when left to the action of the bees. Many flowers on the self-fertilized plants were artificially self-fertilized, and others were allowed to fertilize themselves spontaneously under the net, but they yielded altogether very few seeds. These two lots of seeds – the product of a cross between distinct seedlings, instead of as in the last case between plants multiplied by stolons, and the product of self-fertilized flowers – were allowed to germinate on bare sand, and several equal pairs were planted on opposite sides of two *large* pots. At a very early age the crossed plants showed some superiority over the self-fertilized, which was ever afterwards retained. When the plants were fully grown, the two tallest crossed and the two tallest self-fertilized plants in each pot were measured, as shown in the following table. I regret that from want of time I did not measure all the pairs; but the tallest on each side seemed fairly to represent the average difference between the two lots. /

The average height of the crossed plants is here 20 inches, and that of the self-fertilized 17·12; or as 100 to 86. But this excess of height by no means gives a fair idea of the vast superiority in vigour of the crossed over the self-fertilized plants. The crossed flowered first and produced thirty flower-stems, whilst the self-fertilized produced only fifteen, or half the number. The pots were then bedded out, and the roots probably came out of the holes at the bottom and thus aided their

TABLE XXVIII

No. of pot	Crossed plants (two tallest in each pot)	Self-fertilized plants (two tallest in each pot)
	Inches	Inches
I	26	24
	21	21
II	17	12
	16	11⅛
Total inches	80·0	68·5

growth. Early in the following summer the superiority of the crossed plants, owing to their increase by stolons, over the self-fertilized plants was truly wonderful. In Pot I, and it should be remembered that very large pots had been used, the oval clump of crossed plants was 10 by 4½ inches across, with the tallest stem, as yet young, 5½ inches in height; whilst the clump of self-fertilized plants, on the opposite side of the same pot, was only 3½ by 2½ inches across, with the tallest young stem 4 inches in height. In Pot II, the clump of crossed plants was 18 by 9 inches across, with the tallest young stem 8½ inches in height; whilst the clump of self-fertilized plants on the opposite side of the same pot was 12 by 4½ inches across, with the tallest young stem 6 inches in height. The crossed plants during this season, as during the last, flowered first. Both the crossed and self-fertilized plants being left freely exposed to the visits of bees, manifestly produced much more seed than their grandparents – the plants of the original clump still growing close by in the same garden, and equally left to the action of the bees.

V. ACANTHACEAE

THUNBERGIA ALATA

It appears from Hildebrand's description (*Bot. Zeitung*, 1867, p. 285) that the conspicuous flowers of this plant are adapted for cross-fertilization. Seedlings were twice raised from purchased seed; but during the early summer, when first experimented on, they were extremely sterile, many of the anthers containing hardly any pollen. Nevertheless, during the autumn these same plants spontaneously produced a good many seeds. Twenty-six flowers during the two years were crossed with pollen from a distinct plant, but they yielded only eleven capsules; and these contained very few seeds! Twenty-eight flowers were fertilized with pollen from the same flower, and these yielded only ten capsules, which, however, contained rather more seed than the crossed capsules. Eight pairs of / germinating seeds were planted on opposite sides of five pots; and exactly half the crossed and half the self-fertilized plants exceeded their opponents in height. Two of the self-fertilized plants died young, before they were measured, and their crossed opponents were thrown away. The six remaining pairs grew very unequally, some, both of the crossed and self-fertilized plants, being more than twice as tall as the others. The average height of the crossed plants was 60 inches, and that of the self-fertilized plants 65 inches, or as 100 to 108. A cross, therefore, between distinct individuals here appears to do no good; but this result deduced from so few plants in a very sterile condition and growing very unequally, obviously cannot be trusted. /

CHAPTER IV

CRUCIFERAE, PAPAVERACEAE, RESEDACEAE, ETC.

Brassica oleracea, crossed and self-fertilized plants – Great effect of a cross with a fresh stock on the weight of the offspring – *Iberis umbellata* – *Papaver vagum* – *Eschscholtzia californica*, seedlings from a cross with a fresh stock not more vigorous, but more fertile than the self-fertilized seedlings – *Reseda lutea* and *odorata*, many individuals sterile with their own pollen – *Viola tricolor*, wonderful effects of a cross – *Adonis aestivalis* – *Delphinium consolida* – *Viscaria oculata*, crossed plants hardly taller, but more fertile than the self-fertilized – *Dianthus caryophyllus*, crossed and self-fertilized plants compared for four generations – Great effects of a cross with a fresh stock – Uniform colour of the flowers on the self-fertilized plants – *Hibiscus africanus*.

VI. CRUCIFERAE

BRASSICA OLERACEA

Var. Cattell's Early Barnes Cabbage

The flowers of the common cabbage are adapted, as shown by H. Müller,[1] for cross-fertilization, and should this fail, for self-fertilization. It is well known that the varieties are crossed so largely by insects, that it is impossible to raise pure kinds in the same garden, if more than one kind is in flower at the same time. Cabbages, in one respect, were not well fitted for my experiments, as, after they had formed heads, they were often difficult to measure. The flower-stems also differ much in height; and a poor plant will sometimes throw up a higher stem than that of a fine plant. In the later experiments, the fully-grown plants were cut down and weighed, and then the immense advantage from a cross became manifest.

A single plant of the above variety was covered with a net just before flowering, and was crossed with pollen from another plant of the same variety growing close by; and the seven capsules thus produced con-

[1] *Die Befruchtung*, etc., p. 139.

tained on an average 16·3 seeds, with a / maximum of twenty in one capsule. Some flowers were artificially self-fertilized, but their capsules did not contain so many seeds as those from flowers spontaneously self-fertilized under the net, of which a considerable number were produced. Fourteen of these latter capsules contained on an average 4·1 seeds, with a maximum in one of ten seeds; so that the seeds in the crossed capsules were in number to those in the self-fertilized capsules as 100 to 25. The self-fertilized seeds, fifty-eight of which weighed 3·88 grains, were, however, a little finer than those from the crossed capsules, fifty-eight of which weighed 3·76 grains. When few seeds are produced, these seem often to be better nourished and to be heavier than when many are produced.

The two lots of seeds in an equal state of germination were planted, some on opposite sides of a single pot, and some in the open ground. The young crossed plants in the pot at first exceeded by a little in height the self-fertilized; then equalled them; were then beaten; and lastly were again victorious. The plants, without being disturbed, were turned out of the pot, and planted in the open ground; and after growing for some time, the crossed plants, which were all of nearly the same height, exceeded the self-fertilized ones by 2 inches. When they flowered, the flower-stems of the tallest crossed plant exceeded that of the tallest self-fertilized plant by 6 inches. The other seedlings which were planted in the open ground stood separate, so that they did not compete with one another; nevertheless the crossed plants certainly grew to a rather greater height than the self-fertilized; but no measurements were made. The crossed plants which had been raised in the pot, and those planted in the open ground, all flowered a little before the self-fertilized plants.

Crossed and self-fertilized plants of the second generation. Some flowers on the crossed plants of the last generation were again crossed with pollen from another crossed plant, and produced fine capsules. The flowers on the self-fertilized plants of the last generation were allowed to fertilize themselves spontaneously under a net, and they produced some remarkably fine capsules. The two lots of seeds thus produced germinated on sand, and eight pairs were planted on opposite sides of four pots. These plants were measured to the tips of their leaves on 20 October of the same year, and the eight crossed plants averaged in height 8·4 inches, whilst the self-fertilized averaged 8·53 inches, so that the crossed were a little inferior in height, as 100 to 101·5. By 5 June of the following year these / plants had grown much bulkier, and had begun to form heads. The crossed had now acquired a marked superiority in general appearance, and averaged 8·02 inches in height, whilst the self-fertilized averaged 7·31 inches; or as 100 to 91. The plants were then turned out of their pots and planted undisturbed in the open ground. By 5 August their heads were fully formed, but several had grown so crooked that their heights could hardly be measured with accuracy. The crossed plants, however, were on the whole considerably taller than the self-fertilized. In the following year they flowered; the crossed plants flowering before the self-fertilized in three of the pots, and at the same time in Pot II. The flower-stems were now measured, as shown in Table XXIX.

TABLE XXIX

Measured to tops of flower-stems; o signifies that a flower-stem was not formed

No. of pot	Crossed plants	Self-fertilized plants
	Inches	Inches
I	49²⁄₈	44
	39⁴⁄₈	41
II	37⁴⁄₈	38
	33⁴⁄₈	35⁴⁄₈
III	47	51⅛
	40	41²⁄₈
	42	46⁴⁄₈
IV	43⁶⁄₈	20²⁄₈
	37²⁄₈	33³⁄₈
	o	o
Total in inches	369·75	351·00

The nine flower-stems on the crossed plants here averaged 41·08 inches, and the nine on the self-fertilized plants 39 inches in height, or as 100 to 95. But this small difference, which, moreover, depended almost wholly on one of the self-fertilized plants being only 20 inches high, does not in the least show the vast superiority of the crossed over the self-fertilized plants. Both lots, including the two plants in Pot IV, which did not flower, were now cut down close to the ground and weighed, but / those in Pot II were excluded, for they had been accidentally injured by a fall during transplantation, and one was almost killed. The eight crossed plants weighed 219 ounces, whilst the eight self-fertilized plants weighed only 82 ounces, or as 100 to 37; so that the superiority of the former over the latter in weight was great.

The effects of a cross with a fresh stock. Some flowers on a crossed plant of the last or second generation were fertilized, without being castrated, by pollen taken from a plant of the same variety, but not related to my plants, and brought from a nursery garden (whence my seeds originally came) having a different soil and aspect. The flowers on the self-fertilized plants of the last or second generation (Table XXIX) were allowed to fertilize themselves spontaneously under a net, and yielded plenty of seeds. These latter and the crossed seeds, after germinating on sand, were planted in pairs on the opposite sides of six large pots, which were kept at first in a cool greenhouse. Early in January their heights were measured to the tips of their leaves. The thirteen crossed plants averaged 13·16 inches in height, and the twelve (for one had died) self-fertilized plants averaged 13·7 inches, or as 100 to 104; so that the self-fertilized plants exceeded by a little the crossed plants.

TABLE XXX
Weight of plants after they had formed heads

No. of pot	Crossed plants from pollen of fresh stock	Self-fertilized plants of the third generation
	Ounces	*Ounces*
I	130	18¾
II	74	34¾
III	121	17²⁄₄
IV	127²⁄₄	14
V	90	11²⁄₄
VI	106²⁄₄	46
Total in ounces	649·00	142·25 /

Early in the spring the plants were gradually hardened, and turned out of their pots into the open ground without being disturbed. By the end of August the greater number had formed fine heads, but several grew extremely crooked, from having been drawn up to the light whilst in the greenhouse. As it was scarcely possible to measure their heights, the finest plant on each side of each pot was cut down close to the ground and weighed. In the preceding table we have the result.

The six finest crossed plants average 108·16 ounces, whilst the six finest self-fertilized plants average only 23·7 ounces, or as 100 to 22. This difference shows in the clearest manner the enormous benefit which these plants derived from a cross with another plant belonging to the same subvariety, but to a fresh stock, and grown during at least the three previous generations under somewhat different conditions.

The offspring from a cut-leaved, curled, and variegated white-green cabbage crossed with a cut-leaved, curled, and variegated crimson-green cabbage, compared with the self-fertilized offspring from the two varieties. These trials were made, not for the sake of comparing the growth of the crossed and self-fertilized seedlings, but because I had seen it stated that these varieties would not naturally intercross when growing uncovered and near one another. This statement proved quite erroneous; but the white-green variety was in some degree sterile in my garden, producing little pollen and few seeds. It was therefore no wonder that seedlings raised from the self-fertilized flowers of this variety were greatly exceeded in height by seedlings from a cross between it and the more vigorous crimson-green variety; and nothing more need be said about this experiment.

The seedlings from the reciprocal cross, that is, from the crimson-green variety fertilized with pollen from the white-green variety, offer a somewhat more curious case. A few of these crossed seedlings reverted to a pure variety with their leaves less cut and curled, so that they were altogether in a much more natural state, and these plants grew more vigorously and taller than any of the others. Now it is a strange fact that a much larger number of the self-fertilized seedlings from the crimson-green variety than of the crossed seedlings thus reverted; and as a consequence the self-fertilized seedlings grew taller by 2½ inches on an average than the crossed seedlings, with which they were put into competition. At first, however, the crossed seedlings exceeded the self-fertilized by an average of a quarter / of an inch. We thus see that reversion to a more natural condition acted more powerfully in favouring the ultimate growth of these plants than did a cross; but it should be remembered that the cross was with a semi-sterile variety having a feeble constitution.

IBERIS UMBELLATA
Var. Kermesiana

This variety produced plenty of spontaneously self-fertilized seed under a net. Other plants in pots in the greenhouse were left uncovered, and as I saw small flies visiting the flowers, it seemed probable that they would be intercrossed. Consequently seeds supposed to have been thus crossed and spontaneously self-fertilized seeds were sown on opposite sides of a pot. The self-fertilized seedlings grew from the first quicker than the supposed crossed seedlings, and when both lots were in full flower the former were from 5 to 6 inches higher than the crossed! I record in my notes that the self-fertilized seeds from which these self-fertilized plants were raised were not so well ripened as the crossed; and this may possibly have caused, from premature growth, the great difference in their height, in nearly the same manner as when self-fertilized seeds of other plants were sown a few days before the crossed in the same pot. We have seen a somewhat analogous case with the self-fertilized plants of the eighth generation of Ipomoea, raised from unhealthy parents. It is a curious circumstance, that two other lots of the above seeds were sown in pure sand mixed with burnt earth, and therefore without any organic matter; and here the supposed crossed seedlings grew to double the height of the self-fertilized, before both lots died, as necessarily occurred at an early period. We shall hereafter meet with another case apparently analogous to this of Iberis in the third generation of Petunia.

The above self-fertilized plants were allowed to fertilize themselves again under a net, yielding self-fertilized plants of the second generation, and the supposed crossed plants were crossed by pollen of a distinct plant; but from want of time this was done in a careless manner, namely, by smearing one head of expanded flowers over another. I should have thought that this would have succeeded, and perhaps it did so; but the fact of 108 of the self-fertilized seeds weighing 4·87 grains, whilst the same number of the supposed crossed seeds weighed only 3·57 grains, / does not look like it. Five seedlings from each lot of seeds were raised, and the self-fertilized plants, when fully grown, exceeded in average height by a trifle (viz. 0·4 of an inch) the five probably crossed plants. I have thought it right to give this case and the last, because had the supposed crossed plants proved superior to the self-fertilized in height, I should have assumed without doubt that the former had really been crossed. As it is, I do not know what to conclude.

Being much surprised at the two foregoing trials, I determined to make another, in which there should be no doubt about the crossing. I therefore fertilized with great care (but as usual without castration) twenty-four flowers on the supposed crossed plants of the last generation with pollen from distinct plants, and thus obtained twenty-one capsules. The self-fertilized plants of the last generation were allowed to fertilize themselves again under a net, and the seedlings reared from these seeds formed the third self-fertilized generation. Both lots of seeds, after germinating on bare sand, were planted in pairs on the opposite sides of two pots. All the remaining seeds were sown crowded on opposite sides of a third pot; but as all the self-fertilized seedlings in this latter pot died before they grew to any considerable height, they were not measured. The plants in Pots I and II were measured when between 7 and 8 inches in height, and the crossed exceeded the self-fertilized in average height by 1·57 inches. When fully grown they were again measured to the summits of their flower-heads, with the following result:

TABLE XXXI

No. of pot	*Crossed plants*	*Self-fertilized plants of the third generation*
	Inches	*Inches*
I	18	19
	21	21
	18⅞	19⁴⁄₈
II	19	16⁶⁄₈
	18⁴⁄₈	7⁴⁄₈
	17⁶⁄₈	14⁴⁄₈
	21³⁄₈	16⁴⁄₈
Total in inches	133·88	114·75

The average height of the seven crossed plants is here 19·12 inches, and that of the seven self-fertilized plants 16·39, or as 100 to 86. But as the plants on the self-fertilized side grew very unequally, this ratio cannot be fully trusted, and is probably too high. In both pots a crossed plant flowered before any one of the self-fertilized. These plants were left uncovered in the greenhouse; but from being too much crowded they were not very productive. The seeds from all seven plants of both lots were counted; the crossed produced 206, and the self-fertilized 154; or as 100 to 75.

Cross by a fresh stock. From the doubts caused by the two first trials, in which it was not known with certainty that the plants had been crossed; and from the crossed plants in the last experiment having been put into competition with plants self-fertilized for three generations, which moreover grew very unequally, I resolved to repeat the trial on a larger scale, and in a rather different manner. I obtained seeds of the same crimson variety of *I. umbellata* from another nursery garden, and raised plants from them. Some of these plants were allowed to fertilize themselves spontaneously under a net; others were crossed by pollen taken from plants raised from seed sent me by Dr Durando from Algiers, where the parent-plants had been cultivated for some generations. These latter plants differed in having pale pink instead of crimson flowers, but in no other respect. That the cross had been effective (though the flowers on the crimson mother-plant had *not* been castrated) was well shown when the thirty crossed seedlings flowered, for twenty-four of them produced pale pink flowers, exactly like those of their father; the six others having crimson flowers exactly like those of their mother and like those of all the self-fertilized seedlings. This case offers a good instance of a result which not rarely follows from crossing varieties having differently coloured flowers; namely, that the colours do not blend, but resemble perfectly those either of the father or mother plant. The seeds of both lots, after germinating on sand, were planted on opposite sides of eight pots. When fully grown, the plants were measured to the summits of the flower-heads, as shown in Table XXXII. /

The average height of the thirty crossed plants is here 17·34, and that of the twenty-nine self-fertilized plants (one / having died) 15·51, or as 100 to 89. I am surprised that the difference did not prove somewhat greater, considering that in the last experiment it was as 100 to 86; but this latter ratio, as before explained, was probably too great. It should, however, be observed that in the last experiment (Table XXXI), the crossed plants competed with the plants of the third self-fertilized generation; whilst in the present case, plants derived from a cross with a fresh stock competed with self-fertilized plants of the first generation.

The crossed plants in the present case, as in the last, were more

TABLE XXXII
Iberis umbellata: o *signifies that the plant died*

No. of pot	*Plants from a cross with a fresh stock*	*Plants from spontaneously self-fertilized seeds*
	Inches	*Inches*
I	18 6/8	17 3/8
	17 5/8	16 7/8
	17 6/8	13 1/8
	20 1/8	15 3/8
II	20 2/8	o
	15 7/8	16 6/8
	17	15 2/8
III	19 2/8	13 6/8
	18 1/8	14 2/8
	15 2/8	13 4/8
IV	17 1/8	16 4/8
	18 7/8	14 4/8
	17 5/8	16
	15 6/8	15 3/8
	14 4/8	14 7/8
V	18 1/8	16 4/8
	14 7/8	16 2/8
	16 2/8	14 2/8
	15 5/8	14 2/8
	12 4/8	16 1/8
VI	18 6/8	16 1/8
	18 6/8	15
	17 3/8	15 2/8
VII	18	16 3/8
	16 4/8	14 4/8
	18 2/8	13 5/8
VIII	20 6/8	15 6/8
	17 7/8	16 3/8
	13 5/8	20 2/8
	19 2/8	15 6/8
Total in inches	520·38	449·88

fertile than the self-fertilized, both lots being left uncovered in the greenhouse. The thirty crossed plants produced 103 seed-bearing flower-heads, as well as some heads which yielded no seeds; whereas the twenty-nine self-fertilized plants produced only 81 seed-bearing heads; therefore thirty such plants would have produced 83·7 heads. We thus get the ratio of 100 to 81, for the number of seed-bearing flower-heads produced by the crossed and self-fertilized plants. Moreover, a number of seed-bearing heads from the crossed plants, compared with the same number from the self-fertilized, yielded seeds by weight, in the ratio of 100 to 92. Combining these two elements, viz., the number of seed-bearing heads and the weight of seeds in each head, the productiveness of the crossed to the self-fertilized plants was as 100 to 75.

The crossed and self-fertilized seeds, which remained after the above pairs had been planted (some in a state of germination and some not so), were sown early in the year out of doors in two rows. Many of the self-fertilized seedlings suffered greatly, and a much larger number of them perished than of the crossed. In the autumn the surviving self-fertilized plants were plainly less well-grown than the crossed plants.

VII. PAPAVERACEAE

PAPAVER VAGUM

A subspecies of P. dubium, from the south of France

The poppy does not secrete nectar, but the flowers are highly conspicuous and are visited by many pollen-collecting bees, flies and beetles. The anthers shed their pollen very early, and in the case of *P. rhoeas*, it falls on the circumference of the radiating stigmas, so that this species must often be self-fertilized; but with *P. dubium* the same result does not follow (according to H. Müller, *Die Befruchtung*, p. 128), owing to / the shortness of the stamens, unless the flower happens to stand inclined. The present species, therefore, does not seem so well fitted for self-fertilization as most of the others. Nevertheless *P. vagum* produced plenty of capsules in my garden when insects were excluded, but only late in the season. I may here add that *P. somniferum* produces an abundance of spontaneously self-fertilized capsules, as Professor H. Hoffmann likewise found to be the case.[2] Some species of Papaver cross freely when growing in the same garden, as I have known to be the case with *P. bracteatum* and *orientale*.

[2] *Zur Speciesfrage*, 1875, p. 53.

93

Plants of *Papaver vagum* were raised from seeds sent me from Antibes through the kindness of Dr Bornet. Some little time after the flowers had expanded, several were fertilized with their own pollen, and others (not castrated) with pollen from a distinct individual; but I have reason to believe, from observations subsequently made, that these flowers had been already fertilized by their own pollen, as this process seems to take place soon after their expansion.[3] I raised, however, a few seedlings of both lots, and the self-fertilized rather exceeded the crossed plants in height.

Early in the following year I acted differently, and fertilized seven flowers, very soon after their expansion, with pollen from another plant, and obtained six capsules. From counting the seeds in a medium-sized one, I estimated that the average number in each was at least 120. Four out of twelve capsules, spotaneously self-fertilized at the same time, were found to contain no good seeds; and the remaining eight contained on an average 6·6 seeds per capsule. But it should be observed that later in the season the same plants produced under a net plenty of very fine spontaneously self-fertilized capsules.

The above two lots of seeds, after germinating on sand, were planted in pairs on opposite sides of five pots. The two lots of seedlings, when half an inch in height, and again when 6 inches high, were measured to the tips of their leaves, but presented / no difference. When fully grown, the flower-stalks were measured to the summits of the seed capsules, with the following result in Table XXXIII.

The fifteen crossed plants here average 21·91 inches, and the fifteen self-fertilized plants 19·54 inches in height, or as 100 to 89. These plants did not differ in fertility, as far as could be judged by the number of capsules produced, for there were seventy-five on the crossed side and seventy-four on the self-fertilized side.

---

[3] Mr J. Scott found (*Report on the Experimental Culture of the Opium Poppy*, Calcutta, 1874, p. 47), in the case of *Papaver somniferum*, that if he cut away the stigmatic surface before the flower had expanded, no seeds were produced; but if this was done 'on the second day, or even a few hours after the expansion of the flower on the first day, a partial fertilization had already been effected, and a few good seeds were almost invariably produced.' This proves at how early a period fertilization takes place.

TABLE XXXIII
Papaver vagum

No. of pot	Crossed plants	Self-fertilized plants
	Inches	Inches
I	24²⁄₈	21
	30	26⁵⁄₈
	18⁴⁄₈	16
II	14⁴⁄₈	15³⁄₈
	22	20⅛
	19⁵⁄₈	14⅛
	21⁵⁄₈	16⁴⁄₈
III	26⁶⁄₈	19²⁄₈
	20²⁄₈	13²⁄₈
	20⁶⁄₈	18
IV	25³⁄₈	23²⁄₈
	24²⁄₈	23
V	20	18³⁄₈
	27⁷⁄₈	27
	19	21²⁄₈
Total in inches	328·75	293·13

ESCHSCHOLTZIA CALIFORNICA

This plant is remarkable from the crossed seedlings not exceeding in height or vigour the self-fertilized. On the other hand, a cross greatly increases the productiveness of the flowers on the parent-plants, or, as it would be more correct to say, self-fertilization lessens their productiveness. A cross is indeed sometimes necessary in order that the flowers should produce any seed. Moreover, plants derived from a cross are themselves much more fertile than those raised from self-fertilized flowers / so that the whole advantage of a cross is confined to the reproductive system. It will be necessary for me to give this singular case in considerable detail.

Twelve flowers on some plants in my flower-garden were fertilized with pollen from distinct plants, and produced twelve capsules; but one of these contained no good seed. The seeds of the eleven good capsules weighed 17·4 grains. Eighteen flowers on the same plants were fertilized with their own pollen and produced twelve good capsules, which contained 13·61 grains weight of seed. Therefore an

95

equal number of crossed and self-fertilized capsules would have yielded seed by weight as 100 to 71.[4] If we take into account the fact that a much greater proportion of flowers produced capsules when crossed than when self-fertilized, the relative fertility of the crossed to the self-fertilized flowers was as 100 to 52. Nevertheless these plants, whilst still protected by the net, spontaneously produced a considerable number of self-fertilized capsules.

The seeds of the two lots after germinating on sand were planted in pairs on the opposite sides of four large pots. At first there was no difference in their growth, but ultimately the crossed seedlings exceeded the self-fertilized considerably in height, as shown in the following table. But I believe from / the cases which follow that this

TABLE XXXIV
Eschscholtzia californica

No. of pot	Crossed plants	Self-fertilized plants
	Inches	*Inches*
I	33⅘	25
II	34⅖	35
III	29	27⅖
IV	22	15
Total in inches	118·75	102·25

result was accidental, owing to only a few plants having been measured, and to one of the self-fertilized plants having grown only to a height of 15 inches. The plants had been kept in the greenhouse, and from being drawn up to the light had to be tied to sticks in this and the following trials. They were measured to the summits of their flower-stems.

The four crossed plants here average 29·68 inches, and the four self-fertilized 25·56 in height; or as 100 to 86. The remaining seeds were sown in a large pot in which a Cineraria had long been growing; and in this case again the two crossed plants on the one side greatly exceeded in height the two self-fertilized plants on the opposite side.

---

[4] Professor Hildebrand experimented on plants in Germany on a larger scale than I did, and found them much more self-sterile. Eighteen capsules, produced by cross-fertilization, contained on an average eighty-five seeds, whilst fourteen capsules from self-fertilized flowers contained on an average only nine seeds; that is, as 100 to 11: *Jahrb. für Wissen. Botanik*, vol. vii, p. 467.

The plants in the above four pots from having been kept in the greenhouse did not produce on this or any other similar occasion many capsules; but the flowers on the crossed plants when again crossed were much more productive than the flowers on the self-fertilized plants when again self-fertilized. These plants after seeding were cut down and kept in the greenhouse; and in the following year, when grown again, their relative heights were reversed, as the self-fertilized plants in three out of the four pots were now taller than and flowered before the crossed plants.

Crossed and self-fertilized plants of the second generation. The fact just given with respect to the growth of the cut-down plants made me doubtful about my first trial, so I determined to make another on a larger scale with crossed and self-fertilized seedlings raised from the crossed and self-fertilized plants on the last generation. Eleven pairs were raised and grown in competition in the usual manner; and now the result was different, for the two lots were nearly equal during their whole growth. It would therefore be superfluous to give a table of their heights. When fully grown and measured, the crossed averaged 32·47, and the self-fertilized 32·81 inches in height; or as 100 to 101. There was no great difference in the number of flowers and capsules produced by the two lots when both were left freely exposed to the visits of insects.

Plants raised from Brazilian seed. Fritz Müller sent me from South Brazil seeds of plants which were there absolutely sterile when fertilized with pollen from the same plant, but were perfectly fertile when fertilized with pollen from any other plant. The plants raised by me in England from these / seeds were examined by Professor Asa Gray, and pronounced to belong to *E. californica*, with which they were identical in general appearance. Two of these plants were covered by a net, and were found not to be so completely self-sterile as in Brazil. But I shall recur to this subject in another part of this work. Here it will suffice to state that eight flowers on these two plants, fertilized with pollen from another plant under the net, produced eight fine capsules, each containing on an average, about eighty seeds. Eight flowers on these same plants, fertilized with their own pollen, produced seven capsules, which contained on an average only twelve seeds, with a maximum in one of sixteen seeds. Therefore the cross-fertilized capsules, compared with the self-fertilized, yielded seeds in the ratio of about 100 to 15.

These plants of Brazilian parentage differed also in a marked manner from the English plants in producing extremely few spontaneously self-fertilized capsules under a net.

Crossed and self-fertilized seeds from the above plants, after germinating on bare sand, were planted in pairs on the opposite sides of five large pots. The seedlings thus raised were the grandchildren of the plants which grew in Brazil; the parents having been grown in England. As the grand-parents in Brazil absolutely require cross-fertilization in order to yield any seeds, I expected that self-fertilization would have proved very injurious to these seedlings, and that the crossed ones would have been greatly superior in height and vigour to those raised from self-fertilized flowers. But the result showed that my anticipation was erroneous; for as in the last experiment with plants of the English stock, so in the present one, the self-fertilized plants exceeded the crossed by a little in height. It will be sufficient to state that the fourteen crossed plants averaged 44·64, and the fourteen self-fertilized 45·12 inches in height; or as 100 to 101.

The effects of a cross with a fresh stock. I now tried a different experiment. Eight flowers on the self-fertilized plants of the last experiment (i.e., grandchildren of the plants which grew in Brazil) were again fertilized with pollen from the same plant, and produced five capsules, containing on an average 27·4 seeds, with a maximum in one of forty-two seeds. The seedlings raised from these seeds formed the second *self-fertilized* generation of the Brazilian stock.

Eight flowers on one of the crossed plants of the last experiment / were crossed with pollen from another grandchild, and produced five capsules. These contained on an average 31·6 seeds, with a maximum in one of forty-nine seeds. The seedlings raised from these seeds may be called the *intercrossed*.

Lastly, eight other flowers on the crossed plants of the last experiment were fertilized with pollen from a plant of the English stock, growing in my garden, and which must have been exposed during many previous generations to very different conditions from those to which the Brazilian progenitors of the mother-plant had been subjected. These eight flowers produced only four capsules, containing on an average 63·2 seeds, with a maximum in one of ninety. The plants raised from these seeds may be called the *English-crossed*. As far as the above averages can be trusted from so few capsules, the English-crossed capsules contained twice as many seeds as the intercrossed, and rather more than twice as many as the self-fertilized capsules. The plants which yielded these capsules were grown in pots in the greenhouse, so that their absolute productiveness must not be compared with that of plants growing out of doors.

The above three lots of seeds, viz., the self-fertilized, intercrossed, and English-crossed, were planted in an equal state of germination (having been as usual sown on bare sand) in nine large pots, each divided into three parts by superficial partitions. Many of the self-fertilized seeds germinated before those of the two crossed lots, and these were of course rejected. The seedlings thus raised are the great-grandchildren of the plants which grew in Brazil. When they were from 2 to 4 inches in height, the three lots were equal. They were measured when four-fifths grown, and again when fully grown, and as their relative heights were almost exactly the same at these two ages, I will give only the last measurements. The average height of the nineteen English-crossed plants was 45·92 inches; that of the eighteen intercrossed plants (for one died), 43·38; and that of the nineteen self-fertilized plants, 50·3 inches. So that we have the following ratios in height:

The English-crossed to the self-fertilized plants,	as 100 to 109
The English-crossed to the intercrossed plants,	as 100 to 94
The intercrossed to the self-fertilized plants,	as 100 to 116

After the seed-capsules had been gathered, all these plants were cut down close to the ground and weighed. The nineteen English-crossed plants weighed 18·25 ounces; the intercrossed / plants (with their weight calculated as if there had been nineteen) weighed 18·2 ounces; and the nineteen self-fertilized plants, 21·5 ounces. We have therefore for the weights of the three lots of plants the following ratios:

The English-crossed to the self-fertilized plants,	as 100 to 118
The English-crossed to the intercrossed plants,	as 100 to 100
The intercrossed to the self-fertilized plants,	as 100 to 118

We thus see that in weight, as in height, the self-fertilized plants had a decided advantage over the English-crossed and intercrossed plants.

The remaining seeds of the three kinds, whether or not in a state of germination, were sown in three long parallel rows in the open ground; and here again the self-fertilized seedlings exceeded in height by between 2 and 3 inches the seedlings in the two other rows, which were of nearly equal heights. The three rows were left unprotected throughout the winter, and all the plants were killed, with the exception of two of the self-fertilized; so that as far as this little bit of evidence goes, some of the self-fertilized plants were more hardy than any of the crossed plants of either lot.

We thus see that the self-fertilized plants which were grown in the nine pots were superior in height (as 116 to 100), and in weight (as 118 to 100), and apparently in hardiness, to the intercrossed plants derived from a cross between the grandchildren of the Brazilian stock. The superiority is here much more strongly marked than in the second trial with the plants of the English stock, in which the self-fertilized were to the crossed in height as 101 to 100. It is a far more remarkable fact – if we bear in mind the effects of crossing plants with pollen from a fresh stock in the cases of Ipomoea, Mimulus, Brassica, and Iberis – that the self-fertilized plants exceeded in height (as 109 to 100), and in weight (as 118 to 100), the offspring of the Brazilian stock crosed by the English stock; the two stocks having been long subjected to widely different conditions.

If we now turn to the fertility of the three lots of plants we find a very different result. I may premise that in five out of the nine pots the first plant which flowered was one of the English-crossed; in four of the pots it was a self-fertilized plant; and in not one did an intercrossed plant flower first; so that these latter plants were beaten in this respect, as in so many other ways. The three closely adjoining rows of plants growing in / the open ground flowered profusely, and the flowers were incessantly visited by bees, and certainly thus intercrossed. The manner in which several plants in the previous experiments continued to be almost sterile as long as they were covered by a net, but set a multitude of capsules immediately that they were uncovered, proves how effectually the bees carry pollen from plant to plant. My gardener gathered, at three successive times, an equal number of ripe capsules from the plants of the three lots, until he had collected forty-five from each lot. It is not possible to judge from external appearance whether or not a capsule contains any good seeds; so that I opened all the capsules. Of the forty-five from the English-crossed plants, four were empty; of those from the intercrossed, five were empty; and of those from the self-fertilized, nine were empty. The seeds were counted in twenty-one capsules taken by chance out of each lot, and the average number of seeds in the capsules from the English-crossed plants was 67; from the intercrossed, 6; and from the self-fertilized, 48·52. It therefore follows that:

	Seeds
The forty-five capsules (the four empty ones included) from the English-crossed plants contained	2747
The forty-five capsules (the five empty ones included) from the intercrossed plants contained	2240
The forty-five capsules (the nine empty ones included) from the self-fertilized plants contained	1746·7

The reader should remember that these capsules are the product of cross-fertilization, effected by the bees; and that the difference in the number of the contained seeds must depend on the constitution of the plants; that is, on whether they were derived from a cross with a distinct stock, or from a cross between plants of the same stock, or from self-fertilization. From the above facts we obtain the following ratios:

Number of seeds contained in an equal number of naturally fertilized capsules produced:

By the English-crossed and self-fertilized plants,	as 100 to 63
By the English-crossed and intercrossed plants,	as 100 to 81
By the intercrossed and self-fertilized plants,	as 100 to 78

But to have ascertained the productiveness of the three lots of plants, it would have been necessary to know how many capsules were produced by the same number of plants. The / three long rows, however, were not of quite equal lengths, and the plants were much crowded, so that it would have been extremely difficult to have ascertained how many capsules were produced by them, even if I had been willing to undertake so laborious a task as to collect and count all the capsules. But this was feasible with the plants grown in pots in the greenhouse; and although these were much less fertile than those growing out of doors, their relative fertility appeared, after carefully observing them, to be the same. The nineteen plants of the English-crossed stock in the pots produced altogether 240 capsules; the intercrossed plants (calculated as nineteen) produced 137·22 capsules; and the nineteen self-fertilized plants, 152 capsules. Now, knowing the number of seeds contained in forty-five capsules of each lot, it is easy to calculate the relative numbers of seeds produced by an equal number of the plants of the three lots.

Number of seeds produced by an equal number of naturally-fertilized plants.

Seeds
Plants of English-crossed and self-fertilized parentage   as 100 to 40
Plants of the English-crossed and intercrossed parentage as 100 to 45
Plants of the intercrossed and self-fertilized parentage   as 100 to 89

The superiority in productiveness of the intercrossed plants (that is, the product of a cross between the grandchildren of the plants which grew in Brazil) over the self-fertilized, small as it is, is wholly due to the larger average number of seeds contained in the capsules; for the intercrossed plants produced fewer capsules in the greenhouse than did the self-fertilized plants. The great superiority in productiveness of the English-crossed over the self-fertilized plants is shown by the larger number of capsules produced, the larger average number of contained seeds, and the smaller number of empty capsules. As the English-crossed and intercrossed plants were the offspring of crosses in every previous generation (as must have been the case from the flowers being sterile with their own pollen), we may conclude that the great superiority in productiveness of the English-crossed over the intercrossed plants is due to the two parents of the former having been long subjected to different conditions. /

The English-crossed plants, though so superior in productiveness, were, as we have seen, decidedly inferior in height and weight to the self-fertilized, and only equal to, or hardly superior to, the intercrossed plants. Therefore, the whole advantage of a cross with a distinct stock is here confined to productiveness, and I have met with no similar case.

VIII. RESEDACEAE

RESEDA LUTEA

Seeds collected from wild plants growing in this neighbourhood were sown in the kitchen-garden; and several of the seedlings thus raised were covered with a net. Of these, some were found (as will hereafter be more fully described) to be absolutely sterile when left to fertilize themselves spontaneously, although plenty of pollen fell on their stigmas; and they were equally sterile when artificially and repeatedly fertilized with their own pollen; whilst other plants produced a few spontaneously self-fertilized capsules. The remaining plants were left uncovered, and as pollen was carried from plant to plant by the hive and humble-bees which incessantly visit the flowers, they produced an abundance of capsules. Of the necessity of pollen being carried from one plant to another, I had ample evidence in the case of this species and of *R. odorata*; for those plants, which set no seeds or very few as long as they were protected from insects, became loaded with capsules immediately that they were uncovered.

Seeds from the flowers spontaneously self-fertilized under the net, and from flowers naturally crossed by the bees, were sown on opposite sides of five large pots. The seedlings were thinned as soon as they appeared above ground, so that an equal number were left on the two sides. After a time the pots were plunged into the open ground. The same number of plants of crossed and self-fertilized parentage were measured up to the summits of their flower-stems, with the result given in the following table (XXXV). Those which did not produce flower-stems were not measured.

TABLE XXXV
Reseda latea, in pots

No. of pot	Crossed plants	Self-fertilized plants
	Inches	*Inches*
I	21	$12\frac{7}{8}$
	$14\frac{2}{8}$	16
	$19\frac{1}{8}$	$11\frac{7}{8}$
	7	$15\frac{2}{8}$
	$15\frac{1}{8}$	$19\frac{1}{8}$
II	$20\frac{4}{8}$	$12\frac{4}{8}$
	$17\frac{3}{8}$	$16\frac{2}{8}$
	$23\frac{7}{8}$	$16\frac{2}{8}$
	$17\frac{1}{8}$	$13\frac{3}{8}$
	$20\frac{6}{8}$	$13\frac{5}{8}$
III	$16\frac{1}{8}$	$14\frac{4}{8}$
	$17\frac{6}{8}$	$19\frac{4}{8}$
	$16\frac{2}{8}$	$20\frac{7}{8}$
	10	$7\frac{7}{8}$
	10	$17\frac{6}{8}$
IV	$22\frac{1}{8}$	9
	19	$11\frac{4}{8}$
	$18\frac{7}{8}$	11
	$16\frac{4}{8}$	16
	$19\frac{2}{8}$	$16\frac{3}{8}$
V	$25\frac{2}{8}$	$14\frac{6}{8}$
	22	16
	$8\frac{6}{8}$	$14\frac{3}{8}$
	$14\frac{2}{8}$	$14\frac{2}{8}$
Total in inches	412·25	350·88

The average height of the twenty-four crossed plants is here 17·17 inches, and that of the same number of self-fertilized plants 14·61; or as 100 to 85. Of the crossed plants all but five flowered, whilst several of the self-fertilized did not do so. The above pairs, whilst still in flower, but with some capsules already formed, were afterwards cut down and weighed. The crossed / weighed 90·5 ounces; and an equal number of the self-fertilized only 19 ounces, or as 100 to 21; and this is an astonishing difference.

Seeds of the same two lots were also sown in two adjoining rows in the open ground. There were twenty crossed plants in the one row and thirty-two self-fertilized plants in the other row, so that the experiment was not quite fair; but not so unfair as it at first appears, for the plants in the same row were not crowded so much as seriously to interfere with each other's growth, and the ground was bare on the outside of both rows. / These plants were better nourished than those in the pots and grew to a greater height. The eight tallest plants in each row were measured in the same manner as before, with the following result:

TABLE XXXVI

Reseda lutea, growing in the open ground

Crossed plants	Self-fertilized plants
Inches	*Inches*
28	33²⁄₈
27³⁄₈	23
27⁵⁄₈	21⁵⁄₈
28⁶⁄₈	20⁴⁄₈
29⁷⁄₈	21⁵⁄₈
26⁶⁄₈	22
26²⁄₈	21²⁄₈
30¹⁄₈	21⁷⁄₈
224·75	185·13

The average height of the crossed plants, whilst in full flower, was here 28·09, and that of the self-fertilized 23·14 inches; or as 100 to 82. It is a singular fact that the tallest plant in the two rows, was one of the self-fertilized. The self-fertilized plants had smaller and paler green leaves than the crossed. All the plants in the two rows were afterwards cut down and weighed. The twenty crossed plants weighed 65 ounces, and twenty self-fertilized (by calculation from the actual weight of the thirty-two self-fertilized plants) weighed 26·25 ounces; or as 100 to 40. Therefore the crossed plants did not exceed in weight the self-fertilized plants in nearly so great a degree as those growing in the pots, owing probably to the latter having been subjected to more severe mutual competition. On the other hand, they exceeded the self-fertilized in height in a slightly greater degree.

RESEDA ODORATA

Plants of the common mignonette were raised from purchased seed, and several of them were placed under separate nets. Of these some became loaded with spontaneously self-fertilized capsules; others produced a few, and others not a single one. It must not be supposed that these latter plants produced no seed / because their stigmas did not receive any pollen, for they were repeatedly fertilized with pollen from the same plant with no effect; but they were perfectly fertile with pollen from any other plant. Spontaneously self-fertilized seeds were saved from one of the highly self-fertile plants, and other seeds were collected from the plants growing outside the nets, which had been crossed by the bees. These seeds after germinating on sand were planted in pairs on the opposite sides of five pots. The plants were trained up sticks, and measured to the summits of their leafy stems – the flower-stems not being included. We here have the result:

TABLE XXXVII
Reseda odorata (seedlings from a highly self-fertile plant)

No. of pot	Crossed plants	Self-fertilized plants
	Inches	Inches
I	$20\frac{7}{8}$	$22\frac{4}{8}$
	$34\frac{7}{8}$	$28\frac{5}{8}$
	$26\frac{6}{8}$	$23\frac{2}{8}$
	$32\frac{6}{8}$	$30\frac{4}{8}$
II	$34\frac{3}{8}$	$28\frac{5}{8}$
	$34\frac{5}{8}$	$30\frac{5}{8}$
	$11\frac{6}{8}$	23
	$33\frac{3}{8}$	$30\frac{1}{8}$
III	$17\frac{7}{8}$	$4\frac{4}{8}$
	27	25
	$30\frac{1}{8}$	$26\frac{3}{8}$
	$30\frac{2}{8}$	$25\frac{1}{8}$

TABLE XXXVII – *continued*
Reseda odorata (seedlings from a highly self-fertile plant)

No. of pot	*Crossed plants*	*Self-fertilized plants*
	Inches	*Inches*
IV	21 5/8	22 6/8
	28	25 4/8
	32 5/8	15 1/8
	32 3/8	24 6/8
V	21	11 6/8
	25 2/8	19 7/8
	26 6/8	10 4/8
Total in inches	522·25	428·50

The average height of the nineteen crossed plants is here 27·48, and that of the nineteen self-fertilized 22·55 inches; or as 100 to 82. All these plants were cut down in the early autumn / and weighed: the crossed weighed 11·5 ounces, and the self-fertilized 7·75 ounces, or as 100 to 67. These two lots having been left freely exposed to the visits of insects, did not present any difference to the eye in the number of seed-capsules which they produced.

The remainder of the same two lots of seeds were sown in two adjoining rows in the open ground; so that the plants were exposed to only moderate competition. The eight tallest on each side were measured, as shown in the following table:

TABLE XXXVIII
Reseda odorata, growing in the open ground

	Crossed plants	*Self-fertilized plants*
	Inches	*Inches*
	24 4/8	26 5/8
	27 2/8	25 7/8
	24	25
	26 6/8	28 3/8
	25	29 7/8
	26 2/8	25 7/8
	27 2/8	26 7/8
	25 1/8	28 2/8
Total in inches	206·13	216·75

The average height of the eight crossed plants is 25·76, and that of the eight self-fertilized 27·09; or as 100 to 105.

We here have the anomalous result of the self-fertilized plants being a little taller than the crossed; of which fact I can offer no explanation. It is of course possible, but not probable, that the labels may have been interchanged by accident.

Another experiment was now tried: all the self-fertilized capsules, though very few in number, were gathered from one of the semi-self-sterile plants under a net; and as several flowers on this same plant had been fertilized with pollen from a distinct individual, crossed seeds were thus obtained. I expected that the seedlings from this semi-self-sterile plant would have profited in a higher degree from a cross, than did the seedlings from the fully self-fertile plants. But my anticipation was quite wrong, for they profited in a less degree. An analogous result followed in the case of Eschscholtzia, in which the offspring of the plants of Brazilian parentage (which were partially self-sterile) did not / profit more from a cross, than did the plants of the far more self-

TABLE XXXIX
Reseda odorata (seedlings from a semi-self-sterile plant)

No. of pot	Crossed plants	Self-fertilized plants
	Inches	Inches
I	33$\frac{4}{8}$	31
	30$\frac{6}{8}$	28
	29$\frac{6}{8}$	13$\frac{2}{8}$
	20	32
II	22	21$\frac{6}{8}$
	33$\frac{4}{8}$	26$\frac{6}{8}$
	31$\frac{2}{8}$	25$\frac{2}{8}$
	32$\frac{4}{8}$	30$\frac{4}{8}$
III	30$\frac{1}{8}$	17$\frac{2}{8}$
	32$\frac{1}{8}$	29$\frac{6}{8}$
	31$\frac{4}{8}$	24$\frac{6}{8}$
	32$\frac{2}{8}$	34$\frac{2}{8}$
IV	19$\frac{1}{8}$	20$\frac{6}{8}$
	30$\frac{1}{8}$	32$\frac{6}{8}$
	24$\frac{3}{8}$	31$\frac{4}{8}$
	30$\frac{6}{8}$	36$\frac{6}{8}$
V	34$\frac{6}{8}$	24$\frac{5}{8}$
	37$\frac{1}{8}$	34
	31$\frac{2}{8}$	22$\frac{2}{8}$
	33	37$\frac{1}{8}$
Total in inches	599·75	554·25

fertile English stock. The above two lots of crossed and self-fertilized seeds from the same plant of *Reseda odorata*, after germinating on sand, were planted on opposite sides of five pots, and measured as in the last case, with the result as shown in Table XXXIX.

The average height of the twenty crossed plants is here 29·98, and that of the twenty self-fertilized 27·71 inches; or as 100 to 92. These plants were then cut down and weighed; and the crossed in this case exceeded the self-fertilized in weight by a mere trifle, viz., in the ratio of 100 to 99. The two lots, left freely exposed to insects, seemed to be equally fertile.

The remainder of the seed was sown in two adjoining rows in / the open ground; and the eight tallest plants in each row were measured, with the following result:

TABLE XL

Reseda odorata (seedlings from a semi-self-sterile plant, planted in the open ground)

	Crossed plants	*Self-fertilized plants*
	Inches	*Inches*
	28 2/8	22 3/8
	22 4/8	24 3/8
	25 7/8	23 4/8
	25 3/8	21 4/8
	29 4/8	22 5/8
	27 1/8	27 3/8
	22 4/8	27 3/8
	26 2/8	19 2/8
Total in inches	207.38	188.38

The average height of the eight crossed plants is here 25·92, and that of the eight self-fertilized plants 23·54 inches; or as 100 to 90.

IX. VIOLACEAE

VIOLA TRICOLOR

Whilst the flowers of the common cultivated heartsease are young, the anthers shed their pollen into a little semi-cylindrical passage, formed by the basal portion of the lower petal, and surrounded by papillae. The pollen thus collected lies close beneath the stigma, but can seldom gain access into its cavity, except by the aid of insects, which pass their proboscides down this passage into the nectary.[5] Consequently when I covered up a large plant of a cultivated variety, it set only eighteen capsules, and most of these contained very few good seeds – several from only one to three; whereas an equally fine / uncovered plant of the same variety, growing close by, produced 105 fine capsules. The few flowers which produce capsules when insects are excluded, are perhaps fertilized by the curling inwards of the petals as they wither, for by this means pollen-grains adhering to the papillae might be inserted into the cavity of the stigma. But it is more probable that their fertilization is effected, as Mr Bennet suggests, by Thrips and certain minute beetles which haunt the flowers, and which cannot be excluded by any net. Humble-bees are the usual fertilizers; but I have more than once seen flies (*Rhingia rostrata*) at work, with the under sides of their bodies, heads and legs dusted with pollen; and having marked the flowers which they visited, I found them after a few days fertilized.[6] It is curious for how long a time the flowers of the heartsease and of some other plants may be watched without an insect being seen to visit them. During the summer of 1841, I observed many times daily for more than a fortnight some large clumps of heartsease growing in my garden, before I saw a single humble-bee at work. During another summer I did the same, but at last saw some dark-coloured humble-bees visiting on three successive days almost every flower in several clumps; and almost all these flowers quickly withered and produced fine capsules. I presume that a certain state of the atmosphere is

[5] The flowers of this plant have been fully described by Sprengel, Hildebrand, Delpino, and H. Müller. The latter author sums up all the previous observations in his *Befruchtung der Blumen*, and in *Nature*, 20 November, 1873, p. 44. See also Mr A. W. Bennett, in *Nature*, 15 May, 1873, p. 50; and some remarks by Mr Kitchener, ibid., p. 143. The facts which follow on the effects of covering up a plant of *V. tricolor* have been quoted by Sir J. Lubbock in his *British Wild Flowers*, etc., p. 62.

[6] I should add that this fly apparently did not suck the nectar, but was attracted by the papillae which surround the stigma. H. Müller also saw a small bee, an Andrena, which could not reach the nectar, repeatedly inserting its proboscis beneath the stigma, where the papillae are situated; so that these papillae must be in some way attractive to insects. A writer asserts (*Zoologist*, vol. iii–iv, p. 1225) that a moth (Plusia) frequently visits the flowers of the pansy. Hive-bees do not ordinarily visit them, but a case has been recorded (*Gardeners' Chronicle*, 1844, p. 374) of these bees doing so. H. Müller has also seen the hive-bee at work, but only on the wild small-flowered form. He gives a list (*Nature*, 1873, p. 45) of all the insects which he has seen visiting both the large and small-flowered forms. From his account, I suspect that the flowers of plants in a state of nature are visited more frequently by insects than those of the cultivated varieties. He has seen several butterflies sucking the flowers of wild plants, and this I have never observed in gardens, though I have watched the flowers during many years.

necessary for the secretion of nectar, and that as soon as this occurs the insects discover the fact by the odour emitted, and immediately frequent the flowers.

As the flowers require the aid of insects for their complete / fertilization, and as they are not visited by insects nearly so often as most other nectar-secreting flowers, we can understand the remarkable fact discovered by H. Müller and described by him in *Nature*, namely, that this species exists under two forms. One of these bears conspicuous flowers, which, as we have seen, require the aid of insects, and are adapted to be cross-fertilized by them; whilst the other form has much smaller and less conspicuously coloured flowers, which are constructed on a slightly different plan, favouring self-fertilization, and are thus adapted to ensure the propagation of the species. The self-fertile form, however, is occasionally visited, and may be crossed by insects, though this is rather doubtful.

In my first experiments on *Viola tricolor* I was unsuccessful in raising seedlings, and obtained only one full-grown crossed and self-fertilized plant. The former was 12½ inches and the latter 8 inches in height. On the following year several flowers on a fresh plant were crossed with pollen from another plant, which was known to be a distinct seedling; an to this point it is important to attend. Several other flowers on the same plant were fertilized with their own pollen. The average number of seeds in the ten crossed capsules was 18·7, and in the twelve self-fertilized capsules 12·83; or as 100 to 69. These seeds, after germinating on bare sand, were planted in pairs on the opposite sides of five pots. They were first measured when about a third of their full size, and the crossed plants then averaged 3·87 inches, and the self-fertilized only 2·00 inches in height; or as 100 to 52. They were kept in the greenhouse, and did not grow vigorously. Whilst in flower they were again measured to the summits of their stems (see Table XLI), with the following result:

The average height of the fourteen crossed plants is here 5·58 inches, and that of the fourteen self-fertilized 2·37; or as 100 to 42. In four out of the five pots, a crossed plant flowered before any one of the self-fertilized; as likewise occurred with the pair raised during the previous year. These plants without being disturbed were now turned out of their pots and planted in the open ground, so as to form five separate clumps. Early in the following summer (1869) they flowered profusely, and being visited by humble-bees set many capsules, which were carefully collected from all the plants on both sides. The crossed plants produced 167 capsules, and the self-fertilized only 17; or as 100 to 10. So that the crossed plants were more than twice the / height of the self-fertilized, generally flowered first, and produced ten times as many naturally fertilized capsules.

TABLE XLI
Viola tricolor

No. of pot	Crossed plants	Self-fertilized plants
	Inches	Inches
I	8²⁄₈	0²⁄₈
	7¹⁄₈	2⁴⁄₈
	5	1²⁄₈
II	5	6
	4	4
	4⁴⁄₈	3¹⁄₈
III	9⁴⁄₈	3¹⁄₈
	3³⁄₈	1⁷⁄₈
	8⁴⁄₈	0⁵⁄₈
IV	4⁷⁄₈	2¹⁄₈
	4²⁄₈	1⁶⁄₈
	4	2¹⁄₈
V	6	3
	3³⁄₈	1⁴⁄₈
Total in inches	78·13	33·25

By the early part of the summer of 1870 the crossed plants in all the five clumps had grown and spread so much more than the self-fertilized, that any comparison between them was superfluous. The crossed plants were covered with a sheet of bloom, whilst only a single self-fertilized plant, which was much finer than any of its brethren, flowered. The crossed and self-fertilized plants had now grown all matted together on the respective sides of the superficial partitions still separating them; and in the clump which included the finest self-fertilized plant, I estimated that the surface covered by the crossed plants was about nine times as large as that covered by the self-fertilized plants. The extraordinary superiority of the crossed over the self-fertilized plants in all five clumps, was no doubt due to the crossed plants at first having had a decided advantage over the self-fertilized, and then robbing them more and more of their food during the succeeding seasons. But we should remember / that the same result would follow in a state of nature even to a greater degree; for my plants grew in ground kept clear of weeds, so that the self-fertilized had to compete only with the crossed plants; whereas the whole surface of the ground is naturally covered with various kinds of plants, all of which have to struggle together for existence.

The ensuing winter was very severe, and in the following spring (1871) the plants were again examined. All the self-fertilized were now dead, with the exception of a single branch on one plant, which bore on its summit a minute rosette of leaves about as large as a pea. On the other hand, all the crossed plants without exception were growing vigorously. So that the self-fertilized plants, besides their inferiority in other respects, were more tender.

Another experiment was now tried for the sake of ascertaining how far the superiority of the crossed plants, or to speak more correctly, the inferiority of the self-fertilized plants, would be transmitted to their offspring. The one crossed and one self-fertilized plant, which were first raised, had been turned out of their pot and planted in the open ground. Both produced an abundance of very fine capsules, from which fact we may safely conclude that they had been cross-fertilized by insects. Seeds from both, after germinating on sand, were planted in pairs on the opposite sides of three pots. The naturally

TABLE XLII

Viola tricolor: seedlings from crossed and self-fertilized plants, the parents of both sets having been left to be naturally fertilized

No. of pot	*Naturally crossed plants from artificially crossed plants*	*Naturally crossed plants from self-fertilized plants*
	Inches	*Inches*
I	12 1/8	9 6/8
	11 6/8	8 3/8
II	13 2/8	9 6/8
	10	11 4/8
III	14 4/8	11 1/8
	13 6/8	11 3/8
Total in inches	75·38	61·88

crossed seedlings / derived from the crossed plants flowered in all three pots before the naturally crossed seedlings derived from the self-fertilized plants. When both lots were in full flower, the two tallest plants on each side of each pot were measured, and the result is shown in the preceding table.

The average height of the six tallest plants derived from the crossed plants is 12·56 inches; and that of the six tallest plants derived from the self-fertilized plants is 10·31 inches; or as 100 to 82. We here see a considerable difference in height between the two sets, though very far from equalling that in the previous trials between the offspring from crossed and self-fertilized flowers. This difference must be attributed to the latter set of plants having inherited a weak constitution from their parents, the offspring of self-fertilized flowers; notwithstanding that the parents themselves had been freely intercrossed with other plants by the aid of insects.

X. RANUNCULACEAE

ADONIS AESTIVALIS

The results of my experiments on this plant are hardly worth giving, as I remark in my notes made at the time, 'seedlings, from some unknown cause, all miserably unhealthy'. Nor did they ever become healthy; yet I feel bound to give the present case, as it is opposed to the general results at which I have arrived. Fifteen flowers were crossed and all produced fruit, containing on an average 32·5 seeds; nineteen flowers were fertilized with their own pollen, and they likewise all yielded fruit, containing a rather larger average of 34·5 seeds; or as 100 to 106. Seedlings were raised from these seeds. In one of the pots all the self-fertilized plants died whilst quite young; in the two others, the measurements were as follows:

TABLE XLIII
Adonis aestivalis

No. of pot	Crossed plants	Self-fertilized plants
	Inches	Inches
I	14	13⁴⁄₈
	13⁴⁄₈	13⁴⁄₈
II	16²⁄₈	15²⁄₈
	13²⁄₈	15
Total in inches	57·00	57·25 /

The average height of the four crossed plants is 14·25, and that of the four self-fertilized plants 14·31; or as 100 to 100·4; so that they were in fact of equal height. According to Professor H. Hoffmann,[7] this plant is proterandrous; nevertheless it yields plenty of seeds when protected from insects.

DELPHINIUM CONSOLIDA

It has been said in the case of this plant, as of so many others, that the flowers are fertilized in the bud, and that distinct plants or varieties can never naturally intercross.[8] But this is an error, as we may infer, firstly from the flowers being proterandrous – the mature stamens bending up, one after the other, into the passage which leads to the nectary, and afterwards the mature pistils bending in the same direction; secondly, from the number of humble-bees which visit the flowers;[9] and thirdly, from the greater fertility of the flowers when crossed with pollen from a distinct plant than when spontaneously self-fertilized. In the year 1863 I enclosed a large branch in a net, and crossed five flowers with pollen from a distinct plant; these yielded capsules containing on an average 35·2 very fine seeds, with a maximum of forty-two in one capsule. Thirty-two other flowers on the same branch produced twenty-eight spontaneously self-fertilized capsules, containing on an average 17·2 seeds, with a maximum in one of thirty-six seeds. But six of these capsules were very poor, yielding only from one to five seeds; if these are excluded, the remaining twenty-two capsules give an average of 20·9 seeds, though many of these seeds were small. The fairest ratio, therefore, for the number of seeds produced by a cross and by spontaneous self-fertilization is as 100 to 59. These seeds were not sown, as I had too many other experiments in progress.

In the summer of 1867, which was a very unfavourable one, I again crossed several flowers under a net with pollen from a distinct plant, and fertilized other flowers on the same plant with their own pollen. The former yielded a much larger proportion of capsules than the latter; and many of the seeds in the self-fertilized capsules, though numerous, were so poor that an equal number of seeds from the crossed and self-fertilized capsules / were in weight as 100 to 45. The two lots were allowed to germinate on sand, and pairs were planted on the opposite sides of four pots. When nearly two-thirds grown they were measured, as shown in the following table:

[7] *Zur Speciesfrage*, 1875, p. 11.
[8] Decaisne, *Comptes-Rendus*, July, 1863, p. 5.
[9] Their structure is described by H. Müller, *Befruchtung*, etc., p. 122.

TABLE XLIV
Delphinium consolida

No. of pot	Crossed plants	Self-fertilized plants
	Inches	Inches
I	11	11
II	19	16⅞
	16⅞	11⅛
III	26	22
IV	9⅛	8⅞
	8	6⅛
Total in inches	89·75	75·50

The six crossed plants here average 14·95, and the six self-fertilized 12·50 inches in height; or as 100 to 84. When fully grown they were again measured, but from want of time only a single plant on each side was measured; so that I have thought it best to give the earlier measurements. At the later period the three tallest crossed plants still exceeded considerably in height the three tallest self-fertilized, but not in quite so great a degree as before. The pots were left uncovered in the greenhouse, but whether the flowers were intercrossed by bees or self-fertilized I do not know. The six crossed plants produced 282 mature and immature capsules, whilst the six self-fertilized plants produced only 159; or as 100 to 56. So that the crossed plants were very much more productive than the self-fertilized.

XI. CARYOPHYLLACEAE

VISCARIA OCULATA

Twelve flowers were crossed with pollen from another plant, and yielded ten capsules, containing by weight 5·77 grains of seeds. Eighteen flowers were fertilized with their own pollen and yielded twelve capsules, containing by weight 2·63 grains. Therefore the seeds from an equal number of crossed and self-fertilized / flowers would have been in weight as 100 to 38. I had previously selected a medium-sized capsule from each lot, and counted the seeds in both; the crossed one contained 284, and the self-fertilized one 126 seeds; or as 100 to 44. These seeds were sown on opposite sides of three pots, and several seedlings raised; but only the tallest flower-stem of one plant on each

115

side was measured. The three on the crossed side averaged 32·5 inches, and the three on the self-fertilized side 34 inches in height; or as 100 to 104. But this trial was on much too small a scale to be trusted; the plants also grew so unequally that one of the three flower-stems on the crossed plants was very nearly twice as tall as that on one of the others; and one of the three flower-stems on the self-fertilized plants exceeded in an equal degree one of the others.

In the following year the experiment was repeated on a larger scale: ten flowers were crossed on a new set of plants and yielded ten capsules containing by weight 6·54 grains of seed. Eighteen spontaneously self-fertilized capsules were gathered, of which two contained no seed; the other sixteen contained by weight 6·07 grains of seed. Therefore the weight of seed from an equal number of crossed and spontaneously self-fertilized flowers (instead of artificially fertilized as in the previous case) was as 100 to 58.

The seeds after germinating on sand were planted in pairs on the opposite sides of four pots, with all the remaining seeds sown crowded in the opposite sides of a fifth pot; in this latter pot only the tallest plant on each side was measured. Until the seedlings had grown about 5 inches in height no difference could be perceived in the two lots. Both lots flowered at nearly the same time. When they had almost done flowering, the tallest flower-stem on each plant was measured, as shown in the following table (XLV).

The fifteen crossed plants here average 34·5, and the fifteen self-fertilized 33·55 inches in height; or as 100 to 97. So that the excess of height of the crossed plants is quite insignificant. In productiveness, however, the difference was much more plainly marked. All the capsules were gathered from both lots of plants (except from the crowded and unproductive ones in Pot V), and at the close of the season the few remaining flowers were added in. The fourteen crossed plants produced 381, whilst the fourteen self-fertilized plants produced only 293 capsules and flowers, or as 100 to 77. /

DIANTHUS CARYOPHYLLUS

The common carnation is strongly proterandrous, and therefore depends to a large extent upon insects for fertilization. I have seen only humble-bees visiting the flowers, but I dare say other insects likewise do

TABLE XLV
Viscaria oculata

No. of pot	*Crossed plants*	*Self-fertilized plants*
	Inches	*Inches*
I	19	32 3/8
	33	38
	41	38
	41	28 7/8
II	37 4/8	36
	36 4/8	32 3/8
	38	35 6/8
III	44 4/8	36
	39 4/8	20 7/8
	39	30 5/8
IV	30 2/8	36
	31	39
	33 1/8	29
	24	38 4/8
V Crowded	30 2/8	32
Total in inches	517·63	503·38

so. It is notorious that if pure seed is desired, the greatest care is necessary[10] to prevent the varieties which grow in the same garden from intercrossing. The pollen is generally shed and lost before the two stigmas in the same flower diverge and are ready to be fertilized. I was therefore often forced to use for self-fertilization pollen from the same plant instead of from the same flower. But on two occasions, when I attended to this point, I was not able to detect any marked difference in the number of seeds produced by these two forms of self-fertilization. /

Several single-flowered carnations were planted in good soil, and were all covered with a net. Eight flowers were crossed with pollen from a distinct plant and yielded six capsules, containing on an average 88·6 seeds, with a maximum in one of 112 seeds. Eight other flowers were self-fertilized in the manner above described, and yielded seven capsules containing on an average 82 seeds, with a maximum in one of 112 seeds. So that there was very little difference in the number of seeds produced by cross-fertilization and self-fertilization, viz., as

[10] *Gardeners' Chronicle*, 1847, p. 268

100 to 92. As these plants were covered by a net, they produced spontaneously only a few capsules containing any seeds, and these few may perhaps be attributed to the action of Thrips and other minute insects which haunt the flowers. A large majority of the spontaneously self-fertilized capsules produced by several plants contained no seeds, or only a single one. Excluding these latter capsules, I counted the seeds in eighteen of the finest ones, and these contained on an average 18 seeds. One of the plants was spontaneously self-fertile in a higher degree than any of the others. On another occasion a single covered-up plant produced spontaneously eighteen capsules, but only two of these contained any seed, namely 10 and 15.

Crossed and self-fertilized plants of the first generation. The many seeds obtained from the above crossed and artificially self-fertilized flowers were sown out of doors, and two large beds of seedlings, closely adjoining one another, thus raised. This was the first plant on which I experimented, and I had not then formed any regular scheme of operation. When the two lots were in full flower, I measured roughly a large number of plants, but record only that the crossed were on an average fully 4 inches taller than the self-fertilized. Judging from subsequent measurements, we may assume that the crossed plants were about 28 inches, and the self-fertilized about 24 inches in height; and this will give us a ratio of 100 to 86. Out of a large number of plants four of the crossed ones flowered before any one of the self-fertilized plants.

Thirty flowers on these crossed plants of the first generation were again crossed with pollen from a distinct plant of the same lot, and yielded twenty-nine capsules, containing on an average 55·62 seeds, with a maximum in one of 110 seeds.

Thirty flowers on the self-fertilized plants were again self-fertilized; eight of them with pollen from the same flower, and the remainder with pollen from another flower on the same / plant; and these produced twenty-two capsules, containing on an average 35·95 seeds, with a maximum in one of 61 seeds. We thus see, judging by the number of seeds per capsule, that the crossed plants again crossed were more productive than the self-fertilized again self-fertilized, in the ratio of 100 to 65. Both the crossed and self-fertilized plants, from having grown much crowded in the two beds, produced less fine capsules and fewer seeds than did their parents.

Crossed and self-fertilized plants of the second generation. The crossed and self-fertilized seeds from the crossed and self-fertilized plants of the last generation were sown on opposite sides of two pots; but the seedlings were not thinned enough, so that both lots grew irregularly, and most of the self-fertilized plants after a time died from being smothered. My measurements were, therefore, very incomplete. From the first the crossed seedlings appeared the finest, and when they were on an average, by estimation, 5 inches high, the self-fertilized plants were only 4 inches. In both pots the crossed plants flowered first. The two tallest flower-stems on the crossed plants in the two pots were 17 and 16½ inches in height; and the two tallest flower-stems on the self-fertilized plants 10½ and 9 inches; so that their heights were as 100 to 58. But this ratio, deduced from only two pairs, obviously is not in the least trustworthy, and would not have been given had it not been otherwise supported. I state in my notes that the crossed plants were very much more luxuriant than their opponents, and seemed to be twice as bulky. This latter estimate may be believed from the ascertained weights of the two lots in the next generation. Some flowers on these crossed plants were again crossed with pollen from another plant of the same lot, and some flowers on the self-fertilized plants again self-fertilized; and from the seeds thus obtained the plants of the next generation were raised.

Crossed and self-fertilized plants of the third generation. The seeds just alluded to were allowed to germinate on bare sand, and were planted in pairs on the opposite sides of four pots. When the seedlings were in full flower, the tallest stem on each plant was measured to the base of the calyx. The measurements are given in the following table (XLVI). In Pot I the crossed and self-fertilized plants flowered at the same time; but in the other three pots the crossed flowered first. These latter plants also continued flowering much later in the autumn than the self-fertilized. /

The average height of the eight crossed plants is here 28·39 inches, and of the eight self-fertilized 28·21; or as 100 to 99. So that there was no difference in height worth speaking of; but in general vigour and luxuriance there was an astonishing difference, as shown by their weights. After the seed-capsules had been gathered, the eight crossed and the eight self-fertilized plants were cut down and weighed; the former weighed 43 ounces, and the latter only 21 ounces; or as 100 to 49.

TABLE XLVI
Dianthus caryophyllus (third generation)

No. of pot	*Crossed plants*	*Self-fertilized plants*
	Inches	*Inches*
I	28 6/8	30
	27 3/8	26
II	29	30 7/8
	29 4/8	27 4/8
III	28 4/8	31 6/8
	23 4/8	24 5/8
IV	27	30
	33 4/8	25
Total in inches	227·13	225·75

These plants were all kept under a net, so that the capsules which they produced must have been all spontaneously self-fertilized.The eight crossed plants produced twenty-one such capsules, of which only twelve contained any seed, averaging 8·5 per capsule. On the other hand, the eight self-fertilized plants produced no less than thirty-six capsules, of which I examined twenty-five, and, with the exception of three, all contained seeds, averaging 10·63 seeds per capsule. Thus the proportional number of seeds per capsule produced by the plants of crossed origin to those produced by the plants of self-fertilized origin (both lots being spontaneously self-fertilized) was as 100 to 125. This anomalous result is probably due to some of the self-fertilized plants having varied so as to mature their pollen and stigmas more nearly at the same time than is proper to the species; and we have already seen that some plants in the / first experiment differed from the others in being slightly more self-fertile.

The effects of a cross with a fresh stock. Twenty flowers on the self-fertilized plants of the last or third generation, in Table XLVI, were fertilized with their own pollen, but taken from other flowers on the same plants. These produced fifteen capsules, which contained (omitting two with only three and six seeds) on an average 47·23 seeds, with a maximum of seventy in one. The self-fertilized capsules from the self-fertilized plants of the first generation yielded the much lower average of 35·95 seeds; but as these latter plants grew extremely crowded, nothing can be inferred with respect to this difference in their self-fertility. The seedlings raised from the above seeds constitute the plants of the fourth self-fertilized generation in the following table (XLVII).

Twelve flowers on the same plants of the third self-fertilized generation, in Table XLVI, were crossed with pollen from the crossed plants in the same table. These crossed plants had been intercrossed for the three previous generations; and many of them, no doubt, were more or less closely inter-related, but not so closely as in some of the experiments with other species; for several carnation plants had been raised and crossed in the earlier generations. They were not related, or only in a distant degree, to the self-fertilized plants. The parents of both the self-fertilized and crossed plants had been subjected to as nearly as possible the same conditions during the three previous generations. The above twelve flowers produced ten capsules, containing on an average 48·66 seeds, with a maximum in one of seventy-two seeds. The plants raised from these seeds may be called the *intercrossed*.

Lastly, twelve flowers on the same self-fertilized plants of the third generation were crossed with pollen from plants which had been raised from seeds purchased in London. It is almost certain that the plants which produced these seeds had grown under very different conditions to those to which my self-fertilized and crossed plants had been subjected; and they were in no degree related. The above twelve flowers thus crossed all produced capsules, but these contained the low average of 37·41 seeds per capsule, with a maximum in one of sixty-four seeds. It is surprising that this cross with a fresh stock did not give a much higher average number of seeds; for, as we shall immediately see, the plants raised from these seeds, which may / be called the *London-crossed*, benefited greatly by the cross, both in growth and fertility.

The above three lots of seeds were allowed to germinate on bare sand. Many of the London-crossed germinated before the others, and were rejected; and many of the intercrossed later than those of the other two lots. The seeds after thus germinating were planted in ten pots, made tripartite by superficial / divisions; but when only two kinds of seeds germinated at the same time, they were planted on the opposite sides of other pots; and this is indicated by blank spaces in one of the three columns in Table XLVII. An o in the table signifies that the seedling died before it was measured; and a + signifies that the plant did not produce a flower-stem, and therefore was not measured. It deserves notice that no less than eight out of the eighteen

TABLE XLVII
Dianthus caryophyllus

No. of pot	*London-crossed plants*	*Intercrossed plants*	*Self-fertilized plants*
	Inches	*Inches*	*Inches*
I	$39\frac{5}{8}$	$25\frac{1}{8}$	$29\frac{2}{8}$
	$30\frac{7}{8}$	$21\frac{6}{8}$	+
II	$36\frac{2}{8}$		$22\frac{3}{8}$
	0		+
III	$28\frac{5}{8}$	$30\frac{2}{8}$	
	+	$23\frac{1}{8}$	
IV	$34\frac{4}{8}$	$35\frac{5}{8}$	30
	$28\frac{7}{8}$	32	$24\frac{4}{8}$
V	28	$34\frac{4}{8}$	+
	0	$24\frac{2}{8}$	+
VI	$32\frac{5}{8}$	$24\frac{7}{8}$	$30\frac{3}{8}$
	31	26	$24\frac{4}{8}$
VII	$41\frac{7}{8}$	$29\frac{7}{8}$	$27\frac{7}{8}$
	$34\frac{7}{8}$	$26\frac{4}{8}$	27
VIII	$34\frac{5}{8}$	29	$26\frac{6}{8}$
	$28\frac{5}{8}$	0	+
IX	$25\frac{5}{8}$	$28\frac{5}{8}$	+
	0	+	0
X	38	$28\frac{4}{8}$	$22\frac{7}{8}$
	$32\frac{1}{8}$	+	0
Total in inches	525·13	420·00	265·50

self-fertilized plants either died or did not flower; whereas only three out of the eighteen intercrossed, and four out of the twenty London-crossed plants had a decidedly less vigorous appearance than the plants of the other two lots, their leaves being smaller and narrower. In only one pot did a self-fertilized plant flower before one of the two kinds of crossed plants, between which there was no marked difference in the period of flowering. The plants were measured to the base of the calyx, after they had completed their growth, late in the autumn.

The average height of the sixteen London-crossed plants in the preceding table is 32·82 inches; that of the fifteen intercrossed plants, 28 inches; and that of the ten self-fertilized plants, 26·55.

So that in height we have the following ratios:

The London-crossed to the self-fertilized	as 100 to 81
The London-crossed to the intercrossed	as 100 to 85
The Intercrossed to the self-fertilized	as 100 to 95

These three lots of plants, which it should be remembered were all derived on the mother-side from plants of the third self-fertilized generation, fertilized in three different ways, were left exposed to the visits of insects, and their flowers were freely crossed by them. As the capsules of each lot became ripe they were gathered and kept separate, the empty of bad ones being thrown away. But towards the middle of October, when the capsules could no longer ripen, all were gathered and were counted, whether good or bad. The capsules were then crushed, and the seed cleaned by sieves and weighed. For the sake of uniformity the results are given from calculation, as if there had been twenty plants in each lot.

The sixteen London-crossed plants actually produced 286 capsules; therefore twenty such plants would have produced 357·5 capsules; and from the actual weight of the seeds, the twenty plants would have yielded 462 grains weight of seeds. /

The fifteen intercrossed plants actually produced 157 capsules; therefore twenty of them would have produced 209·3 capsules, and the seeds would have weighed 208·48 grains.

The ten self-fertilized plants actually produced 70 capsules; therefore twenty of them would have produced 140 capsules; and the seeds would have weighed 153·2 grains.

From these data we get the following ratios:

Number of capsules produced by an equal number of plants of the three lots

	Number of capsules
The London-crossed to the self-fertilized,	as 100 to 39
The London-crossed to the intercrossed,	as 100 to 45
The Intercrossed to the self-fertilized	as 100 to 67

Weight of seeds produced by an equal number of plants of the three lots

	Weight of seed
The London-crossed to the self-fertilized	as 100 to 33
The London-crossed to the intercrossed,	as 100 to 45
The Intercrossed to the self-fertilized,	as 100 to 73

We thus see how greatly the offspring from the self-fertilized plants of the third generation crossed by a fresh stock, had their fertility increased, whether tested by the number of capsules produced or by the weight of the contained seeds; this latter being the more trustworthy method. Even the offspring from the self-fertilized plants crossed by one of the crossed plants of the same stock, notwithstanding that both lots had been long subjected to the same conditions, had their fertility considerably increased, as tested by the same two methods.

In conclusion it may be well to repeat in reference to the fertility of these three lots of plants, that their flowers were left freely exposed to the visits of insects and were undoubtedly crossed by them, as may be inferred from the large number of good capsules produced. These plants were all the offspring of the same mother-plants, and the strongly marked difference in their fertility must be attributed to the nature of the pollen employed in fertilizing their parents; and the difference in the nature of the pollen must be attributed to the different treatment to which the pollen-bearing parents had been subjected during several previous generations.

Colour of the flowers. The flowers produced by the self-fertilized / plants of the last or fourth generation were as uniform in tint as those of a wild species, being of a pale pink or rose colour. Analogous cases with Mimulus and Ipomoea, after several generations of self-fertilization, have already been given. The flowers of the intercrossed plants of the fourth generation were likewise nearly uniform in colour. On the other hand, the flowers of the London-crossed plants, or those raised from a cross with the fresh stock which bore dark crimson flowers, varied extremely in colour, as might have been expected, and as is the general rule with seedling carnations. It deserves notice that only two or three of the London-crossed plants produced dark crimson flowers like those of their fathers, and only a very few of a pale pink like those of their mothers. The great majority had their petals longitudinally and variously striped with the two colours – the groundwork tint being, however, in some cases darker than that of the mother-plants.

XII. MALVACEAE

HIBISCUS AFRICANUS

Many flowers on this Hibiscus were crossed with pollen from a distinct plant, and many others were self-fertilized. A rather larger proportional number of the crossed than of the self-fertilized flowers yielded capsules, and the crossed capsules contained rather more seeds. The self-fertilized seeds were a little heavier than an equal number of the crossed seeds, but they germinated badly, and I raised only four plants of each lot. In three out of the four pots, the crossed plants flowered first.

TABLE XLVIII
Hibiscus africanus

No. of pot	Crossed plants	Self-fertilized plants
	Inches	Inches
I	13⅜	16⅝
II	14	14
III	8	7
IV	17⅞	20⅞
Total in inches	53.00	57·75 /

The four crossed plants average 13·25, and the four self-fertilized 14·43 inches in height; or as 100 to 109. Here we have the unusual case of self-fertilized plants exceeding the crossed in height; but only four pairs were measured, and these did not grow well or equally. I did not compare the fertility of the two lots. /

CHAPTER V

GERANIACEAE, LEGUMINOSAE, ONAGRACEAE, ETC.

Pelargonium zonale, a cross between plants propagated by cuttings does no good – *Tropaeolum minus* – *Limnanthes douglasii* – *Lupinus luteus* and *pilosus* – *Phaseolus multiflorus* and *vulgaris* – *Lathyrus odoratus*, varieties of, never naturally intercross in England – *Pisum sativum*, varieties of, rarely intercross, but a cross between them highly beneficial – *Sarothamnus scoparius*, wonderful effects of a cross – *Ononis minutissima*, cleistogene flowers of – Summary on the Leguminosae – *Clarkia elegans* – *Bartonia aurea* – *Passiflora gracilis* – *Apium petroselinum* – *Scabiosa atropurpurea* – *Lactuca sativa* – *Specularia speculum* – *Lobelia ramosa*, advantages of a cross during two generations – *Lobelia fulgens* – *Nemophila insignis*, great advantages of a cross – *Borago officinalis* – *Nolana prostrata*.

XIII. GERANIACEAE

PELARGONIUM ZONALE

This plant, as a general rule, is strongly proterandrous,[1] and is therefore adapted for cross-fertilization by the aid of insects. Some flowers on a common scarlet variety were self-fertilized, and other flowers were crossed with pollen from another plant; but no sooner had I done so, than I remembered that these plants had been propagated by cuttings from the same stock, and were therefore parts in a strict sense of the same individual. Nevertheless, having made the cross I resolved to save the seeds, which, after germinating on sand, were planted on the opposite / sides of three pots. In one pot the quasi-

---

[1] Mr J. Denny, a great raiser of new varieties of pelargoniums, after stating that this species is proterandrous, adds (*The Florist and Pomologist*, January, 1872, p. 11) 'there are some varieties, especially those with petals of a pink colour, or which possess a weakly constitution, where the pistil expands as soon as or even before the pollen-bag bursts, and in which also the pistil is frequently short, so when it expands it is smothered as it were by the bursting anthers; these varieties are great seeders, each pip being fertilized by its own pollen. I would instance Christine as an example of this fact.' We have here an interesting case of variability in an important functional point.

crossed plant was very soon and ever afterwards taller and finer than the self-fertilized. In the two other pots the seedlings on both sides were for a time exactly equal; but when the self-fertilized plants were about 10 inches in height, they surpassed their antagonists by a little, and ever afterwards showed a more decided and increasing advantage; so that the self-fertilized plants, taken altogether, were somewhat superior to the quasi-crossed plants. In this case, as in that of the Origanum, if individuals which have been asexually propagated from the same stock, and which have been long subjected to the same conditions, are crossed, no advantage whatever is gained.

Several flowers on another plant of the same variety were fertilized with pollen from the younger flowers on the same plant, so as to avoid using the old and long-shed pollen from the same flower, as I thought that this latter might be less efficient than fresh pollen. Other flowers on the same plant were crossed with fresh pollen from a plant which, although closely similar, was known to have arisen as a distinct seedling. The self-fertilized seeds germinated rather before the others; but as soon as I got equal pairs they were planted on the opposite sides of four pots. When the two lots of seedlings were between 4 and 5 inches in

TABLE XLIX
Pelargonium zonale

No. of pot	Crossed plants	Self-fertilized plants
	Inches	*Inches*
I	$22\frac{3}{8}$	$25\frac{5}{8}$
	$19\frac{6}{8}$	$12\frac{4}{8}$
II	15	$19\frac{6}{8}$
	$12\frac{2}{8}$	$22\frac{3}{8}$
III	$30\frac{5}{8}$	$19\frac{4}{8}$
	$18\frac{4}{8}$	$7\frac{4}{8}$
IV	38	$9\frac{1}{8}$
Total in inches	156·50	116·38

in height they were equal, excepting in Pot IV, in which the crossed plant was much the tallest. When between 11 and 14 inches in height, they were measured to the tips of their uppermost / leaves; the crossed averaged 13·46, and the self-fertilized 11·07 inches in height, or as 100 to 82. Five months later they were again measured in the same manner, and the results are given in the preceding table.

The seven crossed plants now averaged 22·35, and the seven self-fertilized 16·62 inches in height, or as 100 to 74. But from the great inequality of the several plants, the result is less trustworthy than in most other cases. In Pot II the two self-fertilized plants always had an advantage, except whilst quite young, over the two crossed plants.

As I wished to ascertain how these plants would behave during a second growth, they were cut down close to the ground whilst growing freely. The crossed plants now showed their superiority in another way, for only one out of the seven was killed by the operation, whilst three of the self-fertilized plants never recovered. There was, there-fore, no use in keeping any of the plants excepting those in Pots I and III; and in the following year the crossed plants in these two pots showed during their second growth nearly the same relative superiority over the self-fertilized plants as before.

TROPAEOLUM MINUS

The flowers are proterandrous, and are manifestly adapted for cross-fertilization by insects, as shown by Sprengel and Delpino. Twelve flowers on some plants growing out of doors were crossed with pollen from a distinct plant and produced eleven capsules, containing altogether twenty-four good seeds. Eighteen flowers were fertilized with their own pollen and produced only eleven capsules, containing twenty-two good seeds; so that a much larger proportion of the crossed than of the self-fertilized flowers produced capsules, and the crossed capsules contained rather more seed than the self-fertilized in the ratio of 100 to 92. The seeds from the self-fertilized capsules were however the heavier of the two, in the ratio of 100 to 87.

Seeds in an equal state of germination were planted on the opposite sides of four pots, but only the two tallest plants on each side of each pot were measured to the tops of their stems. The pots were placed in the greenhouse, and the plants trained up sticks, so that they ascended to an unusual height. In three of the pots the crossed plants flowered first, but in the fourth at the same time with the self-fertilized. When the seedlings were between 6 and 7 inches in height, the crossed began to / show a slight advantage over their opponents. When grown to a considerable height the eight tallest crossed plants averaged 44·43 and the eight tallest self-fertilized plants 37·34 inches, or as 100 to 84.

When their growth was completed they were again measured, as shown in the following table:

TABLE L
Tropaeolum minus

No. of pot	Crossed plants	Self-fertilized plants
	Inches	Inches
I	65	31
	50	45
II	69	42
	35	45
III	70	50⁴⁄₈
	59⁴⁄₈	55⁴⁄₈
IV	61⁴⁄₈	37⁴⁄₈
	57⁴⁄₈	61⁴⁄₈
Total in inches	467·5	368·0

The eight tallest crossed plants now averaged 58·43, and the eight tallest self-fertilized plants 46 inches in height, or as 100 to 79.

There was also a great difference in the fertility of the two lots which were left uncovered in the greenhouse. On the 17th of September the capsules from all the plants were gathered, and the seeds counted. The crossed plants yielded 243, whilst the same number of self-fertilized plants yielded only 155 seeds, or as 100 to 64.

LIMNANTHES DOUGLASII

Several flowers were crossed and self-fertilized in the usual manner, but there was no marked difference in the number of seeds which they yielded. A vast number of spontaneously self-fertilized capsules were also produced under the net. Seedlings were raised in five pots from the above seeds, and when the crossed were about 3 inches in height they showed a slight advantage over the self-fertilized. When double this height, the / sixteen crossed and sixteen self-fertilized plants were measured to the tips of their leaves; the former averaged 7·3 inches, and the self-fertilized 6·07 inches in height, or as 100 to 83. In all the pots, excepting IV, a crossed plant flowered before any one of the self-fertilized plants. The plants, when fully grown, were again measured to the summits of their ripe capsules, with the following result:

129

TABLE LI
Limnanthes douglasii

No. of pot	Crossed plants	Self-fertilized plants
	Inches	Inches
I	17 7/8	15 1/8
	17 6/8	16 4/8
	13	11
II	20	14 4/8
	22	15 6/8
	21	16 1/8
	18 4/8	17
III	15 6/8	11 4/8
	17 2/8	10 4/8
	14	0
IV	20 4/8	13 4/8
	14	13
	18	12 2/8
V	17	14 2/8
	18 5/8	14 1/8
	14 2/8	12 5/8
Total in inches	279·50	207·75

The sixteen crossed plants now averaged 17·46, and the fifteen (for one had died) self-fertilized plants 13·85 inches in height, or as 100 to 79. Mr Galton considers that a higher ratio would be fairer, viz., 100 to 76. He made a graphical representation of the above measurements, and adds the words 'very good' to the curvature thus formed. Both lots of plants produced an abundance of seed-capsules, and, as far as could be judged by the eye, there was no difference in their fertility. /

XIV. LEGUMINOSAE

In this family I experimented on the following six genera, Lupinus, Phaseolus, Lathyrus, Pisum, Sarothamnus, and Ononis.

LUPINUS LUTEUS[2]

A few flowers were crossed with pollen from a distinct plant, but owing to the unfavourable season only two crossed seeds were produced.

[2] The structure of the flowers of this plant, and their manner of fertilization, have

Nine seeds were saved from flowers spontaneously self-fertilized under a net, on the same plant which yielded the two crossed seeds. One of these crossed seeds was sown in a pot with two self-fertilized seeds on the opposite side; the latter came up between two and three days before the crossed seed. The second crossed seed was sown in like manner with two self-fertilized seeds on the opposite side; these latter also came up about a day before the crossed one. In both pots, therefore, the crossed seedlings from germinating later, were at first completely beaten by the self-fertilized; nevertheless, this state of things was afterwards completely reversed. The seeds were sown late in the autumn, and the pots, which were much too small, were kept in the greenhouse. The plants in consequence grew badly, and the self-fertilized suffered most in both pots. The two crossed plants when in flower during the following spring were 9 inches in height; one of the self-fertilized plants was 8, and the three others only 3 inches in height, being thus mere dwarfs. The two crossed plants produced thirteen pods, whilst the four self-fertilized plants produced only a single one. Some other self-fertilized plants which had been raised separately in larger pots produced several spontaneously self-fertilized pods under a net, and seeds from these were used in the following experiment.

Crossed and self-fertilized plants of the second generation. The / spontaneously self-fertilized seeds just mentioned, and crossed seeds obtained by intercrossing the two crossed plants of the last generation, after germinating on sand, were planted in pairs on the opposite sides of three large pots. When the seedlings were only 4 inches in height, the crossed had a slight advantage over their opponents. When grown to their full height, every one of the crossed plants exceeded its opponent in height. Nevertheless the self-fertilized plants in all three pots flowered before the crossed! The measurements are given in the following table:

---

been described by H. Müller, *Befruchtung*, etc., p. 243. The flowers do not secrete free nectar, and bees generally visit them for their pollen. Mr Farrer, however, remarks (*Nature*, 1872, p. 499) that 'there is a cavity at the back and base of the vexillum, in which I have not been able to find nectar. But the bees, which constantly visit these flowers, certainly go to this cavity for what they want, and not to the staminal tube.'

TABLE LII
Lupinus luteus

No. of pot	*Crossed plants*	*Self-fertilized plants*
	Inches	*Inches*
I	33 2/8	24 4/8
	30 4/8	18 4/8
	30	28
II	29 4/8	26
	30	25
III	30 4/8	28
	31	27 2/8
	31 4/8	24 4/8
Total in inches	246·25	201·75

The eight crossed plants here average 30·78, and the eight self-fertilized 25·21 inches in height; or as 100 to 82. These plants were left uncovered in the greenhouse to set their pods, but they produced very few good ones, perhaps in part owing to few bees visiting them. The crossed plants produced nine pods, containing on an average 3·4 seeds, and the self-fertilized plants seven pods, containing on an average 3 seeds, so that the seeds from an equal number of plants were as 100 to 88.

Two other crossed seedlings, each with two self-fertilized seedlings on the opposite sides of the same large pot, were turned out of their pots early in the season, without being disturbed, into open ground of good quality. They were thus subjected to but little competition with one another, in comparison with the plants in the above three pots. In the autumn / the two crossed plants were about 3 inches taller than the four self-fertilized plants; they looked also more vigorous and produced many more pods.

Two other crossed and self-fertilized seeds of the same lot, after germinating on sand, were planted on the opposite sides of a large pot, in which a Calceolaria had long been growing, and were therefore exposed to unfavourable conditions; the two crossed plants ultimately attained a height of 20½ and 20 inches, whilst the two self-fertilized were only 18 and 9½ inches high.

LUPINUS PILOSUS

From a series of accidents I was again unfortunate in obtaining a sufficient number of crossed seedlings; and the following results not be worth giving, did they not strictly accord with those just given with respect to *L. luteus*. I raised at first only a single crossed seedling, which was placed in competition with two self-fertilized ones on the opposite side of the same pot. These plants, without being disturbed, were soon afterwards turned into the open ground. By the autumn the crossed plant had grown to so large a size that it almost smothered the two self-fertilized plants, which were mere dwarfs; and the latter died without maturing a single pod. Several self-fertilized seeds had been planted at the same time separately in the open ground; and the two tallest of these were 33 and 32 inches, whereas the one crossed plant was 38 inches in height. This latter plant also produced many more pods than did any one of the self-fertilized plants, although growing separately. A few flowers on the one crossed plant were crossed with pollen from one of the self-fertilized plants, for I had no other crossed plant from which to obtain pollen. One of the self-fertilized plants having been covered by a net produced plenty of spontaneously self-fertilized pods.

Crossed and self-fertilized plants of the second generation. From crossed and self-fertilized seeds obtained in the manner just described, I succeeded in raising to maturity only a pair of plants, which were kept in a pot in the greenhouse. The crossed plants grew to a height of 33 inches, and the self-fertilized to that of 26½ inches. The former produced, whilst still kept in the greenhouse, eight pods, containing on an average 2·77 seeds; and the latter only two pods, containing on an average 2·5 seeds. The average height of the two crossed plants of the two / generations taken together was 35·5, and that of the three self-fertilized plants of the same two generations 30·5; or as 100 to 86.[3]

PHASEOLUS MULTIFLORUS

This plant, the scarlet-runner of English gardeners and the *P. coccineus* of Lamarck, originally came from Mexico, as I am informed by Mr

---

[3] We here see that both *Lupinus luteus* and *pilosus* seed freely when insects are excluded; but Mr Swale, of Christchurch, in New Zealand, informs me (see *Gardeners' Chronicle*, 1858, p. 828) that the garden varieties of the lupine are not there visited by any bees, and that they seed less freely than any other introduced leguminous plant, with the exception of red clover. He adds, 'I have, for amusement, during the summer, released the stamens with a pin, and a pod of seed has always rewarded me for my trouble, the adjoining flowers not so served having all proved blind.' I do not know to what species this statement refers.

Bentham. The flowers are so constructed that hive and humble-bees, which visit them incessantly, almost always alight on the left wing-petal, as they can best suck the nectar from this side. Their weight and movements depress the petal, and this causes the stigma to protrude from the spirally-wound keel, and a brush of hairs round the stigma pushes out the pollen before it. The pollen adheres to the head or proboscis of the bee which is at work, and is thus placed either on the stigma of the same flower, or is carried to another flower.[4] Several years ago I covered some plants under a large net, and these produced on one occasion about one-third, and on another occasion about one-eighth, of the number of pods which the same number of uncovered plants growing close alongside produced.[5] This lessened fertility was not caused by any injury from the net, as I moved the wing-petals of several protected flowers, in the same manner as bees do, and these produced remarkably / fine pods. When the net was taken off, the flowers were immediately visited by bees, and it was interesting to observe how quickly the plants became covered with young pods. As the flowers are much frequented by Thrips, the self-fertilization of most of the flowers under the net may have been due to the action of these minute insects. Dr Ogle likewise covered up a large portion of a plant, and 'out of a vast number of blossoms thus protected not a single one produced a pod, while the unprotected blossoms were for the most part fruitful'. Mr Belt gives a more curious case; this plant grows well and flowers in Nicaragua; but as none of the native bees visit the flowers, not a single pod is ever produced.[6]

From the facts now given we may feel nearly sure that individuals of the same variety or of different varieties, if growing near each other and in flower at the same time, would intercross; but I cannot myself

[4] The flowers have been described by Delpino, and in an admirable manner by Mr Farrer in the *Annals and Mag. of Nat. Hist.*, vol. ii (4th series), October, 1868, p. 256. My son Francis has explained (*Nature*, 8 January, 1874, p. 189) the use of one peculiarity in their structure, namely, a little vertical projection on the single free stamen near its base, which seems placed as if to guard the entrance into the two nectar-holes in the staminal sheath. He shows that this projection prevents the bees reaching the nectar, unless they go to the left side of the flower, and it is absolutely necessary for cross-fertilization that they should alight on the left wing-petal.
[5] *Gardeners' Chronicle*, 1857, p. 725, and more especially ibid., 1858, p. 282. Also *Annals and Mag. of Nat. Hist.* (3rd series), vol. ii, 1858, p. 462.
[6] Dr Ogle, *Pop. Science Review*, 1870, p. 168. Mr Belt, *The Naturalist in Nicaragua*, 1874, p. 70. The latter author gives a case (*Nature*, 1875, p. 26) of a late crop of *P. multiflorus* near London, which 'was rendered barren' by the humble-bees cutting, as they frequently do, holes at the bases of the flowers instead of entering them in the proper manner.

advance any direct evidence of such an occurrence, as only a single variety is commonly cultivated in England. I have, however, received an account from the Rev. W. A. Leighton, that plants raised by him from ordinary seed produced seeds differing in an extraordinary manner in colour and shape, leading to the belief that their parents must have been crossed. In France M. Fermond more than once planted close together varieties which ordinarily come true and which bear differently coloured flowers and seeds; and the offspring thus raised varied so greatly that there could hardly be a doubt that they had intercrossed.[7] On the other hand, Professor H. Hoffmann[8] does not believe in the natural crossing of the varieties; for although seedlings raised from two varieties growing close together produced plants which yielded seeds of a mixed character, he found that this likewise occurred with plants separated by a space of from 40 to 150 paces from any other variety; he therefore attributes the mixed character of the seed to spontaneous / variability. But the above distance would be very far from sufficient to prevent intercrossing: cabbages have been known to cross at several times this distance; and the careful Gärtner[9] gives many instances of plants growing at from 600 to 800 yards apart fertilizing one another. Professor Hoffmann even maintains that the flowers of the kidney-bean are specially adapted for self-fertilization. He enclosed several flowers in bags; and as the buds often dropped off, he attributes the partial sterility of these flowers to the injurious effects of the bags, and not to the exclusion of insects. But the only safe method of experimenting is to cover up a whole plant, which then never suffers.

Self-fertilized seeds were obtained by moving up and down in the same manner as bees do the wing-petals of flowers protected by a net; and crossed seeds were obtained by crossing two of the plants under the same net. The seeds after germinating on sand were planted on the opposite sides of two large pots, and equal-sized sticks were given them to twine up. When 8 inches in height, the plants on the two sides were equal. The crossed plants flowered before the self-fertilized in both pots. As soon as one of each pair had grown to the summit of its stick both were measured.

[7] *Fécondation chez les Végétaux*, 1859, pp. 34–40. He adds that M. Villiers has described a spontaneous hybrid, which he calls *P. coccineus hybridus*, in the *Annales de la Soc. R. de Horticulture*, June, 1844.
[8] *Bestimmung des Werthes von Species und Varietät*, 1869, pp. 47–72.
[9] *Kenntniss der Befruchtung*, 1844, pp. 573, 577.

TABLE LIII
Phaseolus multiflorus

No. of pot	Crossed plants	Self-fertilized plants
	Inches	Inches
I	87	84⁶⁄₈
	88	87
	82⁴⁄₈	76
II	90	76⁴⁄₈
	82⁴⁄₈	87⁴⁄₈
Total in inches	430.00	411.75

The average height of the five crossed plants is 86 inches, and that of the five self-fertilized plants 82·35; or as 100 to 96. The pots were kept in the greenhouse, and there was little or no difference in the fertility of the two lots. Therefore as far as these few observations serve, the advantage gained by a cross is very small. /

PHASEOLUS VULGARIS

With respect to this species, I merely ascertained that the flowers were highly fertile when insects were excluded, as indeed must be the case, for the plants are often forced during the winter when no insects are present. Some plants of two varieties (viz., Canterbury and Fulmer's Forcing Bean) were covered with a net, and they seemed to produce as many pods, containing as many beans, as some uncovered plants growing alongside; but neither the pods nor the beans were actually counted. This difference in self-fertility between *P. vulgaris* and *multiflorus* is remarkable, as these two species are so closely related that Linnaeus thought that they formed one. When the varieties of *P. vulgaris* grow near one another in the open ground, they sometimes cross largely, notwithstanding their capacity for self-fertilization. Mr Coe has given me a remarkable instance of this fact with respect to the negro and a white-seeded and a brown-seeded variety, which were all grown together. The diversity of character in the seedlings of the second generation raised by me from his plants was wonderful. I could add other analogous cases, and the fact is well known to gardeners.[10]

[10] I have given Mr Coe's case in the *Gardeners' Chronicle*, 1858, p. 829. See also for another case, ibid., p. 845.

LATHYRUS ODORATUS

Almost everyone who has studied the structure of papilionaceous flowers has been convinced that they are specially adapted for cross-fertilization, although many of the species are likewise capable of self-fertilization. The case therefore of *Lathyrus odoratus* of the sweetpea is curious, for in this country it seems invariably to fertilize itself. I conclude that this is so, as five varieties, differing greatly in the colour of their flowers but in no other respect, are commonly sold and come true; yet on enquiry from two great raisers of seed for sale, I find that they take no precautions to insure purity – the five varieties being habitually grown close together.[11] I have myself purposely made similar trials with the same result. Although the varieties always come true, yet, as we shall presently see, one / of the five well-known varieties occasionally gives birth to another, which exhibits all its usual characters. Owing to this curious fact, and to the darker-coloured varieties being the most productive, these increase, to the exclusion of the others, as I was informed by the late Mr Masters, if there be no selection.

In order to ascertain what would be the effect of crossing two varieties, some flowers on the Purple sweetpea, which has a dark reddish-purple standard-petal with violet-coloured wing-petals and keel, were castrated whilst very young, and were fertilized with pollen of the Painted Lady. This latter variety has a pale cherry-coloured standard, with almost white wings and keel. On two occasions I raised from a flower thus crossed plants perfectly resembling both parent-forms; but the greater number resembled the paternal variety. So perfect was the resemblance, that I should have suspected some mistake in the label, had not the plants, which were at first identical in appearance with the father or Painted Lady, later in the season produced flowers blotched and streaked with dark purple. This is an interesting example of partial reversion in the same individual plants as it grows older. The purple-flowered plants were thrown away, as they might possibly have been the product of the accidental self-fertilization of the mother-plant, owing to the castration not having been effectual. But the plants which resembled in the colour of their

[11] See Mr W. Earley in *Nature*, 1872, p. 242, to the same effect. He once, however, saw bees visiting the flowers, and supposed that on this occasion they would have been intercrossed.

flowers the paternal variety or Painted Lady were preserved, and their seeds saved. Next summer many plants were raised from these seeds, and they generally resembled their grandfather the Painted Lady, but most of them had their wing-petals streaked and stained with dark pink; and a few had pale purple wings with the standard of a darker crimson than is natural to the Painted Lady, so that they formed a new subvariety. Among these plants a single one appeared having purple flowers like those of the grandmother, but with the petals slightly streaked with a paler tint: this was thrown away. Seeds were again saved from the foregoing plants, and the seedlings thus raised still resembled the Painted Lady, or great-grandfather; but they now varied much, the standard petal varying from pale to dark red, in a few instances with blotches of white; and the wing-petals varied from nearly white to purple, the keel being in all nearly white.

As no variability of this kind can be detected in plants raised from seeds, the parents of which have grown during many successive generations in close proximity, we may infer that they / cannot have intercrossed. What does occasionally occur is that in a row of plants raised from seeds of one variety, another variety true of its kind appears; for instance, in a long row of Scarlets (the seeds of which had been carefully gathered from Scarlets for the sake of this experiment) two Purples and one Painted Lady appeared. Seeds from these three aberrant plants were saved and sown in separate beds. The seedlings from both the Purples were chiefly Purples, but with some Painted Ladies and some Scarlets. The seedlings from the aberrant Painted Lady were chiefly Painted Ladies with some Scarlets. Each variety, whatever its parentage may have been, retained all its characters perfect, and there was no streaking or blotching of the colours, as in the foregoing plants of crossed origin. Another variety, however, is often sold, which is striped and blotched with dark purple; and this is probably of crossed origin, for I found, as well as Mr Masters, that it did not transmit its characters at all truly.

From the evidence now given, we may conclude that the varieties of the sweetpea rarely or never intercross in this country; and this is a highly remarkable fact, considering, firstly, the general structure of the flowers; secondly, the large quantity of pollen produced, far more than is requisite for self-fertilization; and thirdly, the occasional visits of insects. That insects should sometimes fail to cross-fertilize the flowers is intelligible, for I have thrice seen humble-bees of two kinds, as well as hive-bees, sucking the nectar, and they did not depress the keel-petals to as to expose the anthers and stigma; they were therefore quite inefficient for fertilizing the flowers. One of these bees, namely, *Bombus laipdarius*, stood on one side at the base of the standard and inserted its proboscis beneath the single separate stamen, as I afterwards ascertained by opening the flower and finding this stamen prised up. Bees are forced to act in this manner from the slit in the staminal tube being closely covered by the broad membranous margin of the single stamen, and from the tube not being perforated by nectar-passages. On the other hand, in the three British species of Lathyrus which I have examined, and in the allied genus Vicia, two nectar-passages are present. Therefore British bees might well be puzzled how to act in the case of the sweetpea. I may add that the staminal tube of another exotic species, *Lathyrus grandiflorus*, is not perforated by nectar-passages, and this species has rarely set any pods in my garden, unless the wing-petals were moved up and / down, in the same manner as bees ought to do; and then pods were generally formed, but from some cause often dropped off afterwards. One of my sons caught an elephant sphinx-moth whilst visiting the flowers of the sweetpea, but this insect would not depress the wing-petals and keel. On the other hand, I have seen on one occasion hive-bees, and two or three occasions the *Megachile willughbiella* in the act of depressing the keel; and these bees had the under sides of their bodies thickly covered with pollen, and could not thus fail to carry pollen from one flower to the stigma of another. Why then do not the varieties occasionally intercross, though this would not often happen, as insects so rarely act in an efficient manner? The fact cannot, as it appears, be explained by the flowers being self-fertilized at a very early age; for although nectar is sometimes secreted and pollen adheres to the viscid stigma before the flowers are fully expanded, yet in five young flowers which were examined by me the pollen-tubes were not exserted. Whatever the cause may be, we may conclude, that in England the varieties never or very rarely intercross. But it does not follow from this, that they would not be crossed by the aid of other and larger insects in their native country, which in botanical works is said to be the south of Europe and the East Indies. Accordingly I wrote to Professor Delpino, in Florence, and he informs me 'that it is the fixed opinion of gardeners there that the varieties do intercross, and that they cannot be preserved pure unless they are sown separately'.

It follows also from the foregoing facts that the several varieties of the sweetpea must have propagated themselves in England by self-fertilization for very many generations, since the time when each new variety first appeared. From the analogy of the plants of Mimulus and Ipomoea, which had been self-fertilized for several generations, and from trials previously made with the common pea, which is in nearly the same state as the sweetpea, it appeared to me very improbable that a cross between the individuals of the same variety would benefit the offspring. A cross of this kind was therefore not tried, which I now regret. But some flowers of the Painted Lady, castrated at an early age, were fertilized with pollen from the Purple sweetpea; and it should be remembered that these varieties differ in nothing except in the colour of their flowers. The cross was manifestly effectual (though only two seeds were obtained), as was shown by the two seedlings, when they flowered, closely resembling / their father, the Purple pea, excepting that they were a little lighter coloured, with their keels slightly streaked with pale purple. Seeds from flowers spontaneously self-fertilized under a net were at the same time saved from the same mother-plant, the Painted Lady. These seeds unfortunately did not germinate on sand at the same time with the crossed seeds, so that they could not be planted simultaneously. One of the two crossed seeds in a state of germination was planted in a pot (No. I) in which a self-fertilized seed in the same state had been planted four days before, so that this latter seedling had a great advantage over the crossed one. In Pot II the other crossed seed was planted two days before a self-fertilized one; so that here the crossed seedling had a considerable advantage over the self-fertilized one. But this crossed seedling had its summit gnawed off by a slug, and was in consequence for a time quite beaten by the self-fertilized plant. Nevertheless I allowed it to remain, and so great was its constitutional vigour that it ultimately beat its uninjured self-fertilized rival. When all four plants were almost fully grown they were measured, as here shown:

TABLE LIV
Lathyrus odoratus

No. of pot	Crossed plants	Self-fertilized plants
	Inches	Inches
I	80	64⅛
II	78⅛	63
Total in inches	158·5	127·5

The two crossed plants here average 79·25, and the two self-fertilized 63·75 inches in height, or as 100 to 80. Six flowers on these two crossed plants were reciprocally crossed with pollen from the other plant, and the six pods thus produced contained on an average six peas, with a maximum in one of seven. Eighteen spontaneously self-fertilized pods from the Painted Lady, which, as already stated, had no doubt been self-fertilized for many previous generations, contained on an average only 3·93 peas, with a maximum in one of five peas; so that the number of peas in the crossed and self-fertilized pods was as 100 to 65. / The self-fertilized peas were, however, quite as heavy as those from the crossed pods. From these two lots of seeds, the plants of the next generation were raised.

Plants of the second generation. Many of the self-fertilized peas just referred to germinated on sand before any of the crossed ones, and were rejected. As soon as I got equal pairs, they were planted on the opposite sides of two large pots, which were kept in the greenhouse. The seedlings thus raised were the grandchildren of the Painted Lady, which was first crossed by the Purple variety. When the two lots were from 4 to 6 inches in height there was no difference between them. Nor was there any marked difference in the period of their flowering. When fully grown they were measured, as follows:

TABLE LV
Lathyrus odoratus (second generation)

No. of pot	Seedlings from plants crossed during the two previous generations	Seedlings from plants self-fertilized during many previous generations
	Inches	Inches
I	72⁴⁄₈	57⁴⁄₈
	71	67
	52²⁄₈	56²⁄₈
II	81⁴⁄₈	66²⁄₈
	45²⁄₈	38⁷⁄₈
	55	46
Total in inches	377·50	331·86

The average height of the six crossed plants is here 62·91, and that of the six self-fertilized 55·31 inches; or as 100 to 88. There was not

141

much difference in the fertility of the two lots; the crossed plants having produced in the greenhouse thirty-five pods, and the self-fertilized thirty-two pods.

Seeds were saved from the self-fertilized flowers on these two lots of plants, for the sake of ascertaining whether the seedlings thus raised would inherit any difference in growth or vigour. It must therefore be understood that both lots in the following trial are plants of self-fertilized parentage; but that in the one lot the plants were the children of plants which had been crossed during two previous generations, having been before that self-fertilized / for many generations; and that in the other lot they were the children of plants which had not been crossed for very many previous generations. The seeds germinated on sand and were planted in pairs on the opposite sides of four pots. They were measured, when fully grown, with the following result:

TABLE LVI
Lathyrus odoratus

No. of pot	Self-fertilized plants from crossed plants	Self-fertilized plants from self-fertilized plants
	Inches	Inches
I	72	65
	72	61⅛
II	58	64
	68	68⅜
	72⅛	56⅛
III	81	60⅜
IV	77⅛	76⅛
Total in inches	501	452

The average height of the seven self-fertilized plants, the offspring of crossed plants, is 71·57, and that of the seven self-fertilized plants, the offspring of self-fertilized plants, is 64·57; or as 100 to 90. The self-fertilized plants from the self-fertilized produced rather more pods – viz., thirty-six – than the self-fertilized plants from the crossed, for these produced only thirty-one pods.

A few seeds of the same two lots were sown in the opposite corners of a large box in which a Brugmansia had long been growing, and in which the soil was so exhausted that seeds of *Ipomoea purpurea* would hardly vegetate; yet the two plants of the sweetpea which were raised flourished well. For a long time the self-fertilized plant from the self-fertilized beat the self-fertilized plant from the crossed plant; the former flowered first, and was at one time 77½ inches, whilst the latter was only 68½ in height; but ultimately the plant from the previous cross showed its superiority and attained a height of 108½ inches, whilst the other was only 95 inches. I also sowed some of the same two / lots of seeds in poor soil in a shady place in a shrubbery. Here again the self-fertilized plants from the self-fertilized for a long time exceeded considerably in height those from the previously crossed plants; and this may probably be attributed, in the present as in the last case, to these seeds having germinated rather sooner than those from the crossed plants; but at the close of the season the tallest of the self-fertilized plants from the crossed plants was 30 inches, whilst the tallest of the self-fertilized from the self-fertilized was 29⅜ inches in height.

From the various facts now given we see that plants derived from a cross between two varieties of the sweetpea, which differ in no respect except in the colour of their flowers, exceed considerably in height the offspring from self-fertilized plants, both in the first and second generations. The crossed plants also transmit their superiority in height and vigour to their self-fertilized offspring.

PISUM SATIVUM

The common pea is perfectly fertile when its flowers are protected from the visits of insects; I ascertained this with two or three different varieties, as did Dr Ogle with another. But the flowers are likewise adapted for cross-fertilization; Mr Farrer specifies[12] the following points, namely: 'The open blossom displaying itself in the most attractive and convenient position for insects; the conspicuous vexillum; the wings forming an alighting place; the attachment of the wings to the keel, by which any body pressing on the former must press down the latter; the staminal tube enclosing nectar, and affording by means of its partially free stamen with apertures on each side of its base an open passage to an insect seeking the nectar; the moist and sticky

[12] *Nature*, 10 October, 1872, p. 479. H. Müller gives an elaborate description of the flowers, *Befruchtung*, etc., p. 247.

pollen placed just where it will be swept out of the apex of the keel against the entering insect; the stiff elastic style so placed that on a pressure being applied to the keel it will be pushed upwards out of the keel; the hairs on the style placed on that side of the style only on which there is space for the pollen, and in such a direction as to sweep it out; and the stigma so placed as to meet an entering insect – all these become correlated parts of one elaborate mechanism, if we / suppose that the fertilization of these flowers is effected by the carriage of pollen from one to the other.' Notwithstanding these manifest provisions for cross-fertilization, varieties which have been cultivated for very many successive generations in close proximity, although flowering at the same time, remain pure. I have elsewhere[13] given evidence on this head, and if required could give more. There can hardly be a doubt that some of Knight's varieties, which were originally produced by an artificial cross and were very vigorous, lasted for at least sixty years, and during all these years were self-fertilized; for had it been otherwise, they would not have kept true, as the several varieties are generally grown near together. Most of the varieties, however, endure for a shorter period; and this may be in part due to their weakness of constitution from long-continued self-fertilization.

It is remarkable, considering that the flowers secrete much nectar and afford much pollen, how seldom they are visited by insects either in England or, as H. Müller remarks, in North Germany. I have observed the flowers for the last thirty years, and in all this time have only thrice seen bees of the proper kind at work (one of them being *Bombus muscorum*), such as were sufficiently powerful to depress the keel, so as to get the undersides of their bodies dusted with pollen. These bees visited several flowers, and could hardly have failed to cross-fertilize them. Hive-bees and other small kinds sometimes collect pollen from old and already fertilized flowers, but this is of no account. The rarity of the visits of efficient bees to this exotic plant is, I believe, the chief cause of the varieties so seldom intercrossing. That a cross does occasionally take place, as might be expected from what has just been stated, is certain, from the recorded cases of the direct action of the pollen of one variety on the seed-coats of another.[14] The late Mr Masters, who particularly attended to the raising of new varieties of peas, was convinced that some of them had originated from accidental

[13] *Variation of Animals and Plants under Domestication*, chap. ix, 2nd edit., vol. i, p. 348.
[14] *Var. under Domestication*, chap. xi, 2nd edit., vol. i, p. 428.

crosses. But as such crosses are rare, the old varieties would not often be thus deteriorated, more especially as plants departing from the proper type are generally rejected by those who collect seed for sale. There is another cause which probably tends to render cross-fertilization rare, / namely, the early age at which the pollen-tubes are exserted; eight flowers not fully expanded were examined, and in seven of these the pollen-tubes were in this state; but they had not as yet penetrated the stigma. Although so few insects visit the flowers of the pea in this country or in North Germany, and although the anthers seem here to open abnormally soon, it does not follow that the species in its native country would be thus circumstanced.

Owing to the varieties having been self-fertilized for many generations, and to their having been subjected in each generation to nearly the same conditions (as will be explained in a future chapter), I did not expect that a cross between two such plants would benefit the offspring; and so it proved on trial. In 1867 I covered up several plants of the Early Emperor pea, which was not then a very new variety, so that it must already have been propagated by self-fertilization for at least a dozen generations. Some flowers were crossed with pollen from a distinct plant growing in the same row, and others were allowed to fertilize themselves under a net. The two lots of seeds thus obtained were sown on opposite sides of two large pots, but only four pairs came up at the same time. The pots were kept in the greenhouse. The seedlings of both lots when between 6 and 7 inches in height were equal. When nearly full-grown they were measured, as in the following table:

TABLE LVII
Pisum sativum

No. of pot	Crossed plants	Self-fertilized plants
	Inches	Inches
I	35	29⅝
II	31⅛	51
	35	45
	37	33
Total in inches	138·50	158·75

The average height of the four crossed plants is here 34·62, and that of the four self-fertilized plants 39·68, or as 100 to 115. So that the crossed plants, far from beating the self-fertilized, were completely beaten by them. /

There can be no doubt that the result would have been widely different, if any two varieties out of the numberless ones which exist had been crossed. Notwithstanding that both had been self-fertilized for many previous generations, each would almost certainly have possessed its own peculiar constitution; and this degree of differentiation would have been sufficient to make a cross highly beneficial. I have spoken thus confidently of the benefit which would have been derived from crossing any two varieties of the pea from the following facts: Andrew Knight in speaking of the results of crossing reciprocally very tall and short varieties, says,[15] 'I had in this experiment a striking instance of the stimulative effects of crossing the breeds; for the smallest variety, whose height rarely exceeded 2 feet, was increased to 6 feet; whilst the height of the large and luxuriant kind was very little diminished'. Recently Mr Laxton has made numerous crosses, and everyone has been astonished at the vigour and luxuriance of the new varieties which he has thus raised and afterwards fixed by selection. He gave me seed-peas produced from crosses between four distinct kinds; and the plants thus raised were extraordinarily vigorous, being in each case from 1 to 2 or even 3 feet taller than the parent-forms, which were raised at the same time close alongside. But as I did not measure their actual height I cannot give the exact ratio, but it must have been at least as 100 to 75. A similar trial was subsequently made with two other peas from a different cross, and the result was nearly the same. For instance, a crossed seedling between the Maple and Purple-podded pea was planted in poor soil and grew to the extraordinary height of 116 inches; whereas the tallest plant of either parent variety, namely, a Purple-podded pea, was only 70 inches in height; or as 100 to 60.

SAROTHAMNUS SCOPARIUS

Bees incessantly visit the flowers of the common Broom, and these are adapted by a curious mechanism for cross-fertilization. When a bee alights on the wing-petals of a young flower, the keel is slightly opened and the short stamens spring out, which rub their pollen against the abdomen of the bee. If a rather older flower is visited for the first time

---

[15] *Philosophical Transactions*, 1799, p. 200.

(or if the bee exerts great force on a younger flower), the keel opens along its whole length, and the longer as well as the shorter stamens, together / with the much elongated curved pistil, spring forth with violence. The flattened, spoon-like extremity of the pistil rests for a time on the back of the bee, and leaves on it the load of pollen with which it is charged. As soon as the bee flies away, the pistil instantly curls round, so that the stigmatic surface is now upturned and occupies a position, in which it would be rubbed against the abdomen of another bee visiting the same flower. Thus, when the pistil first escapes from the keel, the stigma is rubbed against the back of the bee, dusted with pollen from the longer stamens, either of the same or another flower; and afterwards against the lower surface of the bee dusted with pollen from the shorter stamens, which is often shed a day or two before that from the longer stamens.[16] By this mechanism cross-fertilization is rendered almost inevitable, and we shall immediately see that pollen from a distinct plant is more effective than that from the same flower. I need only add that, according to H. Müller, the flowers do not secrete nectar, and he thinks that bees insert their proboscides only in the hope of finding nectar; but they act in this manner so frequently and for so long a time that I cannot avoid the belief that they obtain something palatable within the flowers.

If the visits of bees are prevented, and if the flowers are not dashed by the wind against any object, the keel never opens, so that the stamens and pistil remain enclosed. Plants thus protected yield very few pods in comparison with those produced by neighbouring uncovered bushes, and sometimes none at all. I fertilized a few flowers on a plant growing almost in a state of nature with pollen from another plant close alongside, and the four crossed capsules contained on an average 9·2 seeds. This large number no doubt was due to the bush being covered up, and thus not exhausted by producing many pods; for fifty pods gathered from an adjoining plant, the flowers of which had been fertilized by the bees, contained an average of only 7·14 seeds. Ninety-three pods spontaneously self-fertilized on a large bush which had been covered up, but had been much agitated by the wind, contained an average of 2·93 seeds. Ten of the finest of these ninety-three / capsules yielded an average of 4·30 seeds, that is less than half the average number in the four artificially crossed capsules. The ratio

[16] These observations have been quoted in an abbreviated form by the Rev. G. Henslow, in the *Journal of Linn. Soc. Bot.*, vol. ix, 1866, p. 358. H. Müller has since published a full and excellent account of the flower in his *Befruchtung*, etc., p. 240.

of 7·14 to 2·93, or as 100 to 41, is probably the fairest for the number of seeds per pod, yielded by naturally-crossed and spontaneously self-fertilized flowers. The crossed seeds compared with an equal number of the spontaneously self-fertilized seeds were heavier, in the ratio of 100 to 88. We thus see that besides the mechanical adaptations for cross-fertilization, the flowers are much more productive with pollen from a distinct plant than with their own pollen.

Eight pairs of the above crossed and self-fertilized seeds, after they had germinated on sand, were planted (1867) on the opposite sides of two large pots. When several of the seedlings were an inch and a half in height, there was no marked difference between the two lots. But even at this early age the leaves of the self-fertilized seedlings were smaller and of not so bright a green as those of the crossed seedlings. The pots were kept in the greenhouse, and as the plants in the following spring (1868) looked unhealthy and had grown but little, they were plunged, still in their pots, into the open ground. The plants all suffered much from the sudden change, especially the self-fertilized, and two of the latter died. The remainder were measured, and I give the measurements in the following table, because I have not seen in any other species so great a difference between the crossed and self-fertilized seedlings at so early an age.

TABLE LVIII

Sarothamnus scoparius (very young plants)

No. of pot	*Crossed plants*	*Self-fertilized plants*
	Inches	*Inches*
I	4 4/8	2 4/8
	6	1 4/8
	2	1
II	2	1 4/8
	2 4/8	1
	0 4/8	0 4/8
Total in inches	17·5	8·0

The six crossed plants here average 2·91, and the six self-fertilized / 1·33 inches in height; so that the former were more than twice as high as the latter, or as 100 to 46.

In the spring of the succeeding year (1869) the three crossed plants in Pot I had all grown to nearly a foot in height, and they had smothered the three little self-fertilized plants so completely that two were dead; and the third, only an inch and a half in height, was dying. It should be remembered that these plants had been bedded out in their pots, so that they were subjected to very severe competition. This pot was now thrown away.

The six plants in Pot II were all alive. One of the self-fertilized was an inch and a quarter taller than any one of the crossed plants; but the other two self-fertilized plants were in a very poor condition. I therefore resolved to leave these plants to struggle together for some years. By the autumn of the same year (1869) the self-fertilized plant which had been victorious was now beaten. The measurements are shown in the following table:

TABLE LIX
Pot II: *Sarothamnus scoparius*

Crossed plants	*Self-fertilized plants*
Inches	*Inches*
15 6/8	13 1/8
9 6/8	3
8 2/8	2 4/8

The same plants were again measured in the autumn of the following year, 1870.

TABLE LX
Pot II: *Sarothamnus scoparius*

Crossed plants	*Self-fertilized plants*
Inches	*Inches*
26 2/8	14 2/8
16 4/8	11 4/8
14	9 6/8
56·75	35·50

The three crossed plants now averaged 18·91, and the three self-fertilized 11·83 inches in height; or as 100 to 63. The three crossed plants in Pot I, as already shown, had beaten the three self-fertilized plants so completely, that any comparison between them was superfluous.

The winter of 1870–71 was severe. In the spring the three crossed plants in Pot II had not even the tips of their shoots in the least injured, whereas all three self-fertilized plants were killed half-way down to the ground; and this shows how much more tender they were. In consequence not one of these latter plants bore a single flower during the ensuing summer of 1871, whilst all three crossed plants flowered.

ONONIS MINUTISSIMA

This plant, of which seeds were sent me from North Italy, produces, besides the ordinary papilionaceous flowers, minute, imperfect, closed or cleistogamic flowers, which can never be cross-fertilized, but are highly self-fertile. Some of the perfect flowers were crossed with pollen from a distinct plant, and six capsules thus produced yielded on an average 3·66 seeds, with a maximum of five in one. Twelve perfect flowers were marked and allowed to fertilize themselves spontaneously under a net, and they yielded eight capsules, containing on an average 2·38 seeds, with a maximum of three seeds in one. So that the crossed and self-fertilized capsules from the perfect flowers yielded seeds in the proportion of 100 to 65. Fifty-three capsules produced by the cleistogamic flowers contained on an average 4·1 seeds, so that these were the most productive of all; and the seeds themselves looked finer even than those from the crossed perfect flowers.

The seeds from the crossed perfect flowers and from the self-fertilized cleistogamic flowers were allowed to germinate on sand; but unfortunately only two pairs germinated at the same time. These were planted on the opposite sides of the same pot, which was kept in the greenhouse. In the summer of the same year, when the seedlings were about 4½ inches in height, the two lots were equal. In the autumn of the following year (1868) the two crossed plants were of exactly the same height, viz., 11⅛ inches, and the two self-fertilized plants 12⅝ and 7⅞ inches; so that one of the self-fertilized exceeded considerably in height all the others. By the autumn of 1869 the two crossed plants had acquired the / supremacy; their height being 16⅜ and 15⅛, whilst that of the two self-fertilized plants was 14⅝ and 11⅛ inches.

By the autumn of 1870, the heights were as follows (Table LXI).

So that the mean height of the two crossed plants was 19·81, and that of the two self-fertilized 17·37 inches; or as 100 to 88. It should be remembered that the two lots were at first equal in height; that one of the self-fertilized plants then had the advantage, the two crossed plants being at last victorious.

TABLE LXI
Ononis minutissima

Crossed plants	Self-fertilized plants
Inches	Inches
20⅜	17⅘
19⅖	17⅖
39·63	34·75

Summary on the Leguminosae

Six genera in this family were experimented on, and the results are in some respect remarkable. The crossed plants of the two species of Lupinus were conspicuously superior to the self-fertilized plants in height and fertility; and when grown under very unfavourable conditions, in vigour. The scarlet-runner (*Phaseolus multiflorus*) is partially sterile if the visits of bees are prevented, and there is reason to believe that varieties growing near one another intercross. The five crossed plants, however, exceeded in height the five self-fertilized only by a little. *Phaseolus vulgaris* is perfectly self-fertile; nevertheless, varieties growing in the same garden sometimes intercross largely. The varieties of *Lathyrus odoratus*, on the other hand, appear never to intercross in this country; and though the flowers are not often visited by efficient insects, I cannot account for this fact, more especially as the varieties are believed to / intercross in North Italy. Plants raised from a cross between two varieties, differing only in the colour of their flowers, grew much taller and were under unfavourable conditions more vigorous than the self-fertilized plants; they also transmitted, when self-fertilized, their superiority to their offspring. The many varieties of the common Pea (*Pisum sativum*), though growing in close proximity, very seldom intercross; and this seems due to the rarity in this country of the visits of bees sufficiently powerful to effect cross-fertilization. A cross between the self-fertilized individuals of the same variety does no good whatever to the offspring; whilst a cross between distinct varieties, though closely allied, does great good, of which we have excellent evidence. The flowers of the Broom (Sarothamnus) are almost sterile if they are not disturbed and if insects are excluded. The pollen from a distinct plant is more effective than that from the same flower in producing seeds. The crossed seedlings have an enormous advantage over the self-fertilized when grown together in close competition. Lastly, only four plants of the *Ononis minutissima* were raised; but as these were observed during their whole growth, the advantage of the crossed over the self-fertilized plants may, I think, be fully trusted.

XV. ONAGRACEAE

CLARKIA ELEGANS

Owing to the season being very unfavourable (1867), few of the flowers which I fertilized formed capsules; twelve crossed flowers produced only four, and eighteen self-fertilized flowers yielded only one capsule. The seeds after germinating on sand were planted in three pots, but all the self-fertilized plants died in one of them. When the two lots were between 4 and 5 inches in height, the crossed began to show a slight superiority over the self-fertilized. When in full flower they were measured, with the following result: /

TABLE LXII
Clarkia elegans

No. of pot	Crossed plants	Self-fertilized plants
	Inches	*Inches*
I	40⅘	33
	35	24
	25	23
II	33⅘	30⅘
Total in inches	134·0	110·5

The average height of the four crossed plants is 33·5, and that of the four self-fertilized plants 27·62 inches, or as 100 to 82. The crossed plants altogether produced 105 and the self-fertilized plants 63 capsules; or as 100 to 60. In both pots a self-fertilized plant flowered before any one of the crossed plants.

XVI. LOASACEAE

BARTONIA AUREA

Some flowers were crossed and self-fertilized in the usual manner during two seasons; but as I reared on the first occasion / only two pairs, the results are given together. On both occasions the crossed capsules contained slightly more seeds than the self-fertilized. During the first year, when the plants were about 7 inches in height, the self-

TABLE LXIII
Bartonia aurea

No. of pot	Crossed plants	Self-fertilized plants
	Inches	Inches
I	31	37
II	18⅝	20⅛
III	19⅞	40⅛
IV	25	35
	36	15⅛
V	31	18
	16	11⅛
VI	20	32⅛
Total in inches	197·0	210·5

fertilized were the tallest, and in the second year the crossed were the tallest. When the two lots were in full flower they were measured, as in the preceding table.

The average height of the eight crossed plants was 24·62, and that of the eight self-fertilized 26·31 inches; or as 100 to 107. So that the self-fertilized had a decided advantage over the crossed. But the plants from some cause never grew well, and finally became so unhealthy that only three crossed and three self-fertilized plants survived to set any capsules, and these were few in number. The two lots seemed to be about equally unproductive.

XVII. PASSIFLORACEAE

PASSIFLORA GRACILIS

This annual species produces spontaneously numerous fruits when insects are excluded, and behaves in this respect very differently from most of the other species in the genus, which are extremely sterile unless fertilized with pollen from a distinct plant.[17] Fourteen fruits from crossed flowers contained on an average 24·14 seeds. Fourteen fruits (two poor ones being rejected), spontaneously self-fertilized under a net, contained on an average 20·58 seeds per fruit; or as 100 to 85. These seeds were sown on the opposite sides of three pots, but only two pairs came up at the same time; and therefore a fair judgement cannot be formed.

[17] *Variation of Animals and Plants under Domestication*, chap. xvii, 2nd edit., vol. ii, p. 118.

TABLE LXIV
Passiflora gracilis

No. of pot	Crossed plants	Self-fertilized plants
	Inches	Inches
I	56	38
II	42	64
Total in inches	98	102

The mean of the two crossed is 49 inches, and that of the two self-fertilized 51 inches; or as 100 to 104. /

XVIII. UMBELLIFERAE

APIUM PETROSELINUM

The Umbelliferae are proterandrous, and can hardly fail to be cross-fertilized by the many flies and small Hymenoptera which visit the flowers.[18] A plant of the common parsley was covered by a net, and it apparently produced as many and as fine spontaneously self-fertilized fruits or seeds as the adjoining uncovered plants. The flowers on the latter were visited by so many insects that they must have received pollen from one another. Some of these two lots of seeds were left on sand, but nearly all the self-fertilized seeds germinated before the others, so that I was forced to throw all away. The remaining seeds were then sown on the opposite sides of four pots. At first the self-fertilized seedlings were a little taller in most of the pots than the naturally crossed seedlings, and this no doubt was due to the self-fertilized seeds having germinated first. But in the autumn all the plants were so equal that it did not seem worth while to measure them. In two of the pots they were absolutely equal; in a third, if there was any difference, it was in favour of the crossed plants, and in a somewhat plainer manner in the fourth pot. But neither side had any substantial advantage over the other; so that in height they may be said to be as 100 to 100.

XIX. DIPSACEAE

SCABIOSA ATRO-PURPUREA

The flowers, which are proterandrous, were fertilized during / the

[18] H. Müller, *Befruchtung*, etc., p. 96. According to M. Mustel (as stated by Godron, *De l'Espèce*, vol. ii, p. 58, 1859), varieties of the carrot growing near each other readily intercross.

TABLE LXV
Scabiosa atro-purpurea

No. of pot	Crossed plants	Self-fertilized plants
	Inches	Inches
I	14	20
II	15	14⁴⁄₈
III	21	14
	18⁴⁄₈	13
Total in inches	68·5	61·5

unfavourable season of 1867, so that I got few seeds, especially from the self-fertilized heads, which were extremely sterile. The crossed and self-fertilized plants raised from these seeds were measured before they were in full flower, as in the preceding table.

The four crossed plants averaged 17·12, and the four self-fertilized 15·37 inches in height; or as 100 to 90. One of the self-fertilized plants in Pot III was killed by an accident, and its fellow pulled up; so that when they were again measured to the summits of their flowers, there were only three on each side; the crossed now averaged in height 32·83, and the self-fertilized 30·16 inches; or as 100 to 92.

XX. COMPOSITAE

LACTUCA SATIVA

Three plants of lettuce[19] (Great London Cos var.) grew close together in my garden; one was covered by a net, and produced self-fertilized seeds, the other two were allowed to be naturally crossed by insects; but the season (1867) was unfavourable, and I did not obtain many seeds. Only one crossed and one self-fertilized plant were raised in Pot I, and their measurements are given in the following table (LXVI). The flowers on this one self-fertilized plant were again self-fertilized under a net, not with pollen from the same floret, but from other florets on the same head. The flowers on the two crossed plants were left to be crossed by insects, but the process was aided by some pollen

[19] The Compositae are well-adapted for cross-fertilization, but a nurseryman on whom I can rely, told me that he had been in the habit of sowing several kinds of lettuce near together for the sake of seed, and had never observed that they became crossed. It is very improbable that all the varieties which were thus cultivated near together flowered at different times; but two which I selected by hazard and sowed near each other did not flower at the same time; and my trial failed.

being occasionally transported by me from plant to plant. These two lots of seeds, after germinating on sand, were planted in pairs on the opposite sides of Pots II and III, which were at first kept in the greenhouse and then turned out of doors. The plants were measured when in full flower. The following table, therefore, includes plants belonging to two generations. When the seedlings of the two lots were only 5 or 6 inches in height they were equal. In Pot III one of the self-fertilized plants died before flowering, as has occurred in so many other cases. /

TABLE LXVI
Lactuca sativa

No. of pot	Crossed plants	Self-fertilized plants
	Inches	Inches
I	27	21⅛
First generation, planted in open ground	25	20
II	29⅛	24
Second generation,	17⅛	10
planted in open ground	12⅛	11
III	14	9⅛
Second generation, kept in the pot	10⅛	0
Total in inches	136	96

The average height of the seven crossed plants is 19·43, and that of the six self-fertilized plants 16 inches; or as 100 to 82.

XXI. CAMPANULACEAE

SPECULARIA SPECULUM

In the closely allied genus, Campanula, in which Specularia was formerly included, the anthers shed at an early period their pollen, and this adheres to the collecting hairs which surround the pistil beneath the stigma; so that without some mechanical aid the flowers cannot be fertilized. For instance, I covered up a plant of *Campanula carpathica*, and it did not produce a single capsule, whilst the surrounding uncovered plants seeded profusely. On the other hand, the present species of Specularia appears to set almost as many capsules when covered up, as when left to the visits of the Diptera, which, as far as I have seen, are the only insects that frequent the flowers.[20] I did not ascertain whether the naturally crossed and spontaneously self-fertilized capsules contained an equal number of seeds, but a comparison of artificially crossed and self-fertilized flowers, / showed that the former were probably the most productive. It appears that this plant is capable of producing a large number of of self-fertilized capsules owing to the petals closing at night, as well as during cold weather. In the act of closing, the margins of the petals became reflexed, and their inwardly projecting midribs then pass between the clefts of the stigma, and in doing so push the pollen from the outside of the pistil on to the stigmatic surfaces.[21]

Twenty flowers were fertilized by me with their own pollen, but owing to the bad season, only six capsules were produced; they contained on an average 21·7 seeds, with a maximum of forty-eight in one. Fourteen flowers were crossed with pollen from another plant, and these produced twelve capsules, containing on an average 30 seeds, with a maximum in one of fifty-seven seeds; so that the crossed seeds were to the self-fertilized from an equal number of capsules as 100 to 72. The former were also heavier than an equal number of self-fertilized seeds, in the ratio of 100 to 86. Thus, whether we judge by the number of

TABLE LXVII
Specularia speculum

No. of pot	Tallest crossed plant in each pot	Tallest self-fertilized plant in each pot
	Inches	Inches
I	18	15⅝
II	17	19
III	22⅛	18
IV	20	23
Total in inches	77·13	75·75

[20] It has long been known that another species of the genus, *Specularia perfoliata*, produces cleistogamic as well as perfect flowers, and the former are of course self-fertile.

[21] Mr Meehan has lately shown (*Proc. Acad. Nat. Sc. Philadelphia*, 16 May, 1876, p. 84) that the closing of the flowers of *Claytonia virginica* and *Ranunculus bulbosus* during the night causes their self-fertilization.

capsules produced from an equal number of flowers, or by the average number of the contained seeds, or the maximum number in any one capsule, or by their weight, crossing does great good in comparison with self-fertilization. The two lots of seeds were / sown on the opposite sides of four pots; but the seedlings were not sufficiently thinned. Only the tallest plant on each side was measured, when fully grown. The measurements are given in the preceding table. In all four pots the crossed plants flowered first. When the seedlings were only about an inch and a half in height both lots were equal.

The four tallest crossed plants averaged 19·28, and the four tallest self-fertilized 18·93 inches in height; or as 100 to 98. So that there was no difference worth speaking of between the two lots in height; though other great advantages are derived, as we have seen, from cross-fertilization. From being grown in pots and kept in the green-house, none of the plants produced any capsules.

LOBELIA RAMOSA[22]
Var. snow-flake

The well-adapted means by which cross-fertilization is ensured in this genus have been described by several authors.[23] The pistil as it slowly increases in length pushes the pollen out of the conjoined anthers, by the aid of a ring of bristles; the two lobes of the stigma being at this time closed and incapable of fertilization. The extrusion of the pollen is also aided by insects, which rub against the little bristles that project from the anthers. The pollen thus pushed out is carried by insects to the older flowers, in which the stigma of the now freely projecting pistil is open and ready to be fertilized. I proved the importance of the gaily-coloured corolla, by cutting off the large lower petal of several

[22] I have adopted the name given to this plant in the *Gardeners' Chronicle*, 1866. Professor T. Dyer, however, informs me that it probably is a white variety of *L. tenuior* of R. Brown, from W. Australia.

[23] See the works of Hildebrand and Delpino. Mr Farrer also has given a remarkably clear description of the mechanism by which cross-fertilization is effected in this genus, in the *Annals and Mag. of Nat. Hist.*, vol. ii (4th series), 1868, p. 260. In the allied genus Isotoma, the curious spike which projects rectangularly from the anthers, and which when shaken causes the pollen to fall on the back of an entering insect, seems to have been developed from a bristle, like one of those which spring from the anthers in some of or all the species of Lobelia, as described by Mr Farrer.

flowers of *Lobelia erinus*; and these flowers were neglected by the hive-bees which were incessantly visiting the other flowers.

A capsule was obtained by crossing a flower of *L. ramosa* / with pollen from another plant, and two other capsules from artificially self-fertilized flowers. The contained seeds wee sown on the opposite sides of four pots. Some of the crossed seedlings which came up before the others had to be pulled up and thrown away. Whilst the plants were very small there was not much difference in height between the two lots; but in Pot III the self-fertilized were for a time the tallest. When in full flower the tallest plant on each side of each pot was measured, and the result is shown in the following table. In all four pots a crossed plant flowered before any one of its opponents.

TABLE LXVIII

Lobelia ramosa (first generation)

No. of pot	*Tallest crossed plant in each pot*	*Tallest self-fertilized plant in each pot*
	Inches	*Inches*
I	22 4/8	17 4/8
II	27 4/8	24
III	16 4/8	15
IV	22 4/8	17
Total in inches	89·0	73·5

The four tallest crossed plants averaged 22·25, and the four tallest self-fertilized 18·37 inches in height; or as 100 to 82. I was surprised to find that the anthers of a good many of these self-fertilized plants did not cohere and did not contain any pollen; and the anthers even of a very few of the crossed plants were in the same condition. Some flowers on the crossed plants were again crossed, four capsules being thus obtained; and some flowers on the self-fertilized plants were again self-fertilized, seven capsules being thus obtained. The seeds from both lots were weighed, and it was calculated that an equal number of capsules would have yielded seed in the proportion by weight of 100 for the crossed to 60 for the self-fertilized capsules. So that the flowers on the crossed plants again crossed were much more fertile than those on the self-fertilized plants again self-fertilized. /

Plants of the second generation. The above two lots of seeds were placed on damp sand, and many of the crossed seeds germinated, as on the last occasion, before the self-fertilized, and were rejected. Three or four pairs in the same state of germination were planted on the opposite sides of two pots; a single pair in a third pot; and all the remaining seeds were sown crowded in a fourth pot. When the seedlings were about one and a half inches in height, they were equal on both sides of the three first pots; but in Pot IV, in which they grew crowded and were thus exposed to severe competition, the crossed were about a third taller than the self-fertilized. In this latter pot, when the crossed averaged 5 inches in height, the self-fertilized were about 4 inches; nor did they look nearly such fine plants. In all four pots the crossed plants flowered some days before the self-fertilized. When in full flower the tallest plant on each side was measured; but before this time the single crossed plant in Pot III, which was taller than its antagonist, had died and was not measured. So that only the tallest plant on each side of three pots was measured, as in the following table:

TABLE LXIX
Lobelia ramosa (second generation)

No. of pot	Tallest crossed plant in each pot	Tallest self-fertilized plant in each pot
	Inches	Inches
I	27⅛	18⅝
II	21	19⅘
IV Crowded	21⅛	19
Total in inches	70	57

The average height of the three tallest crossed plants is here 23·33, and that of the three tallest self-fertilized 19 inches; or as 100 to 81. Besides this difference in height, the crossed plants were much more vigorous and more branched than the self-fertilized plants, and it is unfortunate that they were not weighed. /

LOBELIA FULGENS

This species offers a somewhat perplexing case. In the first generation the self-fertilized plants, though few in number, greatly exceeded the crossed in height; whilst in the second generation, when the trial was made on a much larger scale, the crossed beat the self-fertilized plants. As this species is generally propagated by off-sets, some seedlings were first raised, in order to have distint plants. On one of these plants several flowers were fertilized with their own pollen; and as the pollen is mature and shed long before the stigma of the same flower is ready for fertilization, it was necessary to number each flower and keep its pollen in paper with a corresponding number. By this means well-matured pollen was used for self-fertilization. Several flowers on the same plant were crossed with pollen from a distinct individual, and to obtain this the conjoined anthers of young flowers were roughly squeezed, and as it is naturally protruded very slowly by the growth of the pistil, it is probable that the pollen used by me was hardly mature, certainly less mature than that employed for self-fertilization. I did not at the time think of this source of error, but I now suspect that the growth of the crossed plants was thus injured. Anyhow the trial was not perfectly fair. Opposed to the belief that the pollen used in crossing was not in so good a state as that used for self-fertilization, is the fact that a greater proportional number of the crossed than of the self-fertilized flowers produced capsules; but there was no marked difference in the amount of seed contained in the capsules of the two lots.[24]

As the seeds obtained by the above two methods would not germinate whe left on bare sand, they were sown on the opposite sides of four pots; but I succeeded in raising only a single pair of seedlings of the same age in each pot. The self-fertilized seedlings, when only a few inches in height, were in most of the pots taller than their opponents; and they flowered so much earlier in all the pots, that the height of the flower-stems could be fairly compared only in Pots I and II. /

The mean height of .the flower-stems of the two crossed plants in Pots I and II is here 34·75 inches, and that of the two self-fertilized plants in the same pots 44·25 inches; or as 100 to 127. The self-fertilized plants in Pots III and IV were in every respect very much finer than the crossed plants.

I was so much surprised at this great superiority of the self-fertilized

---

[24] Gärtner has shown that certain plants of *Lobelia fulgens* are quite sterile with pollen from the same plant, though this pollen is efficient on any other individual; but none of the plants on which I experimented, which were kept in the greenhouse, were in this peculiar condition.

TABLE LXX
Lobelia fulgens (first generation)

No. of pot	Height of flower-stems on the crossed plants	Height of flower-stems on the self-fertilized plants
	Inches	Inches
I	33	50
II	36⅛	38⅛
III	21 Not in full flower	43
IV	12 Not in full flower	35⅝

over the crossed plants, that I determined to try how they would behave in one of the pots during a second growth. The two plants, therefore, in Pot I were cut down, and repotted without being disturbed in a much larger pot. In the following year the self-fertilized plant showed even a greater superiority than before; for the two tallest flower-stems produced by the one crossed plant were only 29⅛ and 30⅛ inches in height, whereas the two tallest stems on the one self-fertilized plant were 49⅛ and 49⅝ inches; and this gives a ratio of 100 to 167. Considering all the evidence, there can be no doubt that these self-fertilized plants had a great superiority over the crossed plants.

Crossed and self-fertilized plants of the second generation. I determined on this occasion to avoid the error of using pollen of not quite equal maturity for crossing and self-fertilization; so that I squeezed pollen out of the conjoined anthers of young flowers for both operations. Several flowers on the crossed plant in Pot I in Table LXX were again crossed with pollen from a distinct plant. Several other flowers on the self-fertilized plant / in the same pot were again self-fertilized with pollen from the anthers of other flowers on the *same plant*. Therefore the degree of self-fertilization was not quite so close as in the last generation, in which pollen from the *same flower*, kept in paper, was used. These two lots of seeds were thinly sown on opposite sides of nine pots and the young seedlings were thinned, an equal number of nearly as possible the same age being left on the two sides. In the spring of the following year (1870), when the seedlings had grown to a considerable size, they were measured to the tips of their leaves; and the twenty-three crossed plants averaged 14·04 inches in height, whilst

TABLE LXXI
Lobelia fulgens (second generation)

No. of pot	Crossed plants. Height of flower-stems	Self-fertilized plants. Height of flower-stems
	Inches	Inches
I	$27\frac{3}{8}$	$32\frac{3}{8}$
	26	$26\frac{3}{8}$
	$24\frac{3}{8}$	$25\frac{1}{8}$
	$24\frac{4}{8}$	$26\frac{2}{8}$
II	34	$36\frac{2}{8}$
	$26\frac{6}{8}$	$28\frac{6}{8}$
	$25\frac{1}{8}$	$30\frac{1}{8}$
	26	$32\frac{2}{8}$
III	$40\frac{4}{8}$	$30\frac{4}{8}$
	$37\frac{5}{8}$	$28\frac{2}{8}$
	$32\frac{1}{8}$	23
IV	$34\frac{5}{8}$	$29\frac{4}{8}$
	$32\frac{2}{8}$	$28\frac{3}{8}$
	$29\frac{3}{8}$	26
	$27\frac{1}{8}$	$25\frac{2}{8}$
V	$28\frac{1}{8}$	29
	27	$24\frac{6}{8}$
	$25\frac{3}{8}$	$23\frac{2}{8}$
	$24\frac{3}{8}$	24
VI	$33\frac{5}{8}$	$44\frac{2}{8}$
	32	$37\frac{6}{8}$
	$26\frac{1}{8}$	37
	25	35
VII	$30\frac{6}{8}$	$27\frac{2}{8}$
	$30\frac{3}{8}$	$19\frac{2}{8}$
	$29\frac{2}{8}$	21
VIII	$39\frac{3}{8}$	$23\frac{1}{8}$
	$37\frac{2}{8}$	$23\frac{4}{8}$
	36	$25\frac{4}{8}$
	36	$25\frac{1}{8}$
IX	$33\frac{3}{8}$	$19\frac{3}{8}$
	25	$16\frac{3}{8}$
	$25\frac{3}{8}$	19
	$21\frac{7}{8}$	$18\frac{6}{8}$
Total in inches	1014·00	921·63 /

163

the twenty-three self-fertilized seedlings were 13·54 inches; or as 100 to 96.

In the summer of the same year several of these plants flowered, the crossed and self-fertilized plants flowering almost simultaneously, and all the flower-stems were measured. Those produced by eleven of the crossed plants averaged 30·71 inches, and those by nine of the self-fertilized plants 29·43 inches in height; or as 100 to 96.

The plants in these nine pots, after they had flowered, were repotted without being disturbed in much larger pots; and in the following year, 1871, all flowered freely; but they had grown into such an entangled mass, that the separate plants on each side could no longer be distinguished. Accordingly three or four of the tallest flower-stems on each side of each pot were measured; and the measurements in the preceding table are, I think, more trustworthy than the previous ones, from being more numerous, and from the plants being well established and growing vigorously.

The average height of the thirty-four tallest flower-stems on the twenty-three crossed plants is 29·82 inches, and that of the same number of flower-stems on the same number of self-fertilized plants is 27·10 inches; or as 100 to 91. So that the crossed plants now showed a decided advantage over their self-fertilized opponents.

XXII. POLEMONIACEAE

NEMOPHILA INSIGNIS

Twelve flowers were crossed with pollen from a distinct plant, but produced only six capsules, containing on an average 18·3 seeds. Eighteen flowers were fertilized with their own pollen and produced ten capsules, containing on an average 12·7 / seeds; so that the seeds per capsule were as 100 to 69.[25] The crossed seeds weighed a little less than an equal number of self-fertilized seeds, in the proportion of 100 to 105; but this was clearly due to some of the self-fertilized capsules containing very few seeds, and these were much bulkier than the others, from having been better nourished. A subsequent comparison of the number of seeds in a few capsules did not show so great a superiority on the side of the crossed capsules as in the present case.

The seeds were placed on sand, and after germinating were planted

[25] Several species of Polemoniaceae are known to be proterandrous, but I did not attend to this point in Nemophila. Verlot says (*Des Varriétés*, 1865, p. 66) that varieties growing near one another spontaneously intercross.

in pairs on the opposite sides of five pots, which were kept in the greenhouse. When the seedlings were from 2 to 3 inches in height, most of the crossed had a slight advantage over the self-fertilized. The plants were trained up sticks, and thus grew to a considerable height. In four out of the five pots a crossed plant flowered before any one of the self-fertilized. / The plants were first measured to the tips of their

TABLE LXXII

Nemophila insignis; 0 means that the plant died

No. of pot	*Crossed plants*	*Self-fertilized plants*
	Inches	*Inches*
I	32 4/8	21 2/8
II	34 4/8	23 5/8
III	33 1/8	19
	22 2/8	7 2/8
	29	17 4/8
IV	35 4/8	10 4/8
	33 4/8	27
V	35	0
	38	18 3/8
	36	20 4/8
	37 4/8	34
	32 4/8	0
Total in inches	399·38	199·00

leaves, before they had flowered and when the crossed were under a foot in height. The twelve crossed plants averaged 11·1 inches in height, whilst the twelve self-fertilized were less than half of this height, viz., 5·45; or as 100 to 49. Before the plants had grown to their full height, two of the self-fertilized died, and as I feared that this might happen with others, they were again measured to the tops of their stems, as shown in the preceding table.

The twelve crossed plants now averaged 33·28, and the ten self-fertilized 19·9 inches in height, or as 100 to 60; so that they differed somewhat less than before.

The plants in Pots III and V were placed under a net in the greenhouse, two of the crossed plants in the latter pot being pulled up on account ofthe death of two of the self-fertilized; so that altogether six crossed and six self-fertilized plants were left to fertilize themselves spontaneously. The pots were rather small, and the plants did not produce many capsules. The small size of the self-fertilized plants will largely account for the fewness of the capsules which they produced. The six crossed plants bore 105, and the six self-fertilized only 30 capsules; or as 100 to 29.

The self-fertilized seeds thus obtained from the crossed and self-fertilized plants, after germinating on sand, were planted / on the

TABLE LXXIII
Nemophila insignis

No. of pot	Self-fertilized plants from crossed plants	Self-fertilized plants from self-fertilized plants
	Inches	Inches
I	27	27⁴⁄₈
	14	34²⁄₈
II	17⁶⁄₈	23
	24⁴⁄₈	32
III	16	7
IV	5³⁄₈	7²⁄₈
	5⁴⁄₈	16
Total in inches	110·13	147·00

opposite sides of four small pots, and treated as before. But many of the plants were unhealthy, and their heights were so unequal – some on both sides being five times as tall as the others – that the averages deduced from the measurements in the preceding table are not in the least trustworthy. Nevertheless I have felt bound to give them, as they are opposed to my general conclusions.

The seven self-fertilized plants from the crossed plants here average 15·73, and the seven self-fertilized from the self-fertilized 21 inches in height; or as 100 to 133. Strictly analogous experiments with *Viola tricolor* and *Lathyrus odoratus* gave a very different result.

XXIII. BORAGINACEAE

BORAGO OFFICINALIS

This plant is frequented by a greater number of bees than almost any other one which I have observed. It is strongly proterandrous (H. Müller, *Befruchtung*, etc., p. 267), and the flowers can hardly fail to be cross-fertilized; but should this not occur, they are capable of self-fertilization to a limited extent, as some pollen long remains within the anthers, and is apt to fall on the mature stigma. In the year 1863 I covered up a plant, and examined thirty-five flowers, of which only twelve yielded any seeds; whereas of thirty-five flowers on an exposed plant growing close by, all with the exception of two yielded seeds. The covered-up plant, however, produced altogether twenty-five spontaneously self-fertilized seeds; the exposed plant producing fifty-five seeds, the product, no doubt, of cross-fertilization.

In the year 1868 eighteen flowers on a protected plant were crossed with pollen from a distinct plant, but only seven of these produced fruit; and I suspect that I applied pollen to many of the stigmas before they were mature. These fruits contained on an average 2 seeds, with a maximum in one of three seeds. Twenty-four spontaneously self-fertilized fruits were produced by the same plant, and these contained on an average 1·2 seeds, with a maximum of two in one fruit. So that the fruits from the artificially crossed flowers yielded seeds compared with those from the spontaneously self-fertilized flowers, in the ratio of 100 to 60. But the self-fertilized seeds, as often occurs when few are produced, were heavier than the crossed seeds in the ratio of 100 to 90. /

These two lots of seeds were sown on opposite sides of two large pots; but I succeeded in raising only four pairs of equal age. When the seedlings on both sides were about 8 inches in height they were equal. When in full flower they were measured, as follows:

TABLE LXXIV
Borago officinalis

No. of pot	Crossed plants	Self-fertilized plants
	Inches	Inches
I	19	13⁴⁄8
	21	18⁶⁄8
	16⁴⁄8	20²⁄8
II	26²⁄8	32²⁄8
Total in inches	82·75	84·75

The average height of the four crossed plants is here 20·68, and that of the four self-fertilized 21·18 inches; or as 100 to 102. The self-fertilized plants thus exceeded the crossed in height by a little; but this was entirely due to the tallness of one of the self-fertilized. The crossed plants in both pots flowered before the self-fertilized. Therefore I believe if more plants had been raised, the result would have been different. I regret that I did not attend to the fertility of the two lots.

XXIV. NOLANACEAE

NOLANA PROSTRATA

In some of the flowers the stamens are considerably shorter than the pistil, in others equal to it in length. I suspected, therefore, but erroneously as it proved, that this plant was dimorphic, like Primula, Linum, etc., and in the year 1862 twelve plants, covered by a net in the greenhouse, were subjected to trial. The spontaneously self-fertilized flowers yielded 64 grains weight of seeds, but the product of fourteen artificially crossed flowers is here included, which falsely increases the weight of the self-fertilized seeds. Nine uncovered plants, the flowers of which were eagerly visited by bees for their pollen and were no doubt intercrossed by them, produced 79 grains weight of seeds: therefore twelve plants thus treated would have yielded 105 / grains. Thus the seeds produced by the flowers on an equal number of plants, when crossed by bees, and spontaneously self-fertilized (the product of fourteen artificially crossed flowers being, however, included in the latter) were in weight as 100 to 61.

In the summer of 1867 the trial was repeated; thirty flowers were crossed with pollen from a distinct plant and produced twenty-seven capsules, each containing five seeds. Thirty-two flowers were fertilized with their own pollen, and produced only six capsules, each with five seeds. So that the crossed and self-fertilized capsules contained the same number of seeds, though many more capsules were produced by the cross-fertilized than by the self-fertilized flowers, in the ratio of 100 to 21.

An equal number of seeds of both lots were weighed, and the crossed seeds were to the self-fertilized in weight as 100 to 82. Therefore a cross increases the number of capsules produced and the weight of the seeds, but not the number of seeds in each capsule.

These two lots of seeds, after germinating on sand, were planted on the opposite sides of three pots. The seedlings when from 6 to 7 inches in height were equal. The plants were measured when fully grown, but their heights were so unequal in the several pots, that the result cannot be fully trusted.

TABLE LXXV
Nolana prostrata

No. of pot	Crossed plants	Self-fertilized plants
	Inches	*Inches*
I	8⁴⁄₈	4²⁄₈
	6⁴⁄₈	7⁴⁄₈
II	10⁴⁄₈	14⁴⁄₈
	18	18
III	20²⁄₈	22⁶⁄₈
Total in inches	63·75	67·00

The five crossed plants average 12·75, and the five self-fertilized 13·4 inches in height; or as 100 to 105. /

CHAPTER VI

SOLANACEAE, PRIMULACEAE, POLYGONEAE, ETC.

Petunia violacea, crossed and self-fertilized plants compared for four generations – Effects of a cross with a fresh stock – Uniform colour of the flowers on the self-fertilized plants of the fourth generation – *Nicotiana tabacum,* crossed and self-fertilized plants of equal height – Great effects of a cross with a distinct subvariety on the height, but not on the fertility, of the offspring – *Cyclamen persicum,* crossed seedlings greatly superior to the self-fertilized – *Anagallis collina* – *Primula veris* – Equal-styled variety of *Primula veris,* fertility of, greatly increased by a cross with a fresh stock – *Fagopyrum esculentum* – *Beta vulgaris* – *Canna warscewiczi,* crossed and self-fertilized plants of equal height – *Zea mays* – *Phalaris canariensis.*

XXV. SOLANACEAE

PETUNIA VIOLACEA

Dingy purple variety

The flowers on this plant are so seldom visited during the day by the insects in this country, that I have never seen an instance; but my gardener, on whom I can rely, once saw some humble-bees at work. Mr Meehan says,[1] that in the United States bees bore through the corolla for the nectar, and adds that their 'fertilization is carried on by night-moths'.

In France M. Naudin, after castrating a large number of flowers whilst in bud, left them exposed to the visits of insects, and about a quarter produced capsules;[2] but I am convinced that a much larger proportion of flowers in my garden are cross-fertilized by insects, for protected flowers which had their own pollen placed on the stigmas never yielded nearly a full complement of seed; whilst those left uncovered produced fine capsules, showing that pollen from other plants must have been brought to them, probably by moths. Plants

[1] *Proc. Acad. Nat. Sc. of Philadelphia,* 2 August, 1870, p. 90. Professor Hildebrand also informs me that moths, especially *Sphinx convolvuli,* largely haunt the flowers in Germany. So it is, as I hear from Mr Boulger, with moths in England.

[2] *Annales des Sc. Nat.,* 4th series, *Bot.,* vol. ix, chap. 5.

growing vigorously and flowering in pots in the greenhouse, never yielded a single / capsule; and this may be attributed, at least in chief part, to the exclusion of moths.

Six flowers on a plant covered by a net were crossed with pollen from a distinct plant and produced six capsules, containing by weight 4·44 grains of seed. Six other flowers were fertilized with their own pollen and produced only three capsules, containing only 1·49 grain weight of seed. From this it follows that an equal number of crossed and self-fertilized capsules would have contained seeds by weight as 100 to 67. I should not have thought the proportional contents of so few capsules worth giving, had not nearly the same result been confirmed by several subsequent trials.

Seeds of the two lots were placed on sand, and many of the self-fertilized seeds germinated before the crossed, and were rejected. Several pairs in an equal state of germination were planted on the opposite sides of Pots I and II; but only the tallest plant on each side was measured. Seeds were also sown thickly on the two sides of a large pot (III), the seedlings being afterwards thinned, so that an equal number was left on each side; the three tallest on each side being measured. The pots were kept in the greenhouse, and the plants were trained up sticks. For some time the young crossed plants had no advantage in height over the self-fertilized; but their leaves were larger. When fully grown and in flower the plants were measured, as follows:

TABLE LXXVI
Petunia violacea (first generation)

No. of pot	Crossed plants	Self-fertilized plants
	Inches	Inches
I	30	20⁴⁄₈
II	34⁴⁄₈	27⁴⁄₈
III	34	28⁴⁄₈
	30⁴⁄₈	27⁴⁄₈
	25	26
Total in inches	154	130

The five tallest crossed plants here average 30·8, and the five tallest self-fertilized 26 inches in height, or as 100 to 84. /

Three capsules were obtained by crossing flowers on the above crossed plants, and three other capsules by again self-fertilizing flowers on the self-fertilized plants. One of the latter capsules appeared as fine as any one of the crossed capsules; but the other two contained many imperfect seeds. From these two lots of seeds the plants of the following generation were raised.

Crossed and self-fertilized plants of the second generation. As in the last generation, many of the self-fertilized seeds germinated before the crossed.

Seeds in an equal state of germination were planted on the opposite sides of three pots. The crossed seedlings soon greatly exceeded in height the self-fertilized. In Pot I, when the tallest crossed plant was 10½ inches high, the tallest self-fertilized was only 3½ inches; in Pot II the excess in height of the crossed was not quite so great. The plants were treated as in the last generation, and when fully grown measured as before. In Pot III both the crossed plants were killed at an early age by some animal, so that the self-fertilized had no competitors. Nevertheless these two self-fertilized plants were measured, and are included in the following table. The crossed plants flowered long before their self-fertilized opponents in Pots I and II, and before those growing separately in Pot III.

TABLE LXXVII
Petunia violacea (second generation)

No. of pot	Crossed plants	Self-fertilized plants
	Inches	*Inches*
I	57⅞	13⅜
	36⅝	8
II	44⅛	33⅞
	24	28
III	0	46⅝
	0	28⅛
Total in inches	162·0	157·5

The four crossed plants average 40·5, and the six self-fertilized 26·25 inches in height; or as 100 to 65. But this great inequality is in part accidental, owing to some of the self-fertilized / plants being very short, and to one of the crossed being very tall.

Twelve flowers on these crossed plants were again crossed, and eleven capsules were produced; of these, five were poor and six good;

the latter contained by weight 3·75 grains of seeds. Twelve flowers on the self-fertilized plants were again fertilized with their own pollen and produced no less than twelve capsules, and the six finest of these contained by weight 2·57 grains of seeds. It should however be observed that these latter capsules were produced by the plants in Pot III, which were not exposed to any competition. The seeds in the six fine crossed capsules to those in the six finest self-fertilized capsules were in weight as 100 to 68. From these seeds the plants of the next generation were raised.

Crossed and self-fertilized plants of the third generation. The above seeds were placed on sand, and after germinating were planted in pairs on the opposite sides of four pots; and all the remaining seeds were thickly sown on the two sides of a fifth large pot. The result was surprising, for the self-fertilized seedlings very early in life beat the crossed, and at one time were nearly double their height. At first the case appeared like that of Mimulus, in which after the third generation a tall and highly self-fertile variety appeared. But as in the two succeeding generations the crossed plants resumed their former superiority over the self-fertilized, the case must be looked at as an anomaly. The sole conjecture which I can form is that the self-fertilized / seeds had not been sufficiently ripened, and thus produced weakly plants, which grew at first at an abnormally quick rate, as occurred with seedlings from not well-ripened self-fertilized seeds of Iberis. When the crossed plants were between 3 and 4 inches in height, the six finest in four of the pots were measured to the summits of their stems, and at the same time the six finest of the self-fertilized plants. The

TABLE LXXVIII
Petunia violacea (third generation; plants very young)

No. of pot	Crossed plants	Self-fertilized plants
	Inches	Inches
I	1⁴⁄₈	5⁶⁄₈
	1	4⁴⁄₈
II	5⁷⁄₈	8³⁄₈
	5⁶⁄₈	6⁷⁄₈
III	4	5⁶⁄₈
IV	1⁴⁄₈	5³⁄₈
Total in inches	19·63	36·50

measurements are given in the preceding table (LXXVIII), and it may be here seen that all the self-fertilized plants exceed their opponents in height, whereas when subsequently measured the excess of the self-fertilized depended chiefly on the unusual tallness of two of the plants in Pot II. The crossed plants here average 3·27, and the self-fertilized 6·08 inches in height; or as 100 to 186.

When fully grown they were again measured, as follows:

TABLE LXXIX

Petunia violacea (third generation; plants fully grown)

No. of pot	*Crossed plants*	*Self-fertilized plants*
	Inches	*Inches*
I	41 4/8	40 6/8
	48	39
	36	48
II	36	47
	21	80 2/8
	36 2/8	86 2/8
III	52	46
IV	57	43 6/8
Total in inches	327·75	431·00

The eight crossed plants now averaged 40·96, and the eight self-fertilized plants 53·87 inches in height, or as 100 to 131; and this excess chiefly depended, as already stated, on the unusual tallness of two of the self-fertilized plants in Pot II. The self-fertilized had therefore lost some of their former great superiority over the crossed plants. In three of the pots the self-fertilized plants flowered first; but in Pot III at the same time with the crossed.

The case is rendered the more strange, because the crossed plants in the fifth pot (not included in the two last tables), in / which all the remaining seeds had been thickly sown, were from the first finer plants than the self-fertilized, and had larger leaves. At the period when the two tallest crossed plants in this pot were 6 4/8 and 4 5/8 inches high, the two tallest self-fertilized were only 4 inches. When the two crossed plants were 12 and 10 inches high, the two self-fertilized were only 8 inches. These latter plants, as well as many others on the same side of this pot, never grew any higher, whereas several of the crossed plants grew to the height of two feet! On account of this great superiority of the crossed plants, the plants on neither side of this pot have been included in the two last tables.

Thirty flowers on the crossed plants in Pots I and IV (Table LXXIX) were again crossed, and produced seventeen capsules. Thirty flowers on the self-fertilized plants in the same two pots were again self-fertilized, but produced only seven capsules. The contents of each capsule of both lots were placed in separate watch-glasses, and the seeds from the crossed appeared to the eye to be at least double the number of those from the self-fertilized capsules.

In order to ascertain whether the fertility of the self-fertilized plants had been lessened by the plants having been self-fertilized for the three previous generations, thirty flowers on the crossed plants were fertilized with their own pollen. These yielded only five capsules, and their seeds being placed in separate watch-glasses did not seem more numerous than those from the capsules on the self-fertilized plants self-fertilized for the fourth time. So that as far as can be judged from so few capsules, the self-fertility of the self-fertilized plants had not decreased in comparison with that of the plants which had been intercrossed during the three previous generations. It should, however, be remembered that both lots of plants had been subjected in each generation to almost exactly similar conditions.

Seeds from the crossed plants again crossed, and from the self-fertilized again self-fertilized, produced by the plants in Pot I (Table LXXIX), in which the three self-fertilized plants were on an average only a little taller than the crossed, were used in the following experiment. They were kept separate from two similar lots of seeds produced by the two plants in Pot IV in the same table, in which the crossed plant was much taller than its self-fertilized opponent.

Crossed and self-fertilized plants of the fourth generation / (raised from the plants in Pot I, Table LXXIX). Crossed and self-fertilized seeds from plants of the last generation in Pot I, in Table LXXIX, were placed on sand, and after germinating, were planted in pairs on the opposite sides of four pots. The seedlings when in full flower were measured to the base of the calyx. The remaining seeds were sown crowded on the two sides of Pot V; and the four tallest plants on each side of this pot were measured in the same manner.

The fifteen crossed plants average 46·79, and the fourteen (one having died) self-fertilized plants 32·39 inches in height; or as 100 to 69. So that the crossed plants in this generation had recovered their

TABLE LXXX
Petunia violacea
(fourth generation; raised from plants of the third generation in Pot I, Table LXXIX)

No. of pot	Crossed plants	Self-fertilized plants
	Inches	*Inches*
I	$29\frac{2}{8}$	$30\frac{2}{8}$
	$36\frac{2}{8}$	$34\frac{6}{8}$
	49	$31\frac{3}{8}$
II	$33\frac{3}{8}$	$31\frac{5}{8}$
	$37\frac{3}{8}$	$38\frac{2}{8}$
	$56\frac{4}{8}$	$38\frac{4}{8}$
III	46	$45\frac{1}{8}$
	$67\frac{2}{8}$	45
	$54\frac{3}{8}$	$23\frac{2}{8}$
IV	$51\frac{6}{8}$	34
	$51\frac{7}{8}$	0
V	$49\frac{4}{8}$	$22\frac{3}{8}$
Crowded plants	$46\frac{3}{8}$	$24\frac{2}{8}$
	40	$24\frac{6}{8}$
	53	30
Total in inches	701·88	453·50

wonted superiority over the self-fertilized plants; though the parents of the latter in Pot I, Table LXXIX, were a little taller than their crossed opponents.

Crossed and self-fertilized plants of the fourth generation / (raised from the plants in Pot IV, in Table LXXIX). Two similar lots of seeds, obtained from the plants in Pot IV in Table LXXIX, in which the single crossed plant was at first shorter, but ultimately much taller than its self-fertilized opponent, were treated in every way like their brethren of the same generation in the last experiment. We have in the following Table LXXXI the measurements of the present plants. Although the crossed plants greatly exceeded in height the self-fertilized; yet in three out of the five pots a self-fertilized plant flowered before any one of the crossed; in a fourth pot simultaneously; and in a fifth (viz., Pot II) a crossed plant flowered first.

The thirteen crossed plant here average 44·74, and the thirteen self-fertilized plants 26·87 inches in height; or as 100 to 60. The crossed parents of these plants were much taller, relatively to the self-fertilized parents, than in the last case; and apparently they transmitted some of this superiority to their / crossed offspring. It is unfortunate that I did not turn these plants out of doors, so as to observe their relative fertility, for I compared the pollen from some of the crossed and self-fertilized plants in Pot I, Table LXXXI, and there was a marked

TABLE LXXXI
Petunia violacea
(fourth generation; raised from plants of the third generation in Pot IV, Table LXXIX)

No. of pot	Crossed plants	Self-fertilized plants
	Inches	Inches
I	46	30$^2/_8$
	46	28
II	50$^6/_8$	25
	40$^2/_8$	31$^3/_8$
	37$^3/_8$	22$^4/_8$
III	54$^2/_8$	22$^5/_8$
	61$^1/_8$	26$^6/_8$
	45	32
IV	30	28$^4/_8$
	29$^1/_8$	26
V	37$^4/_8$	40$^2/_8$
Crowded plants	63	18$^5/_8$
	41$^2/_8$	17$^4/_8$
Total in inches	581·63	349·38

difference in its state; that of the crossed plants contained hardly any bad and empty grains, whilst such abounded in the pollen of the self-fertilized plants.

The effects of a cross with a fresh stock. I procured from a garden in Westerham, whence my plants originally came, a fresh plant differing in no respect from mine except in the colour of the flowers, which was a fine purple. But this plant must have been exposed during at least four generations to very different conditions from those to which my plants had been subjected, as these had been grown in pots in the greenhouse. Eight flowers on the self-fertilized plants in Table LXXXI, of the last or fourth self-fertilized generation, were fertilized with pollen from this fresh stock; all eight produced capsules containing together by weight

5·01 grains of seeds. The plants raised from these seeds may be called the *Westerham-crossed.*

Eight flowers on the crossed plants of the last or fourth generation in Table LXXXI were again crossed with pollen from one of the other crossed plants, and produced five capsules, containing by weight 2·07 grains of seeds. The plants raised from these seeds may be called the *intercrossed*; and these form the fifth intercrossed generation.

Eight flowers on the self-fertilized plants of the same generation in Table LXXXI were again self-fertilized, and produced seven capsules, containing by weight 2·1 grains of seeds. The *self-fertilized* plants raised from these seeds form the fifth self-fertilized generation. These latter plants and the intercrossed are comparable in all respects with the crossed and self-fertilized plants of the four previous generations.

From the foregoing data it is easy to calculate that,

	Gr. weight of seed
Ten Westerham-crossed capsules would have contained	6·26
Ten intercrossed capsules would have contained	4·14
Ten self-fertilized capsules would have contained	3·00

We thus get the following ratios: /

Seeds from the Westerham-crossed capsules to those from the capsules of the fifth self-fertilized generation in weight	as 100 to 48
Seeds from the Westerham-crossed capsules to those from the capsules of the fifth intercrossed generation	as 100 to 66
Seeds from the intercrossed to those from the self-fertilized capsules	as 100 to 72

So that a cross with pollen from a fresh stock greatly increased the productiveness of the flowers on plants which had been self-fertilized for the four previous generations, in comparison not only with the flowers on the same plants self-fertilized for the fifth time, but with the flowers on the crossed plants crossed with pollen from another plant of the same old stock for the fifth time.

These three lots of seeds were placed on sand, and were planted in an equal state of germination in seven pots, each made tripartite by three superficial partitions. Some of the remaining seeds, whether or not in a state of germination, were thickly sown in an eighth pot. The pots were kept in the greenhouse, and the plants trained up sticks. They were first measured to the tops of their stems when coming into flower; and the twenty-two Westerham-crossed plants then averaged

25·51 inches; the twenty-three intercrossed plants 30·38; and the twenty-three self-fertilized plants 23·40 inches in height. We thus get the following ratios:

The Westerham-crossed plants in height to the self-fertilized	as 100 to 91
The Westerham-crossed plants in height to the intercrossed	as 100 to 119
The intercrossed plants in height to the self-fertilized	as 100 to 77

These plants were again measured when their growth appeared on a casual inspection to be complete. But in this I was mistaken, for after cutting them down, I found that the summits of the stems of the Westerham-crossed plants were still growing vigorously; whilst the intercrossed had almost, and the self-fertilized had quite completed their growth. Therefore I do not doubt, if the three lots had been left to grow for another month, that the ratios would have been somewhat different / from those deduced from the measurements in the following table (LXXXII).

The twenty-one Westerham-crossed plants now averaged 50·05 inches; the twenty-two intercrossed plants, 54·11 inches; and the twenty-one self-fertilized plants, 33·23 inches in height. We thus get the following ratios: /

The Westerham-crossed plants in height to the self-fertilized	as 100 to 66
The Westerham-crossed plants in height to the intercrossed	as 100 to 108
The intercrossed plants in height to the self-fertilized	as 100 to 61

We here see that the Westerham-crossed (the offspring of plants self-fertilized for four generations and then crossed with a fresh stock) have gained greatly in height, since they were first measured, relatively to the plants self-fertilized for five generations. They were then as 100 to 91, and now as 100 to 66 in height. The intercrossed plants (i.e., those which had been intercrossed for the last five generations) likewise exceed in height the self-fertilized plants, as occurred in all the previous generations with the exception of the abnormal plants of the third generation. On the other hand, the Westerham-crossed plants are exceeded in height by the intercrossed; and this is a surprising fact, judging from most of the other strictly analogous cases. But as the Westerham-crossed plants were still growing vigorously, while the intercrossed had almost ceased to grow, there can hardly be a doubt that if left to grow for another month they would have beaten the intercrossed in height. That they were gaining on them is clear, as when measured before they were as 100 to 119, and now as only 100 to

TABLE LXXXII
Petunia violacea

No. of pot	Westerham-crossed plants from (self-fertilized plants of fourth generation crossed by a fresh stock)	Intercrossed plants (plants of one and the same stock intercrossed for five generations)	Self-fertilized plants (self-fertilized for five generations)
	Inches	Inches	Inches
I	$64\frac{5}{8}$	$57\frac{7}{8}$	$43\frac{6}{8}$
	24	64	$56\frac{3}{8}$
	$51\frac{1}{8}$	$58\frac{6}{8}$	$31\frac{5}{8}$
II	$48\frac{7}{8}$	$59\frac{7}{8}$	$41\frac{5}{8}$
	$54\frac{4}{8}$	$58\frac{7}{8}$	$41\frac{2}{8}$
	$58\frac{1}{8}$	53	$18\frac{7}{8}$
III	62	$52\frac{2}{8}$	$46\frac{6}{8}$
	$53\frac{2}{8}$	$54\frac{6}{8}$	45
	$62\frac{7}{8}$	$61\frac{6}{8}$	$19\frac{4}{8}$
IV	$44\frac{4}{8}$	$58\frac{7}{8}$	$37\frac{5}{8}$
	$49\frac{2}{8}$	$65\frac{2}{8}$	$33\frac{2}{8}$
	..	$59\frac{6}{8}$	$32\frac{7}{8}$
V	$43\frac{1}{8}$	$35\frac{5}{8}$	$41\frac{6}{8}$
	$53\frac{7}{8}$	$34\frac{6}{8}$	$26\frac{5}{8}$
	$53\frac{2}{8}$	$54\frac{6}{8}$	0
VI	$37\frac{4}{8}$	56	$46\frac{5}{8}$
	61	$63\frac{5}{8}$	$29\frac{6}{8}$
	0	$57\frac{7}{8}$	$14\frac{4}{8}$
VII	$59\frac{6}{8}$	51	43
	$43\frac{1}{8}$	$49\frac{6}{8}$	$12\frac{2}{8}$
	$50\frac{5}{8}$	0	0
VIII	$37\frac{7}{8}$	$38\frac{5}{8}$	$21\frac{6}{8}$
Crowded	$37\frac{2}{8}$	$44\frac{5}{8}$	$14\frac{5}{8}$
Total in inches	1051·25	1190·50	697·88

108 in height. The Westerham-crossed plants had also leaves of a darker green, and looked altogether more vigorous than the inter-crossed; and what is much more important, they produced, as we shall presently see, much heavier seed-capsules. So that in fact the offspring from the self-fertilized plants of the fourth generation crossed by a fresh stock were superior to the intercrossed, as well as to the self-fertilized plants of the fifth generation – of which latter fact there could not be the least doubt.

These three lots of plants were cut down close to the ground and weighed. The twenty-one Westerham-crossed plants weighed 32 ounces; the twenty-two intercrossed plants, 34 ounces, and the twenty-one self-fertilized plants 7¼ ounces. The following ratios are calculated for an equal number of plants of each kind. But as the self-fertilized plants were just beginning to wither, their relative weight is here slightly too small; and as the Westerham-crossed were still growing vigorously, their relative weight with time allowed would no doubt have greatly increased. /

The Westerham-crossed plants in weight to the self-fertilized   as 100 to 22
The Westerham-crossed plants in weight to the intercrossed   as 100 to 101
The intercrossed plants in weight to the self-fertilized   as 100 to 22·3

We here see, judging by weight instead of as before by height, that the Westerham-crossed and the intercrossed have an immense advantage over the self-fertilized. The Westeham-crossed are inferior to the intercrossed by a mere trifle; but it is almost certain that if they had been allowed to go on growing for another month, the former would have completely beaten the latter.

As I had an abundance of seeds of the same three lots, from which the foregoing plants had been raised, these were sown in three long parallel and adjoining rows in the open ground, so as to ascertain whether under these circumstances the results would be nearly the same as before. Late in the autumn (13 November) the ten tallest plants were carefully selected out of each row, and their heights measured, with the following result (Table LXXXIII) /.

The ten Westerham-crossed plants here average 36·67 inches in height; the ten intercrossed plants, 38·27 inches; and the ten self-fertilized, 23·31 inches. These three lots of plants were also weighed; the Westerham-crossed plants weighed 28 ounces; the intercrossed, 41 ounces; and the self-fertilized, 14·75 ounces. We thus get the following ratios:

The Westerham-crossed plants in height to the self-fertilized   as 100 to 63
The Westerham-crossed plants in weight to the self-fertilized   as 100 to 53
The Westerham-crossed plants in height to the intercrossed   as 100 to 104
The Westerham-crossed plants in weight to the intercrossed   as 100 to 146
The intercrossed plants in height to the self-fertilized   as 100 to 61
The intercrossed plants in weight to the self-fertilized   as 100 to 36

Here the relative heights of the three lots are nearly the same (within three or four per cent) as with the plants in the pots. In weight there is

TABLE LXXXIII
Petunia violacea (plants growing in the open ground)

Westerham-crossed plants (from self-fertilized plants of the fourth generation crossed by a fresh stock)	Intercrossed plants (plants of one and the same stock intercrossed for five generations)	Self-fertilized plants (self-fertilized for five generations)
Inches	Inches	Inches
34 2/8	38	27 3/8
36 2/8	36 2/8	23
35 2/8	39 5/8	25
32 4/8	37	24 1/8
37	36	22 4/8
36 4/8	41 3/8	23 3/8
40 7/8	37 2/8	21 5/8
37 2/8	40	23 4/8
38 2/8	41 2/8	21 3/8
38 5/8	36	21 2/8
366·75	382·75	233·13

a much greater difference: the Westerham-crossed exceed the self-fertilized by much less than they did before; but the self-fertilized plants ın the pots had become slightly withered, as before stated, and were in consequence unfairly light. The Westerham-crossed plants are here inferior in weight to the intercrossed plants in a much higher degree than in the pots; and this appeared due to their being much less branched, owing to their having germinated in greater numbers and consequently being much crowded. Their leaves were of a brighter green than those of the intercrossed and self-fertilized plants.

Relative fertility of the three lots of plants. None of the plants in pots in the greenhouse ever produced a capsule; and this may be attributed in chief part to the exclusion of moths. Therefore the fertility of the three lots could be judged of only by that of the plants growing out of doors, which from being left uncovered were probably cross-fertilized. The plants in the three rows were exactly of the same age and had been subjected to closely similar conditions, so that any difference in their fertility must be attributed to their different origin; namely, to the / one lot being derived from plants self-fertilized for four generations and then crossed with a fresh stock; to the second lot being derived from plants of the same old stock intercrossed for five generations; and to the third lot being derived from plants self-fertilized for five generations. All the capsules, some nearly mature and some only half-grown, were gathered, counted, and weighed from the ten finest plants in each of the three rows, of which the measurements and weights have already been given. The intercrossed plants, as we have seen, were taller and considerably heavier than the plants of the other two lots, and they produced a greater number of capsules than did even the Westerham-crossed plants; and this may be attributed to the latter having grown more crowded and being in consequence less branched. Therefore the average weight of an equal number of capsules from each lot of plants seems to be the fairest standard of comparison, as their weights will have been determined chiefly by the number of the included seeds. As the intercrossed plants were taller and heavier than the plants of the other two lots, it might have been expected that they would have produced the finest or heaviest capsules; but this was very far from being the case.

The ten tallest Westerham-crossed plants produced 111 ripe and unripe capsules, weighing 121·2 grains. Therefore 100 of such capsules would have weighed 109·18 grains.

The ten tallest intercrossed plants produced 129 capsules, weighing 76·45 grains. Therefore 100 of these capsules would have weighed 59·26 grains.

The ten tallest self-fertilized plants produced only 44 capsules, weighing 22·35 grains. Therefore 100 of these capsules would have weighed 50·79 grains.

From these data we get the following ratios for the fertility of the three lots, as deduced from the relative weights of an equal number of capsules from the finest plants in each lot:

Westerham-crossed plants to self-fertilized plants   as 100 to 46
Westerham-crossed plants to intercrossed plants   as 100 to 54
Intercrossed plants to self-fertilized plants   as 100 to 86

We here see how potent the influence of a cross with pollen from a fresh stock has been on the fertility of plants self-fertilized for four generations, in comparison with plants of the old stock when either intercrossed or self-fertilized for five generations; the / flowers on all these plants having been left to be freely crossed by insects or to fertilize themselves. The Westerham-crossed plants were also much taller and heavier plants than the self-fertilized, both in the pots and open ground; but they were less tall and heavy than the intercrossed plants. This latter result, however, would almost certainly have been reversed, if the plants had been allowed to grow for another month, as the Westerham-crossed were still growing vigorously, whilst the inter-crossed had almost ceased to grow. This case reminds us of the some-what analogous one of Eschscholtzia, in which plants raised from a cross with a fresh stock did not grow higher than the self-fertilized or intercrossed plants, but produced a greater number of seed-capsules, which contained a far larger average number of seeds.

Colour of the flowers on the above three lots of plants. The original mother-plant, from which the five successive self-fertilized generations were raised, bore dingy purple flowers. At no time was any selection prac-tised, and the plants were subjected in each generation to extremely uniform conditions. The result was, as in some previous cases, that the flowers on all the self-fertilized plants, both in the pots and open ground, were absolutely uniform in tint; this being a dull, rather peculiar flesh colour. This uniformity was very striking in the long row of plants growing in the open ground, and these first attracted my attention. I did not notice in which generation the original colour began to change and to become uniform, but I have every reason to believe that the change was gradual. The flowers on the intercrossed plants were mostly of the same tint, but not nearly so uniform as those on the self-fertilized plants, and many of them were pale, approaching almost to white. The flowers on the plants from the cross with the purple-flowered Westerham stock were, as might have been expected, much more purple and not nearly so uniform in tint. The self-fertilized plants were also remarkably uniform in height, as judged by the eye; the intercrossed less so, whilst the Westerham-crossed plants varied much in height.

NICOTIANA TABACUM

This plant offers a curious case. Out of six trials with crossed and self-fertilized plants, belonging to three successive generations, in one alone did the crossed show any marked superiority in height over the self-fertilized; in four of the trials they were / approximately equal; and in one (i.e. the first generation) the self-fertilized plants were greatly superior to the crossed. In no case did the capsules from flowers fertilized with pollen from a distinct plant yield many more, and sometimes they yielded much fewer seeds than the capsules from self-fertilized flowers. But when the flowers of one variety were crossed with pollen from a slightly different variety, which had grown under somewhat different conditions – that is, by a fresh stock – the seedlings derived from this cross exceeded in height and weight those from the self-fertilized flowers in an extraordinary degree.

Twelve flowers on some plants of the common tobacco, raised from purchased seeds, were crossed with pollen from a distinct plant of the same lot, and these produced ten capsules. Twelve flowers on the same plants were fertilized with their own pollen, and produced eleven capsules. The seeds in the ten crossed capsules weighed 31·7 grains, whilst those in ten of the self-fertilized capsules weighed 47·67 grains; or as 100 to 150. The much greater productiveness of the self-fertilized than of the crossed capsules can hardly be attributed to chance, as all the capsules of both lots were very fine and healthy ones.

The seeds were placed on sand, and several pairs in an equal state of germination were planted on the opposite sides of three pots. The remaining seeds were thickly sown on the two sides of Pot IV, so that the plants in this pot were much crowded. The tallest plant on each side of each pot was measured. Whilst the plants were quite young the four tallest crossed plants averaged 7·87 inches, and the four tallest self-fertilized 14·87 inches in height; or as 100 to 189. The heights at this age are given in the two left-hand columns of the following table.

When in full flower the tallest plants on each side were again measured (see the two right-hand columns), with the following result. But I should state that the pots were not large enough, and the plants never grew to their proper hight. The four tallest crossed plants now averaged 18·5, and the four tallest self-fertilized plants 32·75 inches in height; or as 100 to 178. In all four pots a self-fertilized plant flowered before any one of the crossed.

In Pot IV, in which the plants were extremely crowded, the two lots were at first equal; and ultimately the tallest crossed plant exceeded by a trifle the tallest self-fertilized plant. This recalled to my mind an analogous case in the one generation of Petunia, in which the self-fertilized plants were throughout / their growth taller than the crossed in all the pots except in the crowded one. Accordingly another trial was made, and some of the same crossed and self-fertilized seeds of tobacco were sown thickly on opposite sides of two additional pots; the plants being left to grow up much crowded. When they were between 13 and 14 inches in height there was no difference between the two sides, nor was there any marked difference when the plants had grown

TABLE LXXXIV
Nicotiana tabacum (first generation)

| | 20 May, 1868 | | 6 December, 1868 | |
No. of pot	Crossed plants	Self-fertilized plants	Crossed plants	Self-fertilized plants
	Inches	Inches	Inches	Inches
I	15⁴⁄₈	26	40	44
II	3	15	6⁴⁄₈	43
III	8	13⁴⁄₈	16	33
IV	5	5	11⁴⁄₈	11
Crowded				
Total in inches	31·5	59·5	74·0	131·0

as tall as they could; for in one pot the tallest crossed plant was 26½ inches in height, and exceeded by 2 inches the tallest self-fertilized plant, whilst in the other pot, the tallest crossed plant was shorter by 3½ inches than the tallest self-fertilized plant, which was 22 inches in height.

As the plants did not grow to their proper height in the above small pots in Table LXXXIV, four crossed and four self-fertilized plants were raised from the same seed, and were planted in pairs on the opposite sides of four very large pots containing rich soil; so that they were not exposed to at all severe mutual competition. When these plants were in flower I neglected to measure them, but record in my notes that all four self-fertilized plants exceeded in height the four crossed plants by 2 or 3 inches. We have seen that the flowers on the original or parent-plants which were crossed with pollen from a distinct plant yielded much fewer seeds than those fertilzed with their own pollen; and the trial just given, as well as that in Table LXXXIV, show us clearly / that the plants raised from the crossed seeds were inferior in height to those from the self-fertilized seeds; but only when not greatly crowded. When crowded and thus subjected to very severe competition, the crossed and self-fertilized plants were nearly equal in height.

Crossed and self-fertilized plants of the second generation. Twelve flowers on the crossed plants of the last generation growing in the four large pots just mentioned, were crossed with pollen from a crossed plant growing in one of the other pots; and twelve flowers on the self-fertilized plants were fertilized with their own pollen. All these flowers of both lots produced fine capsules. Ten of the crossed capsules contained by weight 38·92 grains of seeds, and ten of the self-fertilized capsules 37·74 grains; or as 100 to 97. Some of these seeds in an equal state of germination were planted in pairs on the opposite sides of five large pots. A good many of the crossed seeds germinated before the self-fertilized, and were of course rejected. The plants thus raised were measured when several of them were in full flower.

TABLE LXXXV
Nicotiana tabacum (second generation)

No. of pot	Crossed plants	Self-fertilized plants
	Inches	Inches
I	14⁴⁄₈	27⁶⁄₈
	78⁴⁄₈	8⁶⁄₈
	9	56
II	60⁴⁄₈	16⁶⁄₈
	44⁶⁄₈	7
	10	50⁴⁄₈
III	57⅛	87 (A)
	1²⁄₈	81²⁄₈ (B)
IV	6⁶⁄₈	19
	31	43²⁄₈
	69⁴⁄₈	4
V	99⁴⁄₈	9⁴⁄₈
	29²⁄₈	3
Total in inches	511·63	413·75 /

The thirteen crossed plants here average 39·35, and the thirteen self-fertilized plants 31·82 inches in height; or as 100 to 81. But it would be a very much fairer plan to exclude all the starved plants of only 10 inches and under in height; and in this case the nine remaining crossed plants average 53·84, and the seven remaining self-fertilized plants 51·78 inches in height, or as 100 to 96; and this difference is so small that the crossed and self-fertilized plants may be considered as of equal heights.

In addition to these plants, three crossed plants were planted separately in three large pots, and three self-fertilized plants in three other large pots, so that they were not exposed to any competition; and now the self-fertilized plants exceeded the crossed in height by a little, for the three crossed averaged 55·91, and the three self-fertilized 59·16 inches; or as 100 to 106.

Crossed and self-fertilized plants of the third generation. As I wished to ascertain, firstly, whether those self-fertilized plants of the last generation, which greatly exceeded in height their crossed opponents, would transmit the same tendency to their offspring, and secondly, whether they possessed the same sexual constitution, I selected for experiment the two self-fertilized plants marked A and B in Pot III in Table LXXXV, as these two / were of nearly equal height, and were greatly

TABLE LXXXVI
Nicotiana tabacum (third generation)
Seedlings from the self-fertilized Plant A in Pot III, Table LXXXV, of the last or second generation

No. of pot	*From self-fertilized plant, crossed by a crossed plant*	*From self-fertilized plant again self-fertilized, forming the third self-fertilized generation*
	Inches	*Inches*
I	100 2/8	98
	91	79
II	110 2/8	59 1/8
	100 4/8	66 6/8
III	104	79 6/8
IV	84 2/8	110 4/8
	76 4/8	64 1/8
Total in inches	666·75	557·25 /

superior to their crossed opponents. Four flowers on each plant were fertilized with their own pollen, and four others on the same plants were crossed with pollen from one of the crossed plants growing in another pot. This plan differs from that before followed, in which seedlings from crossed plants again crossed, have been compared with seedlings from self-fertilized plants again self-fertilized. The seeds from the crossed and self-fertilized capsules of the above two plants were placed in separate watch-glasses and compared, but were not weighed; and in both cases those from the crossed capsules seemed to be rather less numerous than those from the self-fertilized capsules. These seeds were planted in the usual manner, and the heights of the crossed and self-fertilized seedlings, when fully grown, are given in the preceding and following table, LXXXVI and LXXXVII.

The seven crossed plants in the first of these two tables average 95·25, and the seven self-fertilized 79·6 inches in height; or as 100 to 83. In half the pots a crossed plant, and in the other half a self-fertilized plant flowered first.

We now come to the seedlings raised from the other parent-plant B.

TABLE LXXXVII
Nicotiana tabacum (third generation)
Seedlings from the self-fertilized Plant B in Pot III, Table LXXXV,
of the last or second generation

No. of pot	From self-fertilized plant, crossed by a crossed plant	From self-fertilized plant again self-fertilized, forming the third self-fertilized generation
	Inches	Inches
I	87²⁄₈	72⁴⁄₈
	49	14²⁄₈
II	98⁴⁄₈	73
	0	110⁴⁄₈
III	99	106²⁄₈
	15²⁄₈	75⁶⁄₈
IV	97⁶⁄₈	48⁶⁄₈
V	48⁶⁄₈	81²⁄₈
	0	61²⁄₈
Total in inches	495·50	641·75 /

The seven crossed plants (for two of them died) here average 70·78 inches, and the nine self-fertilized plants 71·3 inches in height; or as 100 to barely 101. In four out of these five pots, a self-fertilized plant flowered before any one of the crossed plants. So that, differently from the last case, the self-fertilized plants are in some respects slightly superior to the crossed.

If we now consider the crossed and self-fertilized plants of the three generations, we find an extraordinary diversity in their relative heights. In the first generation, the crossed plants were inferior to the self-fertilized as 100 to 178; and the flowers on the original parent-plants which were crossed with pollen from a distinct plant yielded much fewer seeds than the self-fertilized flowers, in the proportion of

100 to 150. But it is a strange fact that the self-fertilized plants, which were subjected to very severe competition with the crossed, had on two occasions no advantage over them. The inferiority of the crossed plants of this first generation cannot be attributed to the immaturity of the seeds, for I carefully examined them; nor to the seeds being diseased or in any way injured in some one capsule, for the contents of the ten crossed capsules were mingled together and a few taken by chance for sowing. In the second generation the crossed and self-fertilized plants were nearly equal in height. In the third generation, crossed and self-fertilized seeds were obtained from two plants of the previous generation, and the seedlings raised from them differed remarkably in constitution; the crossed in the one case exceeded the self-fertilized in height in the ratio of 100 to 83, and in the other case were almost equal. This difference between the two lots, raised at the same time from two plants growing in the same pot, and treated in every respect alike, as well as the extraordinary superiority of the self-fertilized over the crossed plants in the first generation, considered together, make me believe that some individuals of the present species differ to a certain extent from others in their sexual affinities (to use the term employed by Gärtner), like closely allied species of the same genus. Consequently if two plants which thus differ are crossed, the seedlings suffer and are beaten by those from the self-fertilized flowers, in which the sexual elements are of the same nature. It is known[3] that with our domestic animals / certain individuals are sexually incompatible, and will not produce offspring, although fertile with other individuals. But Kölreuter has recorded a case[4] which bears more closely on our present one, as it shows that in the genus Nicotiana the varieties differ in their sexual affinities. He experimented on five varieties of the common tobacco, and proved that they were varieties by showing that they were perfectly fertile when reciprocally crossed; but one of these varieties, if used either as the father or the mother, was more fertile than any of the others when crossed with a widely distinct species, N. glutinosa. As the different varieties thus differ in their sexual affinities, there is nothing surprising in the individuals of the same variety differing in a like manner to a slight degree.

Taking the plants of the three generations altogether, the crossed

---

[3] I have given evidence on this head in my *Variation of Animals and Plants under Domestication*, chap. xviii, 2nd edit., vol. ii, p. 146.
[4] *Das Geschlecht der Pflanzen, Zweite Fortsetzung*, 1764, pp. 55–60.

show no superiority over the self-fertilized, and I can account for this fact only by supposing that with this species, which is perfectly self-fertile without insect aid, most of the individuals are in the same condition, as those of the same variety of the common pea and of a few other exotic plants, which have been self-fertilized for many generations. In such cases a cross between two individuals does no good; nor does it in any case, unless the individuals differ in general constitution, either from so-called spontaneous variation, or from their progenitors having been subjected to different conditions. I believe that this is the true explanation in the present instance, because, as we shall immediately see, the offspring of plants, which did not profit at all by being crossed with a plant of the same stock, profited to an extraordinary degree by a cross with a slightly different subvariety.

The effects of a cross with a fresh stock. I procured some seed of *N. tabacum* from Kew and raised some plants, which formed a slightly different subvariety from my former plants; as the flowers were a shade pinker, the leaves a little more pointed, and the plants not quite so tall. Therefore the advantage in height which the seedlings gained by this cross cannot be attributed to direct inheritance. Two of the plants of the third self-fertilized generation, growing in Pots II and V in Table LXXXVII, which exceeded in height their crossed opponents (as did their parents in a still higher degree) were self-fertilized with pollen from the Kew plants, that is, by a fresh stock. The seedlings / thus raised may be called the Kew-crossed. Some other flowers on the same two plants were fertilized with their own pollen, and the seedlings thus raised form the fourth self-fertilized generation. The crossed capsules produced by the plant in Pot II, Table LXXXVII, were plainly less fine than the self-fertilized capsules on the same plant. In Pot V the one finest capsule was also a self-fertilized one; but the seeds produced by the two crossed capsules together exceeded in number those produced by the two self-fertilized capsules on the same plant. Therefore as far as the flowers on the parent-plants are concerned, a cross with pollen from a fresh stock did little or no good; and I did not expect that the offspring would have received any benefit, but in this I was completely mistaken.

The crossed and self-fertilized seeds from the two plants were placed on bare sand, and very many of the crossed seeds of both sets germinated before the self-fertilized seeds, and protruded their radicles at a quicker rate. Hence many of the crossed seeds had to be rejected, before pairs in an equal state of germination were obtained for planting on the opposite sides of sixteen large pots. The two series of seedlings raised from the parent-plants in the two Pots II and V were kept separate, and when fully grown were measured to the tips of their highest leaves, as shown in the following double table. But as there was no uniform difference in height between the crossed and self-fertilized seedlings raised from the two plants, their heights have been added together in calculating the averages. I should state that by the accidental fall of a large bush in the greenhouse, several plants in both the series were much injured. These were at once measured together with their opponents and afterwards thrown away. The others were left to grow to their full height, and were measured when in flower.

TABLE LXXXVIII
Nicotiana tabacum
Plants raised from two plants of the third self-fertilized generation in Pots II and V, in Table LXXXVII

From Pot II, Table LXXXVII			From Pot V, Table LXXXVII		
No. of pot	Kew-crossed plants	Plants of the fourth self-fertilized generation	No. of pot	Kew-crossed plants	Plants of the fourth self-fertilized generation
	Inches	Inches		Inches	Inches
I	84⁶⁄₈	68⁴⁄₈	I	77⁶⁄₈	56
	31	5		7²⁄₈	5³⁄₈
II	78⁴⁄₈	51⁴⁄₈	II	55⁴⁄₈	27⁶⁄₈
	48	70		18	7
III	77³⁄₈	12⁶⁄₈	III	76²⁄₈	60⁶⁄₈
	77⁴⁄₈	6⁶⁄₈			
IV	49²⁄₈	29⁴⁄₈	IV	90⁴⁄₈	11⁶⁄₈
	15⁶⁄₈	32		22²⁄₈	4¹⁄₈
V	89	85	V	94²⁄₈	28⁴⁄₈
	17	5³⁄₈			
VI	90	80	VI	78	78⁶⁄₈
VII	84⁴⁄₈	48⁶⁄₈	VII	85⁴⁄₈	61¹⁄₈
	76⁶⁄₈	56⁶⁄₈			
VIII	83⁴⁄₈	84⁴⁄₈	VIII	65⁵⁄₈	78³⁄₈
				72²⁄₈	27¹⁄₈
Total in inches	902·63	636·13	Total in inches	743·13	447·38

This accident accounts for the small height of some of the pairs; but as all the pairs, whether only partly or fully grown, were measured at the same time, the measurements are fair.

The average height of the twenty-six crossed plants in the sixteen pots of the two series is 63·29, and that of the twenty-six self-fertilized plants is 41·67 inches; or as 100 to 66. The superiority of the crossed plants was shown in another way, for in every one of the sixteen pots a crossed plant flowered before a self-fertilized one, with the exception of Pot VI of the second series, in which the plants on the two sides flowered simultaneously. /

Some of the remaining seeds of both series, whether or not in a state of germination, were thickly sown on the opposite sides of two very large pots; and the six highest plants on each side of each pot were measured after they had grown to nearly their full height. But their heights were much less than in the former trials, owing to their extremely crowded condition. Even whilst quite young, the crossed seedlings manifestly had much broader and finer leaves than the self-fertilized seedlings. /

TABLE LXXXIX
Nicotiana tabacum
Plants of the same parentage as those in Table LXXXVIII, but grown extremely crowded in two large pots

From Pot II, Table LXXXVII		From Pot V, Table LXXXVII	
Kew-crossed plants	Plants of the fourth self-fertilized generation	Kew-crossed plants	Plants of the fourth self-fertilized generation
Inches	Inches	Inches	Inches
42⁴⁄₈	22⁴⁄₈	44⁶⁄₈	22⁴⁄₈
34	19²⁄₈	42⁴⁄₈	21
30⁴⁄₈	14²⁄₈	27⁴⁄₈	18
23⁴⁄₈	16	31²⁄₈	15²⁄₈
26⁶⁄₈	13⁴⁄₈	32	13⁵⁄₈
18³⁄₈	16	24⁶⁄₈	14⁶⁄₈
175·63	101·50	202·75	105·13

The twelve tallest crossed plants in the two pots belonging to the two series average here 31·53, and the twelve tallest self-fertilized plants 17·21 inches in height; or as 100 to 54. The plants on both sides, when fully grown, some time after they had been measured, were cut down close to the ground and weighed. The twelve crossed plants weighed

21·25 ounces; and the twelve self-fertilized plants only 7·83 ounces; or in weight as 100 to 37.

The rest of the crossed and self-fertilized seeds from the two parent-plants (the same as in the last experiment) was sown on 1 July in four long parallel and separate rows in good soil in the open ground; so that the seedlings were not subjected to any mutual competition. The summer was wet and unfavourable for their growth. Whilst the seedlings were very small the two crossed rows had a clear advantage over the two self-fertilized rows. When fully grown the twenty tallest crossed plants and the twenty tallest self-fertilized plants were selected and measured on 11 November to the extremities of their leaves, as shown in the following table (XC). Of the twenty crossed plants, twelve had flowered; whilst of the twenty self-fertilized plants, one alone had flowered. /

TABLE XC

Nicotiana tabacum

Plants raised from the same seeds as in the last two experiments, but sown separately in the open ground, so as not to compete together

From Pot II, Table LXXXVII		*From Pot V, Table LXXXVII*	
Kew-crossed plants	*Plants of the fourth self-fertilized generation*	*Kew-crossed plants*	*Plants of the fourth self-fertilized generation*
Inches	*Inches*	*Inches*	*Inches*
42 2/8	22 6/8	54 4/8	34 4/8
54 5/8	37 4/8	51 4/8	38 5/8
39 3/8	34 4/8	45	40 6/8
53 2/8	30	43	43 2/8
49 3/8	28 6/8	43	40
50 3/8	31 2/8	48 6/8	38 2/8
47 1/8	25 4/8	44	35 6/8
57 3/8	26 2/8	48 2/8	39 6/8
37	22 3/8	55 1/8	47 6/8
48	28	63	58 5/8
478·75	286·86	496·13	417·25

The twenty tallest crossed plants here average 48·74, and the twenty tallest self-fertilized 35·2 inches in height; or as 100 to 72. These plants after being measured were cut down close to the ground, and the twenty crossed plants weighed 195·75 ounces, and the twenty self-fertilized plants 123·25 ounces; or as 100 to 63.

In the three preceding tables, LXXXVIII, LXXXIX, and XC, we have the measurements of fifty-six plants derived from two plants of the third self-fertilized generation crossed with pollen from a fresh stock, and of fifty-six plants of the fourth self-fertilized generation derived from the same two plants. These crossed and self-fertilized plants were treated in three different ways, having been put, first, into moderately close competition with one another in pots; secondly, having been subjected to unfavourable conditions and to very severe competition from being greatly crowded in two large pots; and thirdly, having been sown separately in open and good ground, so as not to suffer from any mutual competition. In all these cases the crossed plants in each lot were greatly superior to the self-fertilized. / This was shown in several ways – by the earlier germination of the crossed seeds, by the more rapid growth of the seedlings whilst quite young, by the earlier flowering of the mature plants, as well as by the greater height which they ultimately attained. The superiority of the crossed plants was shown still more plainly when the two lots were weighed; the weight of the crossed plants to that of the self-fertilized in the two crowded pots being as 100 to 37. Better evidence could hardly be desired of the immense advantage derived from a cross with a fresh stock.

XXVI. PRIMULACEAE

CYCLAMEN PERSICUM[5]

Ten flowers crossed with pollen from plants known to be distinct seedlings, yielded nine capsules, containing on an average 34·2 seeds, with a maximum of seventy-seven in one. Ten flowers self-fertilized yielded eight capsules, containing on an average only 13·1 seeds, with a maximum of twenty-five in one. This gives a ratio of 100 to 38 for the average number of seeds per capsule for the crossed and self-fertilized flowers. The flowers hang downwards, and as the stigmas stand close beneath the anthers, it might have been expected that pollen would, have fallen on them, and that they would have been spontaneously self-fertilized; but these covered-up plants did not produce a single capsule. On some other occasions uncovered plants in the same green-house produced plenty of capsules, and I suppose that the flowers had

---

[5] *Cyclamen repandum*, according to Lecoq (*Géographie Botanipue de l'Europe*, vol. viii, 1858, p. 150), is proterandrous, and this I believe to be the case with *C. persicum*.

been visited by bees, which could hardly fail to carry pollen from plant to plant.

The seeds obtained in the manner just described were placed on sand, and after germinating were planted in pairs – three crossed and three self-fertilized plants on the opposite sides of four pots. When the leaves were 2 or 3 inches in length, including the foot-stalks, the seedlings on both sides were equal. In the course of a month of two the crossed plants began to show a slight superiority over the self-fertilized, which steadily increased; and the crossed flowered in all four pots some weeks before, and much more profusely than the self-fertilized. The two tallest flower-stems on the crossed plants in each pot were now measured, and the average height of the eight stems / was 9·49 inches. After a considerable interval of time the self-fertilized plants flowered, and several of their flower-stems (but I forgot to record how many) were roughly measured, and their average height was a little under 7·5 inches; so that the flower-stems on the crossed plants to those on the self-fertilized were at least as 100 to 79. The reason why I did not make more careful measurements of the self-fertilized plants was, that they looked such poor specimens that I determined to have them re-potted in larger pots and in the following year to measure them carefully; but we shall see that this was partly frustrated by so few flower-stems being then produced.

These plants were left uncovered in the greenhouse; and the twelve crossed plants produced forty capsules, whilst the twelve self-fertilized plants produced only five; or as 100 to 12. But this difference does not give a just idea of the relative fertility of the two lots. I counted the seeds in one of the finest capsules on the crossed plants, and it contained seventy-three; whilst the finest of the five capsules produced by the self-fertilized plants contained only thirty-five good seeds. In the other four capsules most of the seeds were barely half as large as those in the crossed capsules. /

In the following year the crossed plants again bore many flowers before the self-fertilized bore a single one. The three tallest flower-stems on the crossed plants in each of the pots were measured, as shown in Table XCI. In Pots I and II the self-fertilized plants did not produce a single flower-stem; in Pot IV only one; and in Pot III six, of which the three tallest were measured.

The average height of the twelve flower-stems on the crossed plants is 9·99, and that of the four flower-stems on the self-fertilized

TABLE XCI

Cyclamen persicum: 0 *implies that no flower-stem was produced*

No. of pot	*Crossed plants*	*Self-fertilized plants*
	Inches	*Inches*
I	10	0
	9 2/8	0
	10 2/8	0
II	9 2/8	0
	10	0
	10 2/8	0
III	9 1/8	8
	9 5/8	6 7/8
	9 5/8	6 6/8
IV	11 1/8	0
	10 5/8	7 7/8
	10 6/8	0
Total in inches	119·88	29·50

plants 7.37 inches; or as 100 to 74. The self-fertilized plants were miserable specimens, whilst the crossed ones looked very vigorous.

ANAGALLIS

Anagallis collina, var. grandiflora (pale red and blue-flowered subvarieties)

First, twenty-five flowers on some plants of the red variety were crossed with pollen from a distinct plant of the same variety, and produced ten capsules; thirty-one flowers were fertilized with their own pollen, and produced eighteen capsules. These plants, which were grown in pots in the greenhouse, were evidently in a very sterile condition, and the seeds in both sets of capsules, especially in the self-fertilized, although numerous, were of so poor a quality that it was very difficult to determine which were good and which bad. But as far as I could judge, the crossed capsules contained on an average 6·3 good seeds, with a maximum in one of thirteen; whilst the self-fertilized contained 6·05 such seeds, with a maximum in one of fourteen.

Secondly, eleven flowers on the red variety were castrated whilst young and fertilized with pollen from the blue variety, and this cross evidently much increased their fertility; for the eleven flowers yielded seven capsules, which contained on an average twice as many good seeds as before, viz., 12·7; with a maximum in two of the capsules of seventeen seeds. Therefore these crossed capsules yielded seeds compared with those in the foregoing self-fertilized capsules, as 100 to 48. These seeds were also conspicuously larger than those from the cross between two individuals of the same red variety, and germinated much more freely. The flowers on most of the plants produced by the cross between the two-coloured varieties (of which several were raised), / took after their mother, and were red-coloured. But on two of the plants the flowers were plainly stained with blue, and to such a degree in one case as to be almost intermediate in tint.

The crossed seeds of the two foregoing kinds and the self-fertilized were sown on the opposite sides of two large pots, and the seedlings were measured when fully grown, as shown in the two following tables:

TABLE XCII
Anagallis collina

Red variety crossed by a distinct plant of the red variety, and red variety self-fertilized

No. of pot	*Crossed plants*	*Self-fertilized plants*
	Inches	*Inches*
I	23 4/8	15 4/8
	21	15 4/8
	17 2/8	14
Total in inches	61·75	45.00

Red variety crossed by blue variety, and red variety self-fertilized

No. of pot	*Crossed plants*	*Self-fertilized plants*
	Inches	*Inches*
	30 4/8	24 4/8
	27 3/8	18 4/8
	25	11 6/8
Total in inches	82·88	54·75
Total of both lots	144·63	99·75

As the plants of the two lots are few in number, they may be run together for the general average; but I may first state that the height of the seedlings from the cross between two individuals of the red variety is to that of the self-fertilized plants of the red variety as 100 to 73; whereas the height of the crossed offspring from the two varieties to the self-fertilized plants of the red variety is as 100 to 66. So that the cross between the two varieties is here seen to be the most advantageous. The average weight of all six crossed plants in the two lots taken together is / 48·20, and that of the six self-fertilized plants 33·25; or as 100 to 69.

These six crossed plants produced spontaneously twenty-six capsules, whilst the six self-fertilized plants produced only two, or as 100 to 8. There is therefore the same extraordinary difference in fertility between the crossed and self-fertilized plants as in the last genus, Cyclamen, which belongs to the same family of the Primulaceae.

PRIMULA VERIS Brit. Flora

(Var. officinalis, Linn.)   The cowslip

Most of the species in this genus are heterostyled or dimorphic; that is, they present two forms – one long-styled with short stamens, and the other short-styled with long stamens.[6] For complete fertilization it is necessary that pollen from the one form should be applied to the stigma of the other form; and this is effected under nature by insects. Such unions, and the seedlings raised from them, I have called legitimate. If one form is fertilized with pollen from the same form, the full complement of seed is not produced; and in the case of some heterostyled genera no seed at all is produced. Such unions, and the seedlings raised from them, I have called illegitimate. These seedlings are often dwarfed and more or less sterile, like hybrids. I possessed some long-styled plants of *P. veris*, which during four successive generations had been produced from illegitimate unions between long-styled plants; they were, moreover, in some degree inter-related, and had been subjected all the time to similar conditions in pots in the greenhouse. As long as they were cultivated in this manner, they grew well and were healthy and fertile. Their fertility even increased in the later generations, as if they were becoming habituated to illegitimate fertilization. Plants of the first illegitimate generation when taken from the greenhouse and planted in moderately good soil out of doors grew well and were healthy; but when those of the two last illegitimate generations were thus treated they became excessively sterile and

---

[6] See my work, *The Different Forms of Flowers on Plants of the same Species*, 1877, or my papers in *Journal of Proc. Linn. Soc.*, vol. vi, 1862, p. 77, and vol. x, 1867, p. 393.

dwarfed, and remained so during the following / year, by which time they ought to have become accustomed to growing out of doors, so that they must have possessed a weak constitution.

Under these circumstances, it seemed advisable to ascertain what would be the effect of legitimately crossing long-styled plants of the fourth illegitimate generation with pollen taken from non-related short-styled plants, growing under different conditions. Accordingly several flowers on plants of the fourth illegitimate generation (i.e., great-great-grandchildren of plants which had been legitimately fertilized), growing vigorously in pots in the greenhouse, were legitimately fertilized with pollen from an almost wild short-styled cowslip, and these flowers yielded some fine capsules. Thirty other flowers on the same illegitimate plants were fertilized with their own pollen, and these yielded seventeen capsules, containing on an average thirty-two seeds. This is a high degree of fertility; higher, I believe, than that which generally obtained with illegitimately fertilized long-styled plants growing out of doors, and higher than that of the previous illegitimate generations, although their flowers were fertilized with pollen taken from a distinct plant of the same form.

These two lots of seeds were sown (for they will not germinate well when placed on bare sand) on the opposite sides of four pots, and the seedlings were thinned, so that an equal number were left on the two sides. For some time there was no marked difference in height between the two lots; and in Pot III, Table XCIII, the self-fertilized plants were rather the tallest. But by the time that they had thrown up young flower-stems, the legitimately crossed plants appeared much the finest, and had greener and larger leaves. The breadth of the largest leaf on each plant was measured, and those on the crossed plants were on an average a quarter of an inch (exactly 0·28 of an inch) broader than those on the self-fertilized plants. The plants, from being too much crowded, produced poor and short flower-stems. The two finest on each side were measured; the eight on the legitimately crossed plants averaged 4·08, and the eight on the illegitimately self-fertilized plants averaged 2·93 inches in height; or as 100 to 72.

These plants after they had flowered were turned out of their pots, and planted in fairly good soil in the open ground. In the following year (1870), when in full flower, the two tallest flower-stems on each side were again measured, as shown in the / following table, which likewise gives the number of flower-stems produced on both sides of all the pots.

TABLE XCIII
Primula veris

	Legitimately crossed plants		*Illegitimately self-fertilized plants*	
No. of pot	*Height in inches*	*No. of flower-stems produced*	*Height in inches*	*No. of flower-stems produced*
I	9 8	16	$2\frac{1}{8}$ $3\frac{4}{8}$	3
II	7 $6\frac{4}{8}$	16	6 $5\frac{4}{8}$	3
III	6 $6\frac{2}{8}$	16	3 $0\frac{4}{8}$	4
IV	$7\frac{3}{8}$ $6\frac{1}{8}$	14	$2\frac{5}{8}$ $2\frac{4}{8}$	5
Total	56·26	62	25·75	15

The average height of the eight tallest flower-stems on the crossed plants is here 7·03 inches, and that of the eight tallest flower-stems on the self-fertilized plants 3·21 inches; or as 100 to 46. We see, also, that the crossed plants bore sixty-two flower-stems; that is, above four times as many as those (viz., fifteen) borne by the self-fertilized plants. The flowers were left exposed to the visits of insects, and as any plants of both forms grew close by, they must have been legitimately and naturally fertilized. Under these circumstances the crossed plants produced 324 capsules, whilst the self-fertilized produced only 16; and these were all produced by a single plant in Pot II, which was much finer than any other self-fertilized plant. Judging by the number of capsules produced, the fertility of an equal number of crossed and self-fertilized plants was as 100 to 5.

In the succeeding year (1871) I did not count all the flower-stems on these plants, but only those which produced capsules containing good seeds. The season was unfavourable, and the crossed plants produced only forty such flower-stems, bearing / 168 good capsules, whilst the self-fertilized plants produced only two such flower-stems, bearing only 6 capsules, half of which were very poor ones. So that the fertility of the two lots, judging by the number of capsules, was as 100 to 3·5.

In considering the great difference in heights and the wonderful difference in fertility between the two sets of plants, we should bear in mind that this is the result of two distinct agencies. The self-fertilized plants were the product of illegitimate fertilization during five successive generations, in all of which, excepting the last, the plants had been fertilized with pollen taken from a distinct individual belonging to the same form, but which was more or less closely related. The plants had also been subjected in each generation to closely similar conditions. This treatment alone, as I know from other observations, would have greatly reduced the size and fertility of the offspring. On the other hand, the crossed plants were the offspring of long-styled plants of the fourth illegitimate generation legitimately crossed with pollen from a short-styled plant, which, as well as its progenitors, had been exposed to very different conditions; and this latter circumstance alone would have given great vigour to the offspring, as we may infer from the several analogous cases already given. How much proportional weight ought to be attributed to these two agencies – the one tending to injure the self-fertilized offspring, and the other to benefit the crossed offspring – cannot be determined. But we shall immediately see that the greater part of the benefit, as far as increased fertility is concerned, must be attributed to the cross having been made with a fresh stock.

PRIMULA VERIS

Equal-styled and red-flowered var.

I have described in my paper 'On the Illegitimate Unions of Dimorphic and Trimorphic Plants' this remarkable variety, which was sent to me from Edinburgh by Mr J. Scott. It possessed a pistil proper to the long-styled form, and stamens proper to the short-styled form; so that it had lost the heterostyled or dimorphic character common to most of the species of the genus, and may be compared with an hermaphrodite form of a bisexual animal. Consequently the pollen and stigma of the same flower are adapted for complete mutual fertilization, instead of its being necessary that pollen should be brought from one / form to another, as in the common cowslip. From the stigma and anthers standing nearly on the same level, the flowers are perfectly self-fertile when insects are excluded. Owing to the fortunate existence of this variety, it is possible to fertilize its flowers in a legitimate manner with their own pollen, and to cross other flowers in a legitimate manner with pollen from another variety or fresh stock. Thus the offspring from both unions can be compared quite fairly, free from any doubt from the injurious effects of an illegitimate union.

The plants on which I experimented had been raised during two successive generations from spontaneously self-fertilized seeds produced by plants under a net; and as the variety is highly self-fertile, its progenitors in Edinburgh may have been self-fertilized during some previous generations. Several flowers on two of my plants were legitimately crossed with pollen from a short-steyled common cowslip growing almost wild in my orchard; so that the cross was between plants which had been subjected to considerably different conditions. Several other flowers on the same two plants were allowed to fertilize themselves under a net; and this union, as already explained, is a legitimate one.

The crossed and self-fertilized seeds thus obtained were sown thickly on the opposite sides of three pots, and the seedlings thinned, so that an equal number were left on the two sides. The seedlings during the first year were nearly equal in height, excepting in Pot III, Table XCIV, in which the self-fertilized plants had a decided advantage. In the autumn the plants were bedded out, in their pots; owing to this circumstance, and to many plants growing in each pot, they did not flourish, and none were very productive in seeds. But the conditions were perfectly equal and fair for both sides. In the following spring I record in my notes that in two of the pots the crossed plants are 'incomparably the finest in general appearance', and in all three pots they flowered before the self-fertilized. When in full flower the tallest flower-stem on each side of each pot was measured, and the number of the flower-stems on both sides counted, as shown in the following table. The plants were left uncovered, and as other plants were growing close by, the flowers no doubt were crossed by insects. When the capsules were ripe they were gathered and counted, and the result is likewise shown in the following table (XCIV). /

The average height of the three tallest flower-stems on the crossed plants is 8·66 inches, and that of the three on the self-fertilized plants 7·33 inches; or as 100 to 85.

All the crossed plants together produced thirty-three flower-stems, whilst the self-fertilized bore only thirteen. The number of the capsules was counted only on the plants in Pots I and II, for the self-fertilized plants in Pot II produced none; therefore those on the crossed plants on the opposite side were not counted. Capsules not containing any good seeds were rejected. The crossed plants in the above two pots produced 206, and the self-fertilized in the same pots only 32 capsules; or as 100 to 15. Judging from the previous

TABLE XCIV
Primula veris (equal-styled, red-flowered variety)

No. of pot	Crossed plants			Self-fertilized plants		
	Height of tallest flower-stem in inches	No. of flower-stems	No. of good capsules	Height of tallest flower-stem in inches	No. of flower-stems	No. of good capsules
I	10	14	163	6⅛	6	6
II	8⅜	12	Several, not counted	5	2	0
III	7⅜	7	43	10⅜	5	26
Totals	26·0	33	206	22·0	13	32

generations, the extreme unproductiveness of the self-fertilized plants in this experiment was wholly due to their having been subjected to unfavourable conditions, and to severe competition with the crossed plants; for had they grown separately in good soil, it is almost certain that they would have produced a large number of capsules. The seeds were counted in twenty capsules from the crossed plants, and they averaged 24·75; whilst in twenty capsules from the self-fertilized plants the average was 17·65; or as 100 to 71. Moreover, the seeds from the self-fertilized plants were not nearly so fine as those from the crossed plants. If we consider together the number of capsules produced and the average number of contained seeds, the fertility of the crossed plants to the self-fertilized plants was 100 to 11. / We thus see what a great effect, as far as fertility is concerned, was produced by a cross between the two varieties, which had been long exposed to different conditions, in comparison with self-fertilization; the fertilization having been in both cases of the legitimate order.

PRIMULA SINENSIS

As the Chinese primrose is a heterostyled or dimorphic plant, like the common cowslip, it might have been expected that the flowers of both forms when illegitimately fertilized with their own pollen or with that from flowers on another plant of the same form, would have yielded less seed than the legitimately crossed flowers; and that the seedlings raised from illegitimately self-fertilized seeds would have been some-what dwarfed and less fertile, in comparison with the seedlings from legitimately crossed seeds. This holds good in relation to the fertility of the flowers; but to my surprise there was no difference in growth between the offspring from a legitimate union between two distinct plants, and from an illegitimate union whether between the flowers on the same plant, or between distinct plants of the same form. But I have shown, in the work lately referred to, that in England this plant is in an abnormal condition, such as, judging from analogous cases, would tend to render a cross between two individuals of no benefit to the offspring. Our plants have been commonly raised from self-fertilized seeds; and the seedlings have generally been subjected to nearly uniform conditions in pots in greenhouses. Moreover, many of the plants are now varying and changing their character, so as to become in a greater or less degree equal-styled, and in consequence highly self-fertile. From the analogy of *P. veris* there can hardly be a doubt that if a plant of *P. sinensis* could have been procured direct from China, and if it had been crossed with one of our English varieties, the offspring would have shown wonderful superiority in height and fertility (though probably not in the beauty of their flowers) over our ordinary plants.

My first experiment consisted in fertilizing many flowers on long-styled and short-styled plants with their own pollen, and other flowers on the same plants with pollen taken from distinct plants belonging to the same form; so that all the unions were illegitimate. There was no uniform and marked difference in / the number of seeds obtained from these two modes of self-fertilization, both of which were illegiti-mate. The two lots of seeds from both forms were sown thickly on opposite sides of four pots, and numerous plants thus raised. But there was no difference in their growth, excepting in one pot, in which the offspring from the illegitimate union of two long-styled plants exceeded in a decided manner in height the offspring of flowers on the same plants fertilized with their own pollen. But in all four pots the plants raised from the union of distinct plants belonging to the same form, flowered before the offspring from the self-fertilized flowers.

Some long-styled and short-styled plants were now raised from purchased seeds, and flowers on both forms were legitimately crossed with pollen from a distinct plant; and other flowers on both forms were illegitimately fertilized with pollen from the flowers on the same plant. The seeds were sown on opposite sides of Pots I to IV in the following table (XCV); a single plant being left on each side. Several flowers on the illegitimate long-styled and short-styled plants described in the last paragraph, were also legitimately and illegitimately fertilized in the manner just described, and their seeds were sown in Pots V to VIII in the same table. As the two sets of seedlings did not differ in any essential manner, their measurements are given in a single table. I should add that the legitimate unions in both cases yielded, as might have been expected, many more seeds than the illegitimate unions.

TABLE XCV
Primula sinensis

No. of pot	Plants from legitimately crossed seeds	Plants from illegitimately self-fertilized seeds
	Inches	*Inches*
I From short-styled mother	$8\frac{2}{8}$	8
II From short-styled mother	$7\frac{4}{8}$	$8\frac{5}{8}$
III From long-styled mother	$9\frac{5}{8}$	$9\frac{3}{8}$
IV From long-styled mother	$8\frac{4}{8}$	$8\frac{2}{8}$
V From illegitimate short-styled mother	$9\frac{3}{8}$	9
VI From illegitimate short-styled mother	$9\frac{7}{8}$	$9\frac{4}{8}$
VII From illegitimate long-styled mother	$8\frac{4}{8}$	$9\frac{4}{8}$
VIII From illegitimate long-styled mother	$10\frac{4}{8}$	10
Total in inches	72·13	72·25

The seedlings whilst half-grown presented no difference in height on the two sides of the several pots. When fully grown they were measured to the tips of their longest leaves, and the result is given in Table XCV.

In six out of the eight pots the legitimately crossed plants exceeded in height by a trifle the illegitimately self-fertilized plants; but the latter exceeded the former in two of the pots in a more strongly marked manner. The average height of the eight legitimately crossed plants is 9·01, and that of the eight illegitimately self-fertilized 9·03 inches; or as 100 to 100·2. The plants on the opposite sides produced, as far as could be judged by the eye, an equal number of flowers. I did not count the capsules or the seeds produced by them; but undoubtedly, judging from many previous observations, the plants derived from the legitimately crossed seeds would have been considerably more fertile than those from the illegitimately self-fertilized seeds. The crossed plants, as in the previous case, flowered before the / self-fertilized plants in all the pots except in Pot II, in which the two sides flowered simultaneously; and this early flowering may, perhaps, be considered as an advantage. /

XXVII. POLYGONEAE

FAGOPYRUM ESCULENTUM

This plant was discovered by Hildebrand to be heterostyled, that is, to present, like the species of Primula, a long-styled and a short-styled form, which are adapted for reciprocal fertilization. Therefore the following comparison of the growth of the crossed and self-fertilized seedlings is not fair, for we do not know whether the difference in their heights may not be wholly due to the illegitimate fertilization of the self-fertilized flowers.

I obtained seeds by legitimately crossing flowers on long-styled and short-styled plants, and by fertilizing other flowers on both forms with pollen from the same plant. Rather more seeds were obtained by the former than by the latter process; and the legitimately crossed seeds were heavier than an equal number of the illegitimately self-fertilized seeds, in the ratio of 100 to 82. Crossed and self-fertilized seeds from the short-styled parents, after germinating on sand, were planted in pairs on the opposite sides of a large pot; and two similar lots of seeds from long-styled parents were planted in a like manner on the opposite sides of two other pots. In all three pots the legitimately crossed seedlings, when a few inches in height, were taller than the self-fertilized; and in all three pots they flowered before them by one or two days. When fully grown they were all cut down close to the ground, and as I was pressed for time, they were placed in a long row, the cut end of one plant touching the tip of another, and the total length of the legitimately crossed plants was 47 ft 7 in, and of the illegitimately self-fertilized plants 32 ft 8 in. Therefore the average height of the fifteen crossed plants in all three pots was 38·06 inches, and that of the fifteen self-fertilized plants 26·13 inches; or as 100 to 69.

XXVIII. CHENOPODIACEAE

BETA VULGARIS

A single plant, no others growing in the same garden, was left to fertilize itself, and the self-fertilized seeds were collected. Seeds were also collected from a plant growing in the midst of a large bed in another garden; and as the incoherent pollen is abundant, the seeds of this plant will almost certainly have been the product of a cross between distinct plants by means of the wind. Some of the two lots of seeds were sown on the opposite sides of two very large pots; and the young seedlings were thinned, so that an equal but considerable number was left on the two sides. These plants were thus subjected to very severe / competition, as well as to poor conditions. The remaining seeds were sown out of doors in good soil in two long and not closely adjoining rows, so that these seedlings were placed under favourable conditions, and were not subjected to any mutual competition. The self-fertilized seeds in the open ground came up very badly; and on removing the soil in two or three places, it was found that many had sprouted under ground and had then died. No such case had been observed before. Owing to the large number of seedlings which thus perished, the surviving self-fertilized plants grew thinly in the row, and thus had an advantage over the crossed plants, which grew very thickly in the other row. The young plants in the two rows were protected by a little straw during the winter, and those in the two large pots were placed in the greenhouse.

There was no difference between the two lots in the pots until the ensuing spring, when they had grown a little, and then some of the crossed plants were finer and taller than any of the self-fertilized. When in full flower their stems were measured, and the measurements are given in the following table:

TABLE XCVI
Beta vulgaris

No. of pot	*Crossed plants*	*Self-fertilized plants*
	Inches	*Inches*
I	34 6/8	36
	30	20 1/8
	33 6/8	32 2/8
	34 4/8	32
II	42 3/8	42 1/8
	33 1/8	26 4/8
	31 2/8	29 2/8
	33	20 2/8
Total in inches	272·75	238·50

The average height of the eight crossed plants is here 34·09, and that of the eight self-fertilized plants 29·81; or as 100 to 87.

With respect to the plants in the open ground, each long row was divided into half, so as to diminish the chance of any accidental advantage in one part of either row; and the four tallest plants in the two halves of the two rows were carefully / selected and measured. The eight tallest crossed plants averaged 30·92, and the eight tallest self-fertilized 30·7 inches in height, or as 100 to 99; so that they were practically equal. But we should bear in mind that the trial was not quite fair, as the self-fertilized plants had a great advantage over the crossed in being much less crowded in their own row, owing to the large number of seeds which had perished under ground after sprouting. Nor were the lots in the two rows subjected to any mutual competition.

XXIX. CANNACEAE

CANNA WARSCEWICZI

In most or all the species belonging to this genus, the pollen is shed before the flower expands, and adheres in a mass to the foliaceous pistil close beneath the stigmatic surface. As the edge of this mass generally touches the edge of the stigma, and as it was ascertained by trials purposely made that a very few pollen-grains suffice for fertilization, the present species and probably all the others of the genus are highly self-fertile. Exceptions occasionally occur in which, from the stamen being slightly shorter than usual, the pollen is deposited a little beneath the stigmatic surface, and such flowers drop off unimpregnated unless they are artificially fertilized. Sometimes, though rarely, the stamen is a little longer than usual, and then the whole stigmatic surface gets thickly covered with pollen. As some pollen is generally deposited in contact with the edge of the stigma, certain authors have concluded that the flowers are invariably self-fertilized. This is an extraordinary conclusion, for it implies that a great amount of pollen is produced for no purpose. On this view, also, the large size of the stigmatic surface is an unintelligible feature in the structure of the flower, as well as the relative position of all the parts, which is such that when insects visit the flowers to suck the copious nectar, they cannot fail to carry pollen from one flower to another.[7] /

According to Delpino, bees eagerly visit the flowers in North Italy, but I have never seen any insect visiting the flowers of the present species in my hothouse, although many plants grew there during several years. Nevertheless these plants produced plenty of seed, as they likewise did when covered by a net; they are therefore fully capable of self-fertilization, and have probably been self-fertilized in this country for many generations. As they are cultivated in pots, and are not exposed to competition with surrounding plants, they have also been subjected for a considerable time to somewhat uniform conditions. This, therefore, is a case exactly parallel with that of the common pea, in which we have no right to expect much or any good from intercrossing plants thus descended and thus treated; and no good did follow, excepting that the cross-fertilized flowers yielded rather more seeds than the self-fertilized. This species was one of the earlier ones on which I experimented, and as I had not then raised any self-fertilized plants for several successive generations under uniform conditions, I did not know or even suspect that such treatment would interfere with the advantages to be gained from a cross. I was therefore much surprised at the crossed plants not growing more vigorously than the self-fertilized, and a large number of plants were raised, notwithstanding that the present species is an extremely troublesome one to experiment on. The seeds, even those which have been long

[7] Delpino has described (*Bot. Zeitung*, 1867, p. 277, and *Scientific Opinion*, 1870, p. 135) the structure of the flowers in this genus, but he was mistaken in thinking that self-fertilization is impossible, at least in the case of the present species. Dr Dickie and Professor Faivre state that the flowers are fertilized in the bud, and that self-fertilization is inevitable. I presume that they were misled by the pollen being deposited at a very early period on the pistil: see *Journal of Linn. Soc. Bot.*, vol. x, p. 55, and *Variabilité des Espèces*, 1868, p. 158.

soaked in water, will not germinate well on bare sand; and those that were sown in pots (which plan I was forced to follow) germinated at very unequal intervals of time; so that it was difficult to get pairs of the same exact age, and many seedlings had to be pulled up and thrown away. My experiments were continued during three successive generations; and in each generation the self-fertilized plants were again self-fertilized, their early progenitors in this country having probably been self-fertilized for many previous generations. In each generation, also, the crossed plants were fertilized with pollen from another crossed plant.

Of the flowers which were crossed in the three generations, taken together, a rather larger proportion yielded capsules than did those which were self-fertilized. The seeds were counted in forty-seven capsules from the crossed flowers, and they contained on an average 9·95 seeds; whereas forty-eight capsules from the self-fertilized flowers contained on an average 8·45 seeds; or as 100 to 85. The seeds from the crossed flowers were / not heavier, on the contrary a little lighter, than those from the self-fertilized flowers, as was thrice ascertained. On one occasion I weighed 200 of the crossed and 106 of the self-fertilized seeds, and the relative weight of an equal number was as 100 for the crossed to 101·5 for the self-fertilized. With other plants, when the seeds from the self-fertilized flowers were heavier than those from the crossed flowers, this appeared to be due generally to fewer having been produced by the self-fertilized flowers, and to their having been in consequence better nourished. But in the present instance the seeds from the crossed capsules were separated into two lots – namely, those from the capsules containing over fourteen seeds, and those from capsules containing under fourteen seeds, and the seeds from the more productive capsules were the heavier of the two; so that the above explanation here fails.

As pollen is deposited at a very early age on the pistil, generally in contact with the stigma, some flowers whilst still in bud were castrated for my first experiment, and were afterwards fertilized with pollen from a distinct plant. Other flowers were fertilized with their own pollen. From the seeds thus obtained, I succeeded in rearing only three pairs of plants of equal age. The three crossed plants averaged 32·79 inches, and the three self-fertilized 2·08 inches in height; so that they were nearly equal, the crossed having a slight advantage. As the same result followed in all three generations, it would be superfluous to give the heights of all the plants, and I will give only the averages.

In order to raise crossed and self-fertilized plants of the second generation, some flowers on the above crossed plants were crossed within twenty-four hours after they had expanded with pollen from a distinct plant; and this interval would probably not be too great to allow of cross-fertilization being effectual. Some flowers on the self-fertilized plants of the last generation were also self-fertilized. From these two lots of seeds, ten crossed and twelve self-fertilized plants of equal ages were raised; and these were measured when fully grown. The crossed averaged 36·98, and the self-fertilized averaged 37·42 inches in height; so that there again the two lots were nearly equal; but the self-fertilized had a slight advantage.

In order to raise plants of the third generation, a better plan was followed, and flowers on the crossed plants of the second generation were selected in which the stamens were too short to / reach the stigmas, so that they could not possibly have been self-fertilized. These flowers were crossed with pollen from a distinct plant. Flowers on the self-fertilized plants of the second generation were again self-fertilized. From the two lots of seeds thus obtained, twenty-one crossed and nineteen self-fertilized plants of equal age, and forming the third generation, were raised in fourteen large pots. They were measured when fully grown, and by an odd chance the average height of the two lots was exactly the same, namely, 35·96 inches; so that neither side had the least advantage over the other. To test this result, all the plants on both sides in ten out of the above fourteen pots were cut down after they had flowered, and in the ensuing year the stems were again measured; and now the crossed plants exceeded by a little (viz., 1·7 inches) the self-fertilized. They were again cut down, and on their flowering for the third time, the self-fertilized plants had a slight advantage (viz., 1·54 inches) over the crossed. Hence the result arrived at with these plants during the previous trials was confirmed, namely, that neither lot had any decided advantage over the other. It may, however, be worth mentioning that the self-fertilized plants showed some tendency to flower before the crossed plants: this occurred with all three pairs of the first generation; and with the cut down plants of the third generation, a self-fertilized plant flowered first in nine out of the twelve pots, whilst in the remaining three pots a crossed plant flowered first.

If we consider all the plants of the three generations taken together, the thirty-four crossed plants average 35·98, and the thirty-four self-fertilized plants 36·39 inches in height; or as 100 to 101. We may therefore conclude that the two lots possessed equal powers of growth; and this I believe to be the result of long-continued self-fertilization, together with exposure to similar conditions in each generation, so that all the individuals had acquired a closely similar constitution.

XXX. GRAMINACEAE

ZEA MAYS

This plant is monoecious, and was selected for trial on this account, no other plant having been experimented on.[8] It is / also anemophilous, or is fertilized by the wind; and of such plants only the common beet has been tried. Some plants were raised in the greenhouse, and were crossed with pollen taken from a distinct plant; and a single plant, growing quite separately in a different part of the house, was allowed to fertilize itself spontaneously. The seeds thus obtained were placed on damp sand, and as they germinated in pairs of equal age were planted on the opposite sides fo four very large pots; nevertheless they were considerably crowded. The pots were kept in the hothouse. The plants were first measured to the tips of their leaves when only between 1 and 2 feet in height, as shown in the following table (XCVII).

The fifteen crossed plants here average 20·19, and the fifteen self-fertilized plants 17·57 inches in height; or as 100 to 87. Mr Galton made a graphical representation, in accordance with the method described in the introductory chapter, of the above / measurements, and adds the words 'very good' to the curves thus formed.

Shortly afterwards one of the crossed plants in Pot I died; another became much diseased and stunted; and the third never grew to its full height. They seemed to have been all injured, probably by some larva gnawing their roots. Therefore all the plants on both sides of this pot were rejected in the subsequent measurements. When the plants were fully grown they were again measured to the tips of the highest leaves, and the eleven crossed plants now averaged 68·1, and the eleven self-fertilized plants 62·34 inches in height; or as 100 to 91. In all four pots a crossed plant flowered before any one of the self-fertilized; but three

[8] Hildebrand remarks that this species seems at first sight adapted to be fertilized by pollen from the same plant, owing to the male flowers standing above the female flowers; but practically it must generally be fertilized by pollen from another plant, as the male flowers usually shed their pollen before the female flowers are mature; *Monatsbericht der K. Akad.*, Berlin, October, 1872, p. 743.

TABLE XCVII
Zea mays

No. of pot	*Crossed plants*	*Self-fertilized plants*
	Inches	*Inches*
I	23 4/8	17 3/8
	12	20 3/8
	21	20
II	22	20
	19 1/8	18 3/8
	21 4/8	18 5/8
III	22 1/8	18 5/8
	20 4/8	15 2/8
	18 2/8	16 4/8
	21 5/8	18
	23 2/8	16 2/8
IV	21	18
	22 1/8	12 6/8
	23	15 4/8
	12	18
Total in inches	302·88	263·63

of the plants did not flower at all. Those that flowered were also measured to the summits of the male flowers: the ten crossed plants averaged 66·51, and the nine self-fertilized plants 61·59 inches in height; or as 100 to 93.

A large number of the same crossed and self-fertilized seeds were sown in the middle of the summer in the open ground in two long rows. Very much fewer of the self-fertilized than of the crossed plants produced flowers; but those that did flower, flowered almost simultaneously. When fully grown the ten tallest plants in each row were selected and measured to the tips of their highest leaves, as well as to the summits of their male flowers. The crossed averaged to the tips of their leaves 54 inches in height, and the self-fertilized 44·65, or as 100 to 83; and to the summits of their male flowers, 53·96 and 43·45 inches; or as 100 to 80.

PHALARIS CANARIENSIS

Hildebrand has shown in the paper referred to under the last species, that this hermaphrodite grass is better adapted for cross-fertilization than for self-fertilization. Several plants were raised in the greenhouse close together, and their flowers were mutually intercrossed. Pollen from a single plant growing quite separately was collected and placed on the stigmas of the same plant. The seeds thus produced were self-fertilized, for they were fertilized with pollen from the same plant, but it will have been a mere chance whether with pollen from the same flowers. Both lots of seeds, after germinating on sand, were planted in pairs on the opposite sides of four pots, which were kept in the greenhouse. When the plants were a little over a / foot in height they were measured, and the crossed plants averaged 13·38, and the self-fertilized 12·29 inches in height; or as 100 to 92.

When in full flower they were again measured to the extremities of their culms, as shown in the following table:

TABLE XCVIII
Phalaris canariensis

No. of pot	Crossed plants	Self-fertilized plants
	Inches	Inches
I	42$\frac{2}{8}$	41$\frac{2}{8}$
	39$\frac{6}{8}$	45$\frac{5}{8}$
II	37	31$\frac{6}{8}$
	49$\frac{4}{8}$	37$\frac{2}{8}$
	29	42$\frac{3}{8}$
	37	34$\frac{7}{8}$
III	37$\frac{6}{8}$	28
	35$\frac{4}{8}$	28
	43	34
IV	40$\frac{2}{8}$	35$\frac{1}{8}$
	37	34$\frac{4}{8}$
Total in inches	428·00	392·63

The eleven crossed plants now averaged 38·9, and the eleven self-fertilized plants 35·69 inches in height; or as 100 to 92, which is the same ratio as before. Differently to what occurred with the maize, the crossed plants did not flower before the self-fertilized; and though both lots flowered very poorly from having been kept in pots in the greenhouse, yet the self-fertilized plants produced twenty-eight flower-heads, whilst the crossed produced only twenty!

Two long rows of the same seeds were sown out of doors, and care was taken that they were sown in nearly equal number; but a far greater number of the crossed than of the self-fertilized seeds yielded plants. The self-fertilized plants were in consequence not so much crowded as the crossed, and thus had an advantage over them. When in full flower, the twelve tallest plants were carefully selected from both rows and measured, as shown in the following table: /

TABLE XCIX
Phalaris canariensis (growing in the open ground)

	Crossed plants, twelve tallest	*Self-fertilized plants, twelve tallest*
	Inches	*Inches*
	34 1/8	35 2/8
	35 7/8	31 1/8
	36	33
	35 5/8	32
	35 5/8	31 5/8
	36 1/8	36
	36 6/8	33
	38 6/8	32
	36 2/8	35 1/8
	35 5/8	33 5/8
	34 1/8	34 2/8
	34 5/8	35
Total in inches	429·5	402·0

The twelve crossed plants here average 35·78, and the twelve self-fertilized 33·5 inches in height; or as 100 to 93. In this case the crossed plants flowered rather before the self-fertilized, and thus differed from those growing in the pots. /

CHAPTER VII

SUMMARY OF THE HEIGHTS AND WEIGHTS OF THE
CROSSED AND SELF-FERTILIZED PLANTS

Number of species and plants measured – Tables given – Preliminary
remarks on the offspring of plants crossed by a fresh stock – Thirteen
cases specially considered – The effects of crossing a self-fertilized plant
either by another self-fertilized plant or by an intercrossed plant of the
old stock – Summary of the results – Preliminary remarks on the crossed
and self-fertilized plants of the same stock – The twenty-six exceptional
cases considered, in which the crossed plants did not exceed greatly in
height the self-fertilized – Most of these cases shown not to be real
exceptions to the rule that cross-fertilization is beneficial – Summary of
results – Relative weights of the crossed and self-fertilized plants.

The details which have been given under the head of each species are
so numerous and so intricate, that it is necessary to tabulate the results.
In Table A, the number of plants of each kind which were raised from
a cross between two individuals of the same stock and from self-
fertilized seeds, together with their mean or average heights at or near
maturity, are given. In the right-hand column, the mean height of the
crossed to that of the self-fertilized plants, the former being taken as
100, is shown. To make this clear, it may be advisable to give an
example. In the first generation of Ipomoea, six plants derived from a
cross between two plants were measured, and their mean height is
86·00 inches; six plants derived from flowers on the same parent-plant
fertilized with their own pollen measured, and their mean height is
65·66 inches. From this it follows, as shown in the right-hand column,
that if the mean height of the crossed plants be taken as 100, that of
the self-fertilized / plants is 76. The same plan is followed with all the
other species.

The crossed and self-fertilized plants were generally grown in pots
in competition with one another, and always under as closely similar
conditions as could be attained. They were, however, sometimes

217

grown in separate rows in the open ground. With several of the species, the crossed plants were again crossed, and the self-fertilized plants again self-fertilized, and thus successive generations were raised and measured, as may be seen in Table A. Owing to this manner of proceeding, the crossed plants became in the later generations of Mimulus are not included, as a new tall variety then prevailed on one side alone, so that a fair comparison between the two sides was no longer possible. With Ipomoea the variety Hero has been excluded for nearly the same reason.

In Table B the relative weights of the crossed and self-fertilized plants, after they had flowered and had been cut down, are given in the few cases in which they were ascertained. The results are, I think, more striking and of greater value as evidence of constitutional vigour than those deduced from the relative heights of the plants.

The most important table is that of C, as it includes the relative heights, weights, and fertility of plants raised from parents crossed by a fresh stock (that is, by non-related plants grown under different conditions), or by a distinct subvariety, in comparison with self-fertilized plants, or in a few cases with plants of the same old stock intercrossed during several generations. The relative fertility of the plants in this and the other tables will be more fully considered in a future chapter. /

TABLE A

Relative heights of plants from parents crossed with pollen from other plants of the same stock, and self-fertilized

NAMES OF PLANTS	Number of the crossed plants measured	Average height of crossed plants in inches	Number of the self-fertilized plants measured	Average height of self-fertilized plants in inches	Average height of the crossed to the self-fertilized plants, the former taken as 100
Ipomoea purpurea — 1st generation	6	86·00	6	65·66	as 100 to 76
Ipomoea purpurea — 2nd generation	6	84·16	6	66·33	„ „ 79
Ipomoea purpurea — 3rd generation	6	77·41	6	52·83	„ „ 68
Ipomoea purpurea — 4th generation	7	69·78	7	60·14	„ „ 86
Ipomoea purpurea — 5th generation	6	82·54	6	62·33	„ „ 75
Ipomoea purpurea — 6th generation	6	87·50	6	63·16	„ „ 72

Relative heights of plants from parents crossed with pollen from other plants of the same stock, and self-fertilized

NAMES OF PLANTS	Number of the crossed plants measured	Average height of crossed plants in inches	Number of the self-fertilized plants measured	Average height of self-fertilized plants in inches	Average height of the crossed to the self-fertilized plants, the former taken as 100
Ipomoea purpurea – 7th generation	9	83·94	9	68·25	as 100 to 81
Ipomoea purpurea – 8th generation	8	113·25	8	96·65	„ „ 85
Ipomoea purpurea – 9th generation	14	81·39	14	64·07	„ „ 79
Ipomoea purpurea – 10th generation	5	93·70	5	50·40	„ „ 54
Number and average height of all the plants of the ten generations	73	85·84	73	66·02	„ „ 77
Mimulus luteus – three first generations, before the new and taller self-fertilized variety appeared	10	8·19	10	5·29	„ „ 65
Digitalis purpurea	16	51·33	8	35·87	„ „ 70
Calceolaria – (common greenhouse variety)	1	19·50	1	15·00	„ „ 77
Linaria vulgaris	3	7·08	3	5·75	„ „ 81
Verbascum thapsus	6	65·34	6	56·50	„ „ 86
Vandellia nummularifolia – crossed and self-fertilized plants, raised from perfect flowers	20	4·30	20	4·27	„ „ 99
Vandellia nummularifolia – crossed and self-fertilized plants, raised from perfect flowers: second trial, plants crowded	24	3·60	24	3·38	„ „ 94
Vandellia nummularifolia – crossed plants raised from perfect flowers, and self-fertilized plants from cleistogene flowers	20	4·30	20	4·06	„ „ 94
Gesneria pendulina	8	32·06	8	29·14	„ „ 90
Salvia coccinea	6	27·85	6	21·16	„ „ 76

TABLE A – *continued*

Relative heights of plants from parents crossed with pollen from other plants of the same stock, and self-fertilized

NAMES OF PLANTS	Number of the crossed plants measured	Average height of crossed plants in inches	Number of the self-fertilized plants measured	Average height of self-fertilized plants in inches	Average height of the crossed to the self-fertilized plants, the former taken as 100
Origanum vulgare	4	20·00	4	17·12	as 100 to 86
Thunbergia alata	6	60·00	6	65·00	„ „ 108
Brassica oleracea	9	41·08	9	39·00	„ „ 95
Iberis umbellata – the self-fertilized plants of the 3rd generation	7	19·12	7	16·39	„ „ 86/
Papaver vagum	15	21·91	15	19·54	„ „ 89
Eschscholtzia californica – English stock, 1st generation	4	29·68	4	25·56	„ „ 86
Eschscholtzia californica – English stock, 2nd generation	11	32·47	11	32·81	„ „ 101
Eschscholtzia californica – Brazilian stock, 1st generation	14	44·64	14	45·12	„ „ 101
Eschscholtzia californica – Brazilian stock, 2nd generation	18	43·38	19	50·30	„ „ 116
Eschscholtzia californica – average height and number of all the plants of Eschscholtzia	47	40·03	48	42·72	„ „ 107
Reseda lutea – grown in pots	24	17·17	24	14·61	„ „ 85
Reseda lutea – grown in open ground	8	28·09	8	23·14	„ „ 82
Reseda odorata – self-fertilized seeds from a highly self-fertile plant, grown in pots	19	27·48	19	22·55	„ „ 82
Reseda odorata – self-fertilized seeds from a highly self-fertile plant, grown in open ground	8	25·76	8	27·09	„ „ 105
Reseda odorata – self-fertilized seeds from a semi-self-sterile plant, grown in pots	20	29·98	20	27·71	„ „ 92

TABLE A – *continued*

Relative heights of plants from parents crossed with pollen from other plants of the same stock, and self-fertilized

NAMES OF PLANTS	Number of the crossed plants measured	Average height of crossed plants in inches	Number of the self-fertilized plants measured	Average height of self-fertilized plants in inches	Average height of the crossed to the self-fertilized plants, the former taken as 100	
Reseda odorata – self-fertilized seeds from a semi-self-sterile plant, grown in open ground	8	25·92	8	23·54	as 100 to	90
Viola tricolor	14	5·58	14	2·37	„ „	42
Adonis aestivalis	4	14·25	4	14·31	„ „	100
Delphinium consolida	6	14·95	6	12·50	„ „	84
Viscaria oculata	15	34·50	15	33·55	„ „	97
Dianthus caryophyllus – open ground, about	6?	28?	6?	24?	„ „	86
Dianthus caryophyllus – 2nd generation, in pots, crowded	2	16·75	2	9·75	„ „	58
Dianthus caryophyllus – 3rd generation, in pots	8	28·39	8	28·21	„ „	99
Dianthus caryophyllus – offspring from plants of the 3rd self-fertilized generation crossed by intercrossed plants of 3rd generation, compared with plants of 4th self-fertilized generation	15	28·00	10	26·55	„ „	95 /
Dianthus caryophyllus – number and average height of all the plants of Dianthus	31	27·37	26	25·18	„ „	92
Hibiscus africanus	4	13·25	4	14·43	„ „	109
Pelargonium zonale	7	22·35	7	16·62	„ „	74
Tropaeolum minus	8	58·43	8	46·00	„ „	79
Limnanthes douglasii	16	17·46	16	13·85	„ „	79
Lupinus luteus – 2nd generation	8	30·78	8	25·21	„ „	82
Lupinus pilosus – plants of two generations	2	35·50	3	30·50	„ „	86
Phaseolus multiflorus	5	86·00	5	82·35	„ „	96
Pisum sativum	4	34·62	4	39·68	„ „	115
Sarothamnus scoparius – small seedlings	6	2·91	6	1·33	„ „	46

TABLE A – continued

Relative heights of plants from parents crossed with pollen from other plants of the same stock, and self-fertilized

NAMES OF PLANTS	Number of the crossed plants measured	Average height of crossed plants in inches	Number of the self-fertilized plants measured	Average height of self-fertilized plants in inches	Average height of the crossed to the self-fertilized plants, the former taken as 100
Sarothamnus scoparius – the three survivors on each side after three year's growth		18·91		11·83	as 100 to 63
Ononis minutissima	2	19·81	2	17·37	,, ,, 88
Clarkia elegans	4	33·50	4	27·62	,, ,, 82
Bartonia aurea	8	24·62	8	26·31	,, ,, 107
Passiflora gracilis	2	49·00	2	51·00	,, ,, 104
Apium petroselinum	?	not measured	?	not measured	,, ,, 100
Scabiosa atro-purpurea	4	17·12	4	15·37	,, ,, 90
Lactuca sativa – plants of two generations	7	19·43	6	16·00	,, ,, 82
Specularia speculum	4	19·28	4	18·93	,, ,, 98
Lobelia ramosa – 1st generation	4	22·25	4	18·37	,, ,, 82
Lobelia ramosa – 2nd generation	3	23·33	3	19·00	,, ,, 81
Lobelia fulgens – 1st generation	2	34·75	2	44·25	,, ,, 127
Lobelia fulgens – 2nd generation	23	29·82	23	27·10	,, ,, 91
Nemophila insignis – half-grown	12	11·10	12	5·45	,, ,, 49
Nemophila insignis – the same fully grown		33·28		19·90	,, ,, 60
Borago officinalis	4	20·68	4	21·18	,, ,, 102
Nolana prostrata	5	12·75	5	13·40	,, ,, 105
Petunia violacea – 1st generation	5	30·80	5	26·00	,, ,, 84
Petunia violacea – 2nd generation	4	40·50	6	26·25	,, ,, 65
Petunia violacea – 3rd generation	8	40·96	8	53·87	,, ,, 131
Petunia violacea – 4th generation	15	46·79	14	32·39	,, ,, 69/
Petunia violacea – 4th generation, from a distinct parent	13	44·74	13	26·87	,, ,, 60

TABLE A – *continued*

Relative heights of plants from parents crossed with pollen from other plants of the same stock, and self-fertilized

NAMES OF PLANTS	Number of the crossed plants measured	Average height of crossed plants in inches	Number of the self-fertilized plants measured	Average height of self-fertilized plants in inches	Average height of the crossed to the self-fertilized plants, the former taken as 100
Petunia violacea – 5th generation	22	54·11	21	33·23	as 100 to 61
Petunia violacea – 5th generation, in open ground	10	38·27	10	23·31	„ „ 61
Petunia violacea – Number and average height of all the plants in pots of Petunia	67	46·53	67	33·12	„ „ 71
Nicotiana tabacum – 1st generation	4	18·50	4	32·75	„ „ 178
Nicotiana tabacum – 2nd generation	9	53·84	7	51·78	„ „ 96
Nicotiana tabacum – 3rd generation	7	95·25	7	79·60	„ „ 83
Nicotiana tabacum – 3rd generation but raised from a distinct plant	7	70·78	9	71·30	„ „ 101
Nicotiana tabacum – number and average height of all the plants of Nicotiana	27	63·73	27	61·31	„ „ 96
Cyclamen persicum	8	9·49	8?	7·50	„ „ 79
Anagallis collina	6	42·20	6	33·35	„ „ 69
Primula sinensis – a dimorphic species	8	9·01	8	9·03	„ „ 100
Fagopyrum esculentum – a dimorphic species	15	38·06	15	26·13	„ „ 69
Beta vulgaris – in pots	8	34·09	8	29·81	„ „ 87
Beta vulgaris – in open ground	8	30·92	8	30·70	„ „ 99
Canna warscewiczi – plants of three generations	34	35·98	34	36·39	„ „ 101
Zea mays – in pots, whilst young, measured to tips of leaves	15	20·19	15	17·57	„ „ 87
Zea mays – when full grown, after the death of some, measured to tips of leaves		68·10		62·34	„ „ 91

TABLE A – *continued*

Relative heights of plants from parents crossed with pollen from other plants of the same stock, and self-fertilized

NAMES OF PLANTS	Number of the crossed plants measured	Average height of crossed plants in inches	Number of the self-fertilized plants measured	Average height of self-fertilized plants in inches	Average height of the crossed to the self-fertilized plants, the former taken as 100
Zea mays – when full grown, after the death of some, measured to tips of flowers		66·51		61·59	as 100 to 93
Zea mays – grown in open ground, measured to tips of leaves	10	54·00	10	44·55	„ „ 83
Zea mays – grown in open ground, measured to tips of flowers		53·96		43·45	„ „ 80
Phalaris canariensis – in pots	11	38·90	11	35·69	„ „ 92
Phalaris canariensis – in open ground	12	35·78	12	33·50	„ „ 93 /

TABLE B

Relative weights of plants from parents crossed with pollen from distinct plants of the same stock, and self-fertilized

NAMES OF PLANTS	Number of crossed plants	Number of self-fertilized plants	Weight of the crossed plants taken as 100
Ipomoea purpurea – plants of the 10th generation	6	6	as 100 to 44
Vandellia nummularifolia – 1st generation	41	41	„ „ 97
Brassica oleracea – 1st generation	9	9	„ „ 37
Eschscholtzia californica – plants of the 2nd generation	19	19	„ „ 118
Reseda lutea – 1st generation, grown in pots	24	24	„ „ 21
Reseda lutea – 1st generation, grown in open ground	8	8	„ „ 40
Reseda odorata – 1st generation, descended from a highly self-fertile plant, grown in pots	19	19	„ „ 67
Reseda odorata – 1st generation, descended from a semi-self-sterile plant, grown in pots	20	20	„ „ 99
Dianthus caryophyllus – plants of the 3rd generation	8	8	„ „ 49
Petunia violacea – plants of the 5th generation, in pots	22	21	„ „ 22
Petunia violacea – plants of the 5th generation, in open ground	10	10	„ „ 36 /

TABLE C

Relative heights, weights, and fertility of plants from parents crossed by a fresh stock, and from parents either self-fertilized or intercrossed with plants of the same stock

NAMES OF PLANTS AND NATURE OF THE EXPERIMENTS	Numbers of the plants from a cross with a fresh stock	Average height in inches and weight	Number of the plants from self-fertilized or intercrossed parents of the same stock	Average height in inches and weight	Height, weight, and fertility of the plants from the cross with a fresh stock taken as 100
Ipomoea purpurea – offspring of plants intercrossed for nine generations and then crossed by a fresh stock, compared with plants of the 10th intercrossed generation	19	84·03	19	65·78	as 100 to 78
Ipomoea purpurea – offspring of plants intercrossed for nine generations and then crossed by a fresh stock, compared with plants of the 10th intercrossed generation, in fertility	..	..	..	..	„ „ 51
Mimulus luteus – offspring of plants self-fertilized for eight generations and then crossed by a fresh stock, compared with plants of the 9th self-fertilized generation	28	21·62	19	10·44	„ „ 52
Mimulus luteus – offspring of plants self-fertilized for eight generations and then crossed by a fresh stock, compared with plants of the 9th self-fertilized generation, in fertility	..	..	..	..	„ „ 3
Mimulus luteus – offspring of plants self-fertilized for eight generations and then crossed by a fresh stock, compared with the offspring of a plant self-fertilized for eight generations, and then intercrossed with another self-fertilized plant of the same generation	28	21·62	27	12·20	„ „ 56/

TABLE C – *continued*

Relative heights, weights, and fertility of plants from parents crossed by a fresh stock, and from parents either self-fertilized or intercrossed with plants of the same stock

NAMES OF PLANTS AND NATURE OF THE EXPERIMENTS	Numbers of the plants from a cross with a fresh stock	Average height in inches and weight	Number of the plants from self-fertilized or intercrossed parents of the same stock	Average height in inches and weight	Height, weight, and fertility of the plants from the cross with a fresh stock taken as 100
Mimulus luteus – offspring of plants self-fertilized for eight generations and then crossed by a fresh stock, compared with the offspring of a plant self-fertilized for eight generations, and then intercrossed with another self-fertilized plant of the same generation, in fertility	..	..	..	..	as 100 to 4
Brassica oleracea – offspring of plants self-fertilized for two generations and then crossed by a fresh stock, compared with plants of the 3rd self-fertilized generation, by weight	6		6		,, ,, 22
Iberis umbellata – offspring from English variety crossed by slightly different Algerine variety, compared with the self-fertilized offspring of the English variety	30	17·34	29	15·51	,, ,, 89
Iberis umbellata – offspring from English variety, crossed by slightly different Algerine variety, compared with the self-fertilized offspring of the English variety, in fertility	..	..	..	..	,, ,, 75
Eschscholtzia californica – offspring of a Brazilian stock crossed by an English stock, compared with plants of the Brazilian stock of the 2nd self-fertilized generation	19	45·92	19	50·30	,, ,, 109

227

TABLE C – *continued*

Relative heights, weights, and fertility of plants from parents crossed by a fresh stock, and from parents either self-fertilized or intercrossed with plants of the same stock

NAMES OF PLANTS AND NATURE OF THE EXPERIMENTS	Numbers of the plants from a cross with a fresh stock	Average height in inches and weight	Number of the plants from self-fertilized or intercrossed parents of the same stock	Average height in inches and weight	Height, weight, and fertility of the plants from the cross with a fresh stock taken as 100
Eschscholtzia californica – offspring of a Brazilian stock crossed by an English stock, compared with plants of the Brazilian stock of the 2nd self-fertilized generation, in weight	..	..	..	..	as 100 to 118/
Eschscholtzia californica – offspring of a Brazilian stock crossed by an English stock, compared with plants of the Brazilian stock of the 2nd self-fertilized generation, in fertility	..	..	..	..	,, ,, 40
Eschscholtzia californica – offspring of a Brazilian stock crossed by an English stock, compared with plants of the Brazilian stock of the 2nd intercrossed generation, in height	19	45·92	18	43·38	,, ,, 94
Eschscholtzia californica – offspring of a Brazilian stock crossed by an English stock, compared with plants of the Brazilian stock of the 2nd intercrossed generation, in weight	..	..	..	..	,, ,, 100
Eschscheltzia californica – offspring of a Brazilian stock crossed by an English stock, compared with plants of the Brazilian stock of the 2nd intercrossed generation, in fertility	..	..	..	..	,, ,, 45

TABLE C – *continued*

Relative heights, weights, and fertility of plants from parents crossed by a fresh stock, and from parents either self-fertilized or intercrossed with plants of the same stock

NAMES OF PLANTS AND NATURE OF THE EXPERIMENTS	Numbers of the plants from a cross with a fresh stock	Average height in inches and weight	Number of the plants from self-fertilized or intercrossed parents of the same stock	Average height in inches and weight	Height, weight, and fertility of the plants from the cross with a fresh stock taken as 100
Dianthus caryophyllus – offspring of plants self-fertilized for three generations and then crossed by a fresh stock, compared with plants of the 4th self-fertilized generation	16	32·82	10	26·55	as 100 to 81
Dianthus caryophyllus – offspring of plants self-fertilized for three generations and then crossed by a fresh stock, compared with plants of the 4th self-fertilized generation, in fertility	..	..	..	..	„ „ 33 /
Dianthus caryophyllus – offspring of plants self-fertilized for three generations and then crossed by a fresh stock, compared with the offspring of plants self-fertilized for three generations and then crossed by plants of the 3rd intercrossed generation	16	32·82	15	28·00	„ „ 85
Dianthus caryophyllus – offspring of plants self-fertilized for three generations and then crossed by a fresh stock, compared with the offspring of plants self-fertilized for three generations and then crossed by plants of the 3rd intercrossed generation, in fertility	..	..	..	..	„ „ 45

229

TABLE C – *continued*

Relative heights, weights, and fertility of plants from parents crossed by a fresh stock, and from parents either self-fertilized or intercrossed with plants of the same stock

NAMES OF PLANTS AND NATURE OF THE EXPERIMENTS	Numbers of the plants from a cross with a fresh stock	Average height in inches and weight	Number of the plants from self-fertilized or intercrossed parents of the same stock	Average height in inches and weight	Height, weight, and fertility of the plants from the cross with a fresh stock taken as 100
Pisum sativum – offspring from a cross between two closely allied varieties, compared with the self-fertilized offspring of one of the varieties, or with intercrossed plants of the same stock	?		?		as 100 to }60 to 75
Lathyrus odoratus – offspring from two varieties, differing only in colour of their flowers, compared with the self-fertilized offspring of one of the varieties: in 1st generation	2	79·25	2	63·75	,, ,, 80
Lathyrus odoratus – offspring from two varieties, differing only in colour of their flowers, compared with the self-fertilized offspring of one of the varieties: in 2nd generation	6	62·91	6	55·31	,, ,, 88/
Petunia violacea – offspring of plants self-fertilized for four generations and then crossed by a fresh stock, compared with plants of the 5th self-fertilized generation, in height	21	50·05	21	33·23	,, ,, 66
Petunia violacea – offspring of plants self-fertilized for four generations and then crossed by a fresh stock, compared with plants of the 5th self-fertilized generation, in weight	..	..	..	..	,, ,, 23

TABLE C – *continued*

Relative heights, weights, and fertility of plants from parents crossed by a fresh stock, and from parents either self-fertilized or intercrossed with plants of the same stock

NAMES OF PLANTS AND NATURE OF THE EXPERIMENTS	Numbers of the plants from a cross with a fresh stock	Average height in inches and weight	Number of the plants from self-fertilized or intercrossed parents of the same stock	Average height in inches and weight	Height, weight, and fertility of the plants from the cross with a fresh stock taken as 100
Petunia violacea – offspring of plants self-fertilized for four generations and then crossed by a fresh stock, compared with plants of the 5th self-fertilized generation, grown in open ground, in height	10	36·67	10	23·31	as 100 to 63
Petunia violacea – offspring of plants self-fertilized for four generations and then crossed by a fresh stock, compared with plants of the 5th self-fertilized generation, grown in open ground, in weight	..	..	..	..	,, ,, 53
Petunia violacea – offspring of plants self-fertilized for four generations and then crossed by a fresh stock, compared with plants of the 5th self-fertilized generation, grown in open ground, in fertility	..	..	..	..	,, ,, 46
Petunia violacea – offspring of plants self-fertilized for four generations and then crossed by a fresh stock, compared with plants of the 5th intercrossed generation, in height	21	50·05	22	54·11	,, ,, 108/
Petunia violacea – offspring of plants self-fertilized for four generations and then crossed by a fresh stock, compared with plants of the 5th intercrossed generation, in weight	..	..	..	..	,, ,, 101

231

TABLE C — *continued*

Relative heights, weights, and fertility of plants from parents crossed by a fresh stock, and from parents either self-fertilized or intercrossed with plants of the same stock

NAMES OF PLANTS AND NATURE OF THE EXPERIMENTS	Numbers of the plants from a cross with a fresh stock	Average height in inches and weight	Number of the plants from self-fertilized or intercrossed parents of the same stock	Average height in inches and weight	Height, weight, and fertility of the plants from the cross with a fresh stock taken as 100
Petunia violacea — offspring of plants self-fertilized for four generations and then crossed by a fresh stock, compared with plants of the 5th intercrossed generation, grown in open ground, in height	10	36·67	10	38·27	as 100 to 104
Petunia violacea – offspring of plants self-fertilized for four generations and then crossed by a fresh stock, compared with plants of the 5th intercrossed generation, grown in open ground, in weight	..	..	..	..	„ „ 146
Petunia violacea – offspring of plants self-fertilized for four generations and then crossed by a fresh stock, compared with plants of the 5th intercrossed generation, grown in open ground, in fertility	..	..	..	..	„ „ 54
Nicotiana tabacum – offspring of plants self-fertilized for three generations and then crossed by a slightly different variety, compared with plants of the 4th self-fertilized generation, grown not much crowded in pots, in height	26	63·29	26	41·67	„ „ 66/

TABLE C – *continued*

Relative heights, weights, and fertility of plants from parents crossed by a fresh stock, and from parents either self-fertilized or intercrossed with plants of the same stock

NAMES OF PLANTS AND NATURE OF THE EXPERIMENTS	Numbers of the plants from a cross with a fresh stock	Average height in inches and weight	Number of the plants from self-fertilized or intercrossed parents of the same stock	Average height in inches and weight	Height, weight, and fertility of the plants from the cross with a fresh stock taken as 100
Nicotiana tabacum – offspring of plants self-fertilized for three generations and then crossed by a slightly different variety, compared with plants of the 4th self-fertilized generation, grown much crowded in pots, in height	12	31·53	12	17·21	as 100 to 54
Nicotiana tabacum – offspring of plants self-fertilized for three generations and then crossed by a slightly different variety, compared with plants of the 4th self-fertilized generation, grown much crowded in pots, in weight	..	..	..	..	„ „ 37
Nicotiana tabacum – offspring of plants self-fertilized for three generations and then crossed by a slightly different variety, compared with plants of the 4th self-fertilized generation, grown in open ground, in height	20	48·74	20	35·20	„ „ 72
Nicotiana tabacum – offspring of plants self-fertilized for three generations and then crossed by a slightly different variety, compared with plants of the 4th self-fertilized generation, grown in open ground, in weight	..	..	..	..	„ „ 63
Anagallis collina – offspring from a red variety crossed by a blue variety, compared with the self-fertilized offspring of the red variety	3	27·62	3	18·21	„ „ 66/

233

TABLE C — *continued*

Relative heights, weights, and fertility of plants from parents crossed by a fresh stock, and from parents either self-fertilized or intercrossed with plants of the same stock

NAMES OF PLANTS AND NATURE OF THE EXPERIMENTS	Numbers of the plants from a cross with a fresh stock	Average height in inches and weight	Number of the plants from self-fertilized or intercrossed parents of the same stock	Average height in inches and weight	Height, weight, and fertility of the plants from the cross with a fresh stock taken as 100
Anagallis collina – offspring from a red variety crossed by a blue variety, compared with the self-fertilized offspring of the red variety, in fertility	..	..	..	..	as 100 to 6
Primula veris – offspring from long-styled plants of the 3rd illegitimate generation, crossed by a fresh stock, compared with plants of the 4th illegitimate and self-fertilized generation	8	7·03	8	3·21	„ „ 46
Primula veris – offspring from long-styled plants of the 3rd illegitimate generation, crossed by a fresh stock, compared with plants of the 4th illegitimate and self-fertilized generation, in fertility	..	..	..	..	„ „ 5
Primula veris – offspring from long-styled plants of the 3rd illegitimate generation, crossed by a fresh stock, compared with plants of the 4th illegitimate and self-fertilized generation, in fertility in following year	..	..	..	..	„ „ 35
Primula veris (equal-styled, red-flowered variety) – offspring from plants self-fertilized for two generations and then crossed by a different variety, compared with plants of the 3rd self-fertilized generation	3	8·66	3	7·33	„ „ 85

TABLE C — *continued*

Relative heights, weights, and fertility of plants from parents crossed by a fresh stock, and from parents either self-fertilized or intercrossed with plants of the same stock

NAMES OF PLANTS AND NATURE OF THE EXPERIMENTS	Numbers of the plants from a cross with a fresh stock	Average height in inches and weight	Number of the plants from self-fertilized or intercrossed parents of the same stock	Average height in inches and weight	Height, weight, and fertility of the plants from the cross with a fresh stock taken as 100
Primula veris (equal-styled, red-flowered variety) – offspring from plants self-fertilized for two generations and then crossed by a different variety, compared with plants of the 3rd self-fertilized generation, in fertility . .	. .	. .	. .	as 100 to 11 /

In these three tables the measurements of fifty-seven species, belonging to fifty-two genera and to thirty great natural families, are given. The species are natives of various parts of the world. The number of crossed plants, including those derived from a cross between plants of the same stock and of two different stocks, amounts to 1,101; and the number of self-fertilized plants (including a few in Table C derived from a cross between plants of the same old stock) is 1,076. Their growth was observed from the germination of the seeds to maturity; and most of them were measured twice and some thrice. The various precautions taken to prevent either lot being unduly favoured, have been described in the introductory chapter. Bearing all these circumstances in mind, it may be admitted that we have a fair basis for judging of the comparative effects of cross-fertilization and of self-fertilization on the growth of the offspring.

It will be the most convenient plan first to consider the results given in Table C, as an opportunity will thus be afforded of incidentally discussing some important points. If the reader will look down the right-hand column of this table, he will see at a glance what an extraordinary advantage in height, weight, and fertility the plants derived from a cross with a fresh stock or with another subvariety have over the self-fertilized plants, as well as over the intercrossed plants of the same old stock. There are only two exceptions to this rule, and these are hardly real ones. In the case of Eschscholtzia, the advantage is confined to fertility. In that of Petunia, though the plants derived from a cross with a fresh stock had an immense superiority in height, weight, and fertility over the self-fertilized plants, they were conquered by the intercrossed plants of the same old stock in height and weight, but not / in fertility. It has, however, been shown that the superiority of these intercrossed plants in height and weight was in all probability not real; for if the two sets had been allowed to grow for another month, it is almost certain that those from a cross with the fresh stock would have been victorious in every way over the intercrossed plants.

Before we consider in detail the several cases given in Table C, some preliminary remarks must be made. There is the clearest evidence, as we shall presently see, that the advantage of a cross depends wholly on the plants differing somewhat in constitution; and that the disadvantages of self-fertilization depend on the two parents, which are combined in the same hermaphrodite flower, having a closely similar constitution. A certain amount of differentiation in the sexual elements seems indispensable for the full fertility of the parents, and for the full vigour of the offspring. All the individuals of the same species, even those produced in a state of nature, differ somewhat, though often very slightly, from one another in external characters and probably in constitution. This obviously holds good between the varieties of the same species, as far as external characters are concerned; and much evidence could be advanced with respect to their generally differing somewhat in constitution. There can hardly be a doubt that the differences of all kinds between the individuals and varieties of the same species depend largely, and as I believe exclusively, on their progenitors having been subjected to different conditions; though the conditions to which the individuals of the same species are exposed in a state of nature often falsely appear to us the same. For instance, the individuals growing together are necessarily exposed to the same climate, and they seem to us at first sight to be subjected to identically / the same conditions; but this can hardly be the case, except under the unusual contingency of each individual being surrounded by other kinds of plants in exactly the same proportional numbers. For the surrounding plants absorb different amounts of various substances from the soil, and thus greatly affect the nourishment and even the life of the individuals of any particular species. These will also be shaded and otherwise affected by the nature of the surrounding plants. Moreover, seeds often lie dormant in the ground, and those which germinate during any one year will often have been matured during very different seasons. Seeds are widely dispersed by various means, and some will occasionally be brought from distant stations, where their parents have grown under somewhat different conditions, and the plants produced from such seeds will intercross with the old residents, thus mingling their constitutional peculiarities in all sorts of proportions.

Plants when first subjected to culture, even in their native country, cannot fail to be exposed to greatly changed conditions of life, more especially from growing in cleared ground, and from not having to compete with many or any surrounding plants. They are thus enabled to absorb whatever they require which the soil may contain. Fresh seeds are often brought from distant gardens, where the parent-plants have been subjected to different conditions. Cultivated plants like those in a state of nature frequently intercross, and will thus mingle their constitutional peculiarities. On the other hand, as long as the individuals of any species are cultivated in the same garden, they will apparently be subjected to more uniform conditions than plants in a state of nature, as the individuals have not to compete / with various surrounding species. The seeds sown at the same time in a garden have generally been matured during the same season and in the same place; and in this respect they differ much from the seeds sown by the hand of nature. Some exotic plants are not frequented by insects in their new home, and therefore are not intercrossed; and this appears to be a highly important factor in the individuals acquiring uniformity of constitution.

In my experiments the greatest care was taken that in each generation all the crossed and self-fertilized plants should be subjected to the same conditions. Not that the conditions were absolutely the same, for the more vigorous individuals will have robbed the weaker ones of nutriment, and likewise of water when the soil in the pots was becoming dry; and both lots at one end of the pot will have received a little more light than those at the other end. In the successive generations, the plants were subjected to somewhat different conditions for the seasons necessarily varied, and they were sometimes raised at different periods of the year. But as they were all kept under glass, they were exposed to far less abrupt and great changes of temperature and moisture than are plants growing out of doors. With respect to the intercrossed plants, their first parents, which were not related, would almost certainly have differed somewhat in constitution; and such constitutional peculiarities would be variously mingled in each succeeding intercrossed generation, being sometimes augmented, but more commonly neutralized in a greater or less degree, and sometimes revived through reversion; just as we know to be the case with the external characters of crossed species and varieties. With the plants which were self-fertilized during the successive generations, this latter important / source of some diversity of constitution will have been wholly eliminated; and the sexual elements produced by the same flower must have been developed under as nearly the same conditions as it is possible to conceive.

In Table C the crossed plants are the offspring of a cross with a fresh stock, or with a distinct variety; and they were put into competition either with self-fertilized plants, or with intercrossed plants of the same old stock. By the term fresh stock I mean a non-related plant, the progenitors of which have been raised during some generations in another garden, and have consequently been exposed to somewhat different conditions. In the case of Nicotiana, Iberis, the red variety of Primula, the common Pea, and perhaps Anagallis, the plants which were crossed may be ranked as distinct varieties or subvarieties of the same species; but with Ipomoea, Mimulus, Dianthus, and Petunia, the plants which were crossed differed exclusively in the tint of their flowers; and as a large proportion of the plants raised from the same lot of purchased seeds thus varied, the differences may be estimated as merely individual. Having made these preliminary remarks, we will now consider in detail the several cases given in Table C, and they are well worthy of full consideration.

(1.) *Ipomoea purpurea.* Plants growing in the same pots, and subjected in each generation to the same conditions, were intercrossed for nine consecutive generations. These intercrossed plants thus became in the later generations more or less closely inter-related. Flowers on the plants of the ninth intercrossed generation were fertilized with pollen taken from a fresh stock, and seedlings thus raised. Other flowers on the same intercrossed plants were fertilized with pollen from another intercrossed plant, producing seedlings of the tenth intercrossed generation. These two sets of / seedlings were grown in competition with one another, and differed greatly in height and fertility. For the offspring from the cross with a fresh stock exceeded in height the intercrossed plants in the ratio of 100 to 78; and this is nearly the same excess which the intercrossed had over the self-fertilized plants in all ten generations taken together, namely, as 100 to 77. The plants raised from the cross with a fresh stock were also greatly superior in fertility to the intercrossed, namely, in the ratio of 100 to 51, as judged by the relative weight of the seed-capsules produced by an equal number of plants of the two sets, both having been left to be naturally fertilized. It should be especially observed that none of the plants of either lot were the product of self-fertilization. On the contrary, the intercrossed plants had certainly been crossed for the last ten generations and probably during all previous generations, as we may infer from the structure of the flowers and from the frequency of the visits of humble-bees. And so it wil have been with the parent-plants of the fresh stock. The whole great difference in height and fertility between the two lots must be attributed to the one being the product of a cross with pollen from a fresh stock, and the other of a cross between plants of the same old stock.

This species offers another interesting case. In the five first generations in which intercrossed and self-fertilized plants were put into competition with one another, every single intercrossed plant beat its self-fertilized antagonist, except in one instance, in which they were equal in height. But in the sixth generation a plant appeared, named by me the Hero, remarkable for its tallness and increased self-fertility, and which transmitted its characters to the next three generations. The children of Hero were again self-fertilized, forming the eighth self-fertilized generation, and were likewise intercrossed one with another; but this cross between plants which had been subjected to the same conditions and had been self-fertilized during the seven previous generations, did not effect the least good; for the intercrossed grandchildren were actually shorter than the self-fertilized grandchildren, in the ratio of 100 to 107. We here see that the mere act of crossing two distinct plants does not by itself benefit the offspring. This case is almost the converse of that in the last paragraph, in which the offspring profited so greatly by a cross with a fresh stock. A similar trial was made with the descendants of Hero in the following generation, and with the same result. But the trial cannot be fully trusted, owing to the extremely unhealthy condition of the plants. Subject to this same serious cause of doubt, even a cross with a fresh stock did not benefit the great-grandchildren of Hero; and if this were really the case, it is the greatest anomaly observed by me in all my experiments.

(2.) *Mimulus luteus.* During the three first generations the intercrossed plants taken together exceeded in height the self-fertilized taken together, in the ratio of 100 to 65, and in fertility in a still higher degree. In the fourth generation a new variety, which grew taller and had whiter and larger flowers than the old varieties, began to prevail, especially among the self-fertilized plants. This variety transmitted its characters with remarkably fidelity, so that all the plants in the later self-fertilized generations belonged to it. These consequently exceeded the intercrossed plants considerably in height. Thus in the seventh generation the intercrossed plants were to the self-fertilized in height as 100 to 137. It is a more remarkable fact that the / self-fertilized plants of the sixth generation had become much more fertile than the intercrossed plants, judging by the number of capsules spontaneously produced, in the ratio of 147 to 100. This variety, which as we have seen appeared amongst the plants of the fourth self-fertilized generation, resembles in almost all its constitutional peculiarities the variety called Hero, which appeared in the sixth self-fertilized generation of Ipomoea. No other such case, with the partial exception of that of Nicotiana, occurred in my experiments, carried on during eleven years.

Two plants of this variety of Mimulus, belonging to the sixth self-fertilized generation, and growing in separate pots, were intercrossed; and some flowers on the same plants were again self-fertilized. From the seeds thus obtained, plants derived from a cross between the self-fertilized plants, and others of the seventh self-fertilized generation, were raised. But this cross did not do the least good, the intercrossed plants being inferior in height to the self-fertilized, in the ratio of 100 to 110. This case is exactly parallel with that given under Ipomoea, of the grandchildren of Hero, and apparently of its great-grandchildren; for the seedlings raised by intercrossing these plants were not in any way superior to those of the corresponding generation raised from the self-fertilized flowers. Therefore in these several cases the crossing of plants, which had been self-fertilized for several generations and which had been cultivated all the time under as nearly as possible the same conditions, was not in the least beneficial.

Another experiment was now tried. First, plants of the eighth self-fertilized generation were again self-fertilized, producing plants of the ninth self-fertilized generation. Secondly, two of the plants of the / eighth self-fertilized generation were intercrossed one with another, as in the experiment above referred to; but this was now effected on plants which had been subjected to two additional generations of self-fertilization. Thirdly, the same plants of the eighth self-fertilized generation were crossed with pollen from plants of a fresh stock brought from a distant garden. Numerous plants were raised from these three sets of seeds, and grown in competition with one another. The plants derived from a cross between the self-fertilized plants exceeded in height by a little the self-fertilized, viz., as 100 to 92; and in fertility in a greater degree, viz., as 100 to 73. I do not know whether this difference in the result, compared with that in the previous case, can be accounted for by the increased deterioration of the self-fertilized plants from two additional generations of self-fertilization, and the consequent advantage of any cross whatever, although merely between the self-fertilized plants. But however this may be, the effects of crossing the self-fertilized plants of the eighth generation with a fresh stock were extremely striking; for the seedlings thus raised were to the self-fertilized of the ninth generation as 100 to 52 in height, and as 100 to 3 in fertility! They were also to the intercrossed plants (derived from crossing two of the self-fertilized plants of the eighth generation) in height as 100 to 56, and in fertility as 100 to 4. Better evidence could hardly be desired of the potent influence of a cross with a fresh stock on plants which had been self-fertilized for eight generations, and had been cultivated all the time under nearly uniform conditions, in comparison with plants self-fertilized for nine generations continuously, or then once intercrossed, namely in the last generation.

(3.) *Brassica oleracea.* Some flowers on cabbage / plants of the second self-fertilized generation were crossed with pollen from a plant of the same variety brought from a distant garden, and other flowers were again self-fertilized. Plants derived from a cross with a fresh stock and plants of the third self-fertilized generation were thus raised. The former were to the self-fertilized in weight as 100 to 22; and this enormous difference must be attributed in part to the beneficial effects of a cross with a fresh stock, and in part to the deteriorating effects of self-fertilization continued during three generations.

(4.) *Iberis umbellata.* Seedlings from a crimson English variety crossed by a pale-coloured variety which had been grown for some generations in Algiers, were to the self-fertilized seedlings from the crimson variety in height as 100 to 89, and as 100 to 75 in fertility. I am surprised that this cross with another variety did not produce a still more strongly marked beneficial effect; for some intercrossed plants of the crimson English variety, put into competition with plants of the same variety self-fertilized during three generations, were in height as 100 to 86, and in fertility as 100 to 75. The slightly greater difference in height in this latter case, may possibly be attributed to

241

the deteriorating effects of self-fertilization carried on for two additional generations.

(5.) *Eschscholtzia californica.* This plant offers an almost unique case, inasmuch as the good effects of a cross or the evil effects of self-fertilization are confined to the reproductive system. Intercrossed and self-fertilized plants of the English stock did not differ in height (nor in weight, as far as was ascertained), in any constant manner; the self-fertilized plants usually having the advantage. So it was with the offspring of plants of the Brazilian stock, tried in the same / manner. The parent-plants, however, of the English stock produced many more seeds when fertilized with pollen from another plant than when self-fertilized; and in Brazil the parent-plants were absolutely sterile unless they were fertilized with pollen from another plant. Intercrossed seedlings, raised in England from the Brazilian stock, compared with self-fertilized seedlings of the corresponding second generation, yielded seeds in number as 100 to 89; both lots of plants being left freely exposed to the visits of insects. If we now turn to the effects of crossing plants of the Brazilian stock with pollen from the English stock – so that plants which had been long exposed to very different conditions were intercrossed – we find that the offspring were, as before, inferior in height and weight to the plants of the Brazilian stock after two generations of self-fertilization, but were superior to them in the most marked manner in the number of seeds produced, namely, as 100 to 40; both lots of plants being left freely exposed to the visits of insects.

In the case of Ipomoea, we have seen that the plants derived from a cross with a fresh stock were superior in height as 100 to 78, and in fertility as 100 to 51, to the plants of the old stock, although these had been intercrossed during the last ten generations. With Eschscholtzia we have a nearly parallel case, but only as far as fertility is concerned, for the plants derived from a cross with a fresh stock were superior in fertility in the ratio of 100 to 45 to the Brazilian plants, which had been artificially intercrossed in England for the two last generations, and which must have been naturally intercrossed by insects during all previous generations in Brazil, where otherwise they are quite sterile.

(6.) *Dianthus caryophyllus.* Plants self-fertilized / for three generations were crossed with pollen from a fresh stock, and their offspring were grown in competition with plants of the fourth self-fertilized generation. The crossed plants thus obtained were to the self-fertilized in height as 100 to 81, and in fertility (both lots being left to be naturally fertilized by insects) as 100 to 33.

These same crossed plants were also to the offspring from the plants of the third self-fertilized generation crossed by the intercrossed plants of the corresponding generation, in height as 100 to 85, and in fertility as 100 to 45.

We thus see what a great advantage the offspring from a cross with a fresh stock had, not only over the self-fertilized plants of the fourth generation, but over the offspring from the self-fertilized plants of the third generation, when crossed by the intercrossed plants of the old stock.

(7.) *Pisum sativum.* It has been shown under the head of this species, that the several varieties in this country almost invariably fertilize themselves, owing to insects rarely visiting the flowers; and as the plants have been long cultivated under nearly similar conditions, we can understand why a cross between two individuals of the same variety does not do the least good to the offspring either in height or fertility. This case is almost exactly parallel with that of Mimulus, or that of the Ipomoea named Hero; for in these two instances, crossing plants which had been self-fertilized for seven generations did not at all benefit the offspring. On the other hand, a cross between two varieties of the pea causes a marked superiority in the growth and vigour of the offspring, over the self-fertilized plants of the same varieties, as shown by two excellent observers. From my own / observations (not made with great care) the offspring from crossed varieties were to self-fertilized plants in height, in one case as 100 to about 75, and in a second case as 100 to 60.

(8.) *Lathyrus odoratus.* The sweetpea is in the same state in regard to self-fertilization as the common pea; and we have seen that seedlings from a cross between two varieties, which differed in no respect except in the colour of their flowers, were to the self-fertilized seedlings from the same mother-plant in height as 100 to 80; and in the second generation as 100 to 88. Unfortunately I did not ascertain whether crossing two plants of the same variety failed to produce any beneficial effect, but I venture to predict such would be the result.

(9.) *Petunia violacea.* The intercrossed plants of the same stock in four out of the five successive generations plainly exceeded in height the self-fertilized plants. The latter in the fourth generation were crossed by a fresh stock, and the seedlings thus obtained were put into competition with the self-fertilized plants of the fifth generation. The crossed plants exceeded the self-fertilized in height in the ratio of 100 to 66, and in weight as 100 to 23; but this difference, though so great, is not much greater than that between the intercrossed plants of the same stock in comparison with the self-fertilized plants of the corresponding generation. This case, therefore, seems at first sight opposed to the rule that a cross with a fresh stock is much more beneficial than a cross between individuals of the same stock. But as with Eschscholtzia, the reproductive system was here chiefly benefited; for the plants raised from the cross with the fresh stock were to the self-fertilized plants in fertility, both lots being naturally fertilized, as 100 to 46, whereas the / intercrossed plants of the same stock were to the self-fertilized plants of the corresponding fifth generation in fertility only as 100 to 86.

Although at the time of measurement the plants raised from the cross with the fresh stock did not exceed in height or weight the intercrossed plants of the old stock (owing to the growth of the former not having been completed, as explained under the head of this species), yet they exceeded the intercrossed plants in fertility in the ratio of 100 to 54. This fact is interesting, as it shows that plants self-fertilized for four generations and then crossed by a fresh stock, yielded seedlings which were nearly twice as fertile as those from plants of the same stock which had been intercrossed for the five previous generations. We here see, as with Eschscholtzia and Dianthus, that the mere act of crossing, independently of the state of the crossed plants, has little efficacy in giving increased fertility to the offspring. The same conclusion holds good, as we have already seen, in the analogous cases of Ipomoea, Mimulus, and Dianthus, with respect to height.

(10.) *Nicotiana tabacum.* My plants were remarkably self-fertile, and the capsules from the self-fertilized flowers apparently yielded more seeds than those which were cross-fertilized. No insects were seen to visit the flowers in the hothouse, and I suspect that the stock on which I experimented had been raised under glass, and had been self-fertilized during several previous generations; if so, we can understand why, in the course of three generations, the crossed seedlings of the same stock did not uniformly exceed in height the self-fertilized seedlings. But the case is complicated by individual plants having different constitutions, so that some of the crossed and self-fertilized seedlings raised at the same time from the same parents behaved differently. / However this may be, plants raised from self-fertilized plants of the third generation crossed by a slightly different subvariety, exceeded greatly in height and weight the self-fertilized plants of the fourth generation; and the trial was made on a large scale. They exceeded them in height when grown in pots, and not much crowded, in the ratio of 100 to 66; and when much crowded, as 100 to 54. These crossed plants, when thus subjected to severe competition, also exceeded the self-fertilized in weight in the ratio of 100 to 37. So it was, but in a less degree (as may be seen in Table C), when the two lots were grown out of doors and not subjected to any mutual competition. Nevertheless, strange as is the fact, the flowers on the mother-plants of the third self-fertilized generation did not yield more seed when they were crossed with pollen from plants of the fresh stock than when they were self-fertilized.

(11.) *Anagallis collina*. Plants raised from a red variety crossed by another plant of the same variety were in height to the self-fertilized plants from the red variety as 100 to 73. When the flowers on the red variety were fertilized with pollen from a closely similar blue-flowered variety, they yielded double the number of seeds to what they did when crossed by pollen from another individual of the same red variety, and the seeds were much finer. The plants raised from this cross between the two varieties were to the self-fertilized seedlings from the red variety, in height as 100 to 66, and in fertility as 100 to 6.

(12.) *Primula veris*. Some flowers on long-styled plants of the third illegitimate generation were legitimately crossed with pollen from a fresh stock, and others were self-fertilized with their own pollen. From the seeds thus produced crossed plants, and self-fertilized / plants of the fourth illegitimate generation, were raised. The former were to the latter in height as 100 to 46, and in fertility during one year as 100 to 5, and as 100 to 3·5 during the next year. In this case, however, we have no means of distinguishing between the evil effects of illegitimate fertilization continued during four generations (that is, by pollen of the same form, but taken from a distinct plant) and strict self-fertilization. But these two processes perhaps do not differ so essentially as at first appears to be the case. In the following experiment any doubt arising from illegitimate fertilization was completely eliminated.

(13.) *Primula veris*. (Equal-styled, red-flowered variety.) Flowers on plants of the second self-fertilized generation were crossed with pollen from a distinct variety or fresh stock, and others were again self-fertilized. Crossed plants and plants of the third self-fertilized generation, all of legitimate origin, were thus raised; and the former was to the latter in height as 100 to 85, and in fertility (as judged by the number of capsules produced, together with the average number of seeds) as 100 to 11.

Summary of the measurements in Table C

This table includes the heights and often the weights of 292 plants derived from a cross with a fresh stock, and of 305 plants, either of self-fertilized origin, or derived from an intercross between plants of the same stock. These 597 plants belong to thirteen species and twelve genera. The various precautions which were taken to ensure a fair comparison have already been stated. If we now look down the right-hand column, in which the mean height, weight, and fertility of the plants derived from a cross with a fresh stock are represented by 100, we shall see / by the other figures how wonderfully superior they are both to the self-fertilized and to the intercrossed plants of the same stock. With respect to height and weight, there are only two exceptions to the rule, namely, with Eschscholtzia and Petunia, and the latter is probably no real exception. Nor do these two species offer an exception in regard to fertility, for the plants derived from the cross with a fresh stock were much more fertile than the self-fertilized plants. The difference between the two sets of plants in the table is generally much greater in fertility than in height or weight. On the other hand, with some of the species, as with Nicotiana, there was no difference in fertility between the two sets, although a great difference in height and weight. Considering all the cases in this table, there can be no doubt that plants profit immensely, though in different ways, by a cross with a fresh stock or with a distinct subvariety. It cannot be maintained that the benefit thus derived is due merely to the plants of the fresh stock being perfectly healthy, whilst those which had been long intercrossed or self-fertilized had become unhealthy; for in most cases there was no appearance of such unhealthiness, and we shall see under Table A that the intercrossed plants of the same stock are generally superior to a certain extent to the self-fertilized – both lots having been subjected to exactly the same conditions and being equally healthy or unhealthy.

We further learn from Table C, that a cross between plants that have been self-fertilized during several successive generations and kept all the time under nearly uniform conditions, does not benefit the offspring in the least or only in a very slight degree. Mimulus and the descendants of Ipomoea named Hero offer instances of this rule. Again, plants self-fertilized / during several generations profit only to a small extent by a cross with intercrossed plants of the same stock (as in the case of Dianthus), in comparison with the effects of a cross by a fresh stock. Plants of the same stock intercrossed during several generations (as with Petunia) were inferior in a marked manner in fertility to those derived from the corresponding self-fertilized plants crossed by a fresh stock. Lastly, certain plants which are regularly intercrossed by insects in a state of nature, and which were artificially crossed in each succeeding generation in the course of my experiments, so that they can never or most rarely have suffered any evil from self-fertilization (as with Eschscholtzia and Ipomoea), nevertheless profited greatly by a cross with a fresh stock. These several cases taken together show us in the clearest manner that it is not the mere crossing of any two individuals which is beneficial to the offspring. The benefit thus derived depends on the plants which are united differing in some manner, and there can hardly be a doubt that it is in the constitution or nature of the sexual elements. Anyhow, it is certain that the differences are not of an external nature, for two plants which resemble each other as closely as the individuals of the same species ever do, profit in the plainest manner when intercrossed, if their progenitors have been exposed during several generations to different conditions. But to this latter subject I shall have to recur in a future chapter.

TABLE A

We will now turn to our first table, which relates to crossed and self-fertilized plants of the same stock. These consist of fifty-four species belonging to thirty natural orders. The total number of crossed plants of which measurements are given is 796, and / of self-fertilized plants 809; that is altogether 1,605 plants. Some of the species were experimented on during several successive generations; and it should be borne in mind that in such cases the crossed plants in each generation were crossed with pollen from another crossed plant, and the flowers on the self-fertilized plants were almost always fertilized with their own pollen, though sometimes with pollen from other flowers on the same plant. The crossed plants thus became more or less closely inter-related in the later generations; and both lots were subjected in each generation to almost absolutely the same conditions, and to nearly the same conditions in the successive generations. It would have been a better plan in some respects if I had always crossed some flowers either on the self-fertilized or intercrossed plants of each generation with pollen from a non-related plant, grown under different conditions, as was done with the plants in Table C; for by this procedure I should have learnt how much the offspring became deteriorated through continued self-fertilization in the successive generations. As the case stands, the self-fertilized plants of the successive generations in Table A were put into competition with and compared with intercrossed plants, which were probably deteriorated in some degree by being more or less inter-related and grown under similar conditions. Nevertheless, had I always followed the plan in Table C, I should not have discovered the important fact that, although a cross between plants which are rather closely related and which had been subjected to closely similar conditions, gives during several generations some advantage to the offspring, yet that after a time the may be intercrossed with no advantage whatever to the offspring. Nor should I have learnt that the self-fertilized plants of the later generations might / be crossed with intercrossed plants of the same stock with little or no advantage, although they profited to an extraordinary degree by a cross with a fresh stock.

With respect to the greater number of the plants in Table A, nothing special need here be said; full particulars may be found under the head of each species by the aid of the Index. The figures in the right-hand column show the mean height of the self-fertilized plants, that of the crossed plants with which they competed being represented by 100. No notice is here taken of the few cases in which crossed and self-fertilized plants were grown in the open ground, so as not to compete together. The table includes, as we have seen, plants belonging to fifty-four species, but as some of these were measured during several successive generations, there are eighty-three cases in which crossed and self-fertilized plants were compared. As in each generation the number of plants which were measured (given in the table) was never very large and sometimes small, whenever in the right-hand column the mean height of the crossed and self-fertilized plants is the same within five per cent, their heights may be considered as practically equal. Of such cases, that is, of self-fertilized plants of which the mean height is expressed by figures between 95 and 105, there are eighteen, either in some one or all the generations. There are eight cases in which the self-fertilized plants exceed the crossed by about five per cent, as shown by the figures in the right-hand column being above 105. Lastly, there are fifty-seven cases in which the crossed plants exceed the self-fertilized in a ratio of at least 100 to 95, and generally in a much higher degree.

If the relative heights of the crossed and self-fertilized plants had been due to mere chance, there would have / been about as many cases of self-fertilized plants exceeding the crossed in height by above five per cent as of the crossed thus exceeding the self-fertilized; but we see that of the latter there are fifty-seven cases, and of the former only eight cases; so that the cases in which the crossed plants exceed in height the self-fertilized in the above proportion are more than seven times as numerous as those in which the self-fertilized exceed the crossed in the same proportion. For our special purpose of comparing the powers of growth of crossed and self-fertilized plants, it may be said that in fifty-seven cases the crossed plants exceeded the self-fertilized by more than five per cent, and that in twenty-six cases (18+8) they did not thus exceed them. But we shall now show that in several of these twenty-six cases the crossed plants had a decided advantage over the self-fertilized in other respects, though not in height; that in other cases the mean heights are not trustworthy, owing to too few plants having been measured, or to their having grown unequally from being unhealthy, or to both causes combined. Nevertheless, as these cases are opposed to my general conclusion I have felt bound to give them. Lastly, the cause of the crossed plants having no advantage over the self-fertilized can be explained in some other cases. Thus a very small residue is left in which the self-fertilized plants appear, as far as my experiments serve, to be really equal or superior to the crossed plants.

We will now consider in some little detail the eighteen cases in which the self-fertilized plants equalled in average height the crossed plants within five per cent; and the eight cases in which the self-fertilized plants exceeded in average height the crossed plants by above five per cent; making altogether twenty-six / cases in which the crossed plants were not taller than the self-fertilized plants in any marked degree.

(1.) *Dianthus caryophyllus* (third generation). This plant was experimented on during four generations, in three of which the crossed plants exceeded in height the self-fertilized generally by much more than five per cent; and we have seen under Table C that the offspring from the plants of the third self-fertilized generation crossed by a fresh stock profited in height and fertility to an extraordinary degree. But in this third generation the crossed plants of the same stock were in height to the self-fertilized only as 100 to 99, that is, they were practically equal. Nevertheless, when the eight crossed and eight self-fertilized plants were cut down and weighed, the former were to the latter in weight as 100 to 49! There can therefore be not the least doubt that the crossed plants of this species are greatly superior in vigour and luxuriance to the self-fertilized; and what was the cause of the self-fertilized plants of the third generation, though so light and thin, growing up so as almost to equal the crossed in height, I cannot explain.

(2.) *Lobelia fulgens* (first generation). The crossed plants of this generation were much inferior in height to the self-fertilized, in the proportion of 100 to 127. Although only two pairs were measured, which is obviously much too few to be trusted, yet from other evidence given under the head of this species, it is certain that the self-fertilized plants were very much more vigorous than the crossed. As I used pollen of unequal maturity for crossing and self-fertilizing the parent-plants, it is possible that the great difference in the growth of their offspring may have been due to this cause. In the next generation this source of error was avoided, and many more plants were raised, and now the average height of the twenty-three crossed plants was to that of the twenty-three self-fertilized plants as 100 to 91. We can therefore hardly doubt that a cross is beneficial to this species.

(3.) *Petunia violacea* (third generation). Eight crossed plants were to eight self-fertilized of the third generation in average height as 100 to 131; and at an early age the crossed were inferior even in a still higher degree. But it is a remarkable fact that in one pot in which plants of both lots grew extremely crowded, the crossed were thrice as tall as the self-fertilized. As in the two preceding and two succeeding genera-tions, as well as / with plants raised by a cross with a fresh stock, the crossed greatly exceeded the self-fertilized in height, weight, and fertility (when these two latter points were attended to), the present case must be looked at as an anomaly not affecting the general rule. The most probable explanation is that the plants grew prematurely, owing to the seeds of the last generation not having been well ripened; for I have observed an analogous case with Iberis. Self-fertilized seedlings of this latter plant, which were known to have been produced from seeds not well matured, grew from the first much more quickly than the crossed plants, which were raised from better matured seeds; so that having thus once got a great start they were enabled ever afterwards to retain their advantage. Some of these same seeds of the Iberis were sown on the opposite sides of pots filled with burnt earth and pure sand, not containing any organic matter; and now the young crossed seedlings grew during their short life to double the height of the self-fertilized, in the same manner as occurred with the above two sets of seedlings of Petunia which were much crowded and thus exposed to very unfavourable conditions. We have seen also in the eighth generation of Ipomoea that self-fertilized seedlings raised from unhealthy parents grew at first very much more quickly than the crossed seedlings, so that they were for a long time much taller, though ultimately beaten by them.

(4, 5, 6.) *Eschscholtzia californica.* Four sets of measurements are given in Table A. In one of these the crossed plants exceed the self-fertilized in average height, so that this is not one of the exceptions here to be considered. In two other cases the crosses equalled the self-fertilized in height within five per cent; and in the fourth case the self-fertilized exceeded the crossed by above this limit. We have seen in Table C that the whole advantage of a cross by a fresh stock is confined to the number of seeds produced, and so it was with the disadvantage from self-fertilization with the intercrossed plants of the same stock compared with the self-fertilized, for the former were in fertility to the latter as 100 to 89. The intercrossed plants thus have at least one important advantage over the self-fertilized. Moreover, the flowers on the parent-plants when fertilized with pollen from another individual of the same stock yield far more seeds than when self-fertilized; the flowers in this latter case being often quite sterile. We may therefore conclude that a cross does some good, though it does not give to the crossed seedlings increased powers of growth. /

(7.) *Viscaria oculata.* The average height of the fifteen intercrossed plants to that of the fifteen self-fertilized plants was only as 100 to 97; but the former produced many more capsules than the latter, in the ratio of 100 to 77. Moreover, the flowers on the parent-plants which were crossed and self-fertilized, yielded seeds on one occasion in the proportion of 100 to 38, and on a second occasion in the proportion of 100 to 58. So that there can be no doubt about the beneficial effects of a cross, although the mean height of the crossed plants was only three per cent above that of the self-fertilized plants.

(8.) *Specularia speculum.* Only the four tallest of the crossed and the four tallest of the self-fertilized plants, growing in four pots, were measured; and the former were to the latter in height as 100 to 98. In all four pots a crossed plant flowered before any one of the self-fertilized plants, and this is usually a safe indication of some real superiority in the crossed plants. The flowers on the parent-plants which were crossed with pollen from another plant yielded seeds compared with the self-fertilized flowers in the ratio of 100 to 72. We may therefore draw the same conclusion as in the last case with respect to a cross being decidedly beneficial.

(9.) *Borago officinalis.* Only four crossed and four self-fertilized plants were raised and measured, and the former were to the latter in height as 100 to 102. So small a number of measurements ought never to be trusted; and in the present instance the advantage of the self-fertilized over the crossed plants depended almost entirely on one of the self-fertilized plants having grown to an unusual height. All four crossed plants flowered before their self-fertilized opponents. The cross-fertilized flowers on the parent-plants in comparison with the self-fertilized flowers yielded seeds in the proportion of 100 to 60. So that here again we may draw the same conclusion as in the two last cases.

(10.) *Passiflora gracilis.* Only two crossed and two self-fertilized plants were raised; and the former were to the latter in height as 100 to 104. On the other hand, fruits from the cross-fertilized flowers on the parent-plants contained seeds in number, compared with those from the self-fertilized flowers, in the proportion of 100 to 85.

(11.) *Phaseolus multiflorus.* The five crossed plants were to the five self-fertilized in height as 100 to 96. Although the crossed plants were thus only four per cent taller than the / self-fertilized, they flowered in both pots before them. It is therefore probable that they had some real advantage over the self-fertilized plants.

(12.) *Adonis aestivalis.* The four crossed plants were almost exactly equal in height to the four self-fertilized plants, but as so few plants were measured, and as these were all 'miserably unhealthy', nothing can be inferred with safety with respect to their relative heights.

(13.) *Bartonia aurea.* The eight crossed plants were to the eight self-fertilized in height as 100 to 107. This number of plants, considering the care with which they were raised and compared, ought to have given a trustworthy result. But from some unknown cause they grew very unequally, and they became so unhealthy that only three of the crossed and three of the self-fertilized plants set any seeds, and these few in number. Under these circumstances the mean height of neither lot can be trusted, and the experiment is valueless. The cross-fertilized flowers on the parent-plants yielded rather more seeds than the self-fertilized flowers.

(14.) *Thunbergia alata.* The six crossed plants were to the six self-fertilized in height as 100 to 108. Here the self-fertilized plants seem to have a decided advantage; but both lots grew unequally, some of the plants in both being more than twice as tall as others. The parent-plants also were in an odd semi-sterile condition. Under these circumstances the superiority of the self-fertilized plants cannot be fully trusted.

(15.) *Nolana prostrata.* The five crossed plants were to the five self-fertilized in height as 100 to 105; so that the latter seem here to have a small but decided advantage. On the other hand, the flowers on the parent-plants which were cross-fertilized produced very many more capsules than the self-fertilized flowers, in the ratio of 100 to 21; and the seeds which the former contained were heavier than an equal number from the self-fertilized capsules in the ratio of 100 to 82.

(16.) *Hibiscus africanus.* Only four pairs were raised, and the crossed were to the self-fertilized in height as 100 to 109. Excepting that too few plants were measured, I know of nothing else to cause distrust in the result. The cross-fertilized flowers on the parent-plants were, on the other hand, rather more productive than the self-fertilized flowers.

(17.) *Apium petroselinum.* A few plants (number not recorded) derived from flowers believed to have been crossed by / insects and a few self-fertilized plants were grown on the opposite sides of four pots. They attained to a nearly equal height, the crossed having a very slight advantage.

(18.) *Vandellia nummularifolia.* Twenty crossed plants raised from the seeds of perfect flowers were to twenty self-fertilized plants, likewise raised from the seeds of perfect flowers, in height as 100 to 99. The experiment was repeated, with the sole difference that the plants were allowed to grow more crowded; and now the twenty-four tallest of the crossed plants were to the twenty-four tallest self-fertilized plants in height as 100 to 94, and in weight as 100 to 97. Moreover, a larger number of the crossed than of the self-fertilized plants grew to a moderate height. The above-mentioned twenty crossed plants were also grown in competition with twenty self-fertilized plants raised from the closed or cleistogamic flowers, and their heights were as 100 to 94.

Therefore had it not been for the first trial, in which the crossed plants were to the self-fertilized in height only as 100 to 99, this species might have been classed with those in which the crossed plants exceed the self-fertilized by above five per cent. On the other hand, the crossed plants in the second trial bore fewer capsules, and these contained fewer seeds, than did the self-fertilized plants, all the capsules having been produced by cleistogamic flowers. The whole case therefore must be left doubtful.

(19.) *Pisum sativum* (common pea). Four plants derived from a cross between individuals of the same variety were in height to four self-fertilized plants belonging to the same variety as 100 to 115. Although this cross did no good, we have seen under Table C that a cross between distinct varieties adds greatly to the height and vigour of the offspring; and it was there explained that the fact of a cross between the individuals of the same variety not being beneficial, is almost certainly due to their having been self-fertilized for many generations, and in each generation grown under nearly similar conditions.

(20, 21, 22.) *Canna warscewiczi.* Plants belonging to three generations were observed, and in all of three the crossed were approximately equal to the self-fertilized; the average height of the thirty-four crossed plants being to that of the same number of self-fertilized plants as 100 to 101. Therefore the crossed plants had no advantage over the self-fertilized; and it is probable that the same explanation here holds good as in the case of *Pisum sativum*; for the flowers of this Canna are perfectly / self-fertile, and were never seen to be visited by insects in the hothouse, so as to be crossed by them. This plant, moreover, has been cultivated under glass for several generations in pots, and therefore under nearly uniform conditions. The capsules produced by the cross-fertilized flowers on the above thirty-four crossed plants contained more seeds than did the capsules produced by the self-fertilized flowers on the self-fertilized plants, in the proportion of 100 to 85; so that in this respect crossing was beneficial.

(23.) *Primula sinensis.* The offspring of plants, some of which were legitimately and others illegitimately fertilized with pollen from a distinct plant, were almost exactly of the same height as the offspring of self-fertilized plants; but the former with rare exceptions flowered before the latter. I have shown in my work on heterostyled plants that this species is commonly raised in England from self-fertilized seed, and the plants from having been cultivated in pots have been subjected to nearly uniform conditions. Moreover, many of them are now varying and changing their character, so as to become in a greater or less degree equal-styled, and in consequence highly self-fertile. Therefore I believe that the cause of the crossed plants not exceeding in height the self-fertilized is the same as in the two previous cases of *Pisum sativum* and Canna.

(24, 25, 26.) *Nicotiana tabacum*. Four sets of measurements were made; in one, the self-fertilized plants greatly exceeded in height the crossed, in two others they were approximately equal to the crossed, and in the fourth were beaten by them; but this latter case does not here concern us. The individual plants differ in constitution, so that the descendants of some profit by their parents having been intercrossed, whilst others do not. Taking all three generations together, the twenty-seven crossed plants were in height to the twenty-seven self-fertilized plants as 100 to 96. This excess of height in the crossed plants is so small compared with that displayed by the offspring from the same mother-plants when crossed by a slightly different variety, that we may suspect (as explained under Table C) that most of the individuals belonging to the variety which served as the mother-plants in my experiments, had acquired a nearly similar constitution, so as not to profit by being mutually intercrossed.

Reviewing these twenty-six cases, in which the crossed plants either do not exceed the self-fertilized / by above five per cent in height, or are inferior to them, we may conclude that much the greater number of the cases do not form real exceptions to the rule – that a cross between two plants, unless these have been self-fertilized and exposed to nearly the same conditions for many generations, gives a great advantage of some kind to the offspring. Of the twenty-six cases, at least two, namely, those of Adonis and Bartonia, may be wholly excluded, as the trials were worthless from the extreme unhealthiness of the plants. In twelve other cases (three trials with Eschscholtzia here included) the crossed plants either were superior in height to the self-fertilized in all the other generations excepting the one in question, or they showed their superiority in some different manner, as in weight, fertility, or in flowering first; or again, the cross-fertilized flowers on the mother-plant were much more productive of seed than the self-fertilized.

Deducting these fourteen cases, there remain twelve in which the crossed plants show no well-marked advantage over the self-fertilized. On the other hand, we have seen that there are fifty-seven cases in which the crossed plants exceed the self-fertilized in height by at least five per cent, and generally in a much higher degree. But even in the twelve cases just referred to, the want of any advantage on the crossed side is far from certain: with Thunbergia the parent-plants were in an odd semi-sterile condition, and the offspring grew very unequally; with Hibiscus and Apium much too few plants were raised for the measurements to be trusted, and the cross-fertilized flowers of Hibiscus produced rather more seed than did the self-fertilized; with Vandellia the crossed plants were a little taller and heavier than the self-fertilized, but as they were less fertile the case must be left doubtful. / Lastly, with Pisum, Primula, the three generations of Canna, and the three of Nicotiana (which together complete the twelve cases), a cross between two plants certainly did no good or very little good to the offspring; but we have reason to suspect that this is the result of these plants having been self-fertilized and cultivated under nearly uniform conditions for several generations. The same result followed with the experimental plants of Ipomoea and Mimulus, and to a certain extent with some other species, which had been intentionally treated by me in this manner; yet we know that these species in their normal condition profit greatly by being intercrossed. There is, therefore, not a single case in Table A which affords decisive evidence against the rule that a cross between plants, the progenitors of which have been subjected to somewhat diversified conditions, is beneficial to the offspring. This is a surprising conclusion, for from the analogy of domesticated animals it could not have been anticipated, that the good effects of crossing or the evil effects of self-fertilization would have been perceptible until the plants had been thus treated for several generations.

The results given in Table A may be looked at under another point of view. Hitherto each generation has been considered as a separate case, of which there are eighty-three; and this no doubt is the more correct method of comparing the crossed and self-fertilized plants.

But in those cases in which plants of the same species were observed during several generations, a general average of their heights in all the generations together may be made; and such averages are given in Table A; for instance, under Ipomoea the general average for the plants of all ten generations is as 100 for the crossed, to 77 for the self-fertilized / plants. This having been done in each case in which more than one generation was raised, it is easy to calculate the average of the average heights of the crossed and self-fertilized plants of all the species included in Table A. It should however be observed that as only a few plants of some species, whilst a considerable number of others, were measured, the value of the mean or average heights of the several species is very different. Subject to this source of error, it may be worth while to give the mean of the mean heights of the fifty-four species in Table A; and the result is, calling the mean of the mean heights of the crossed plants 100, that of the self-fertilized plants is 87. But it is a better plan to divide the fifty-four species into three groups, as was done with the previously given eighty-three cases. The first group consists of species of which the mean heights of the self-fertilized plants are within five per cent of 100; so that the crossed and self-fertilized plants are approximately equal; and of such species there are twelve about which nothing need be said, the mean of the mean heights of the self-fertilized being of course very nearly 100, or exactly 99·58. The second group consists of the species, thirty-seven in number, of which the mean heights of the crossed plants exceed that of the self-fertilized plants by more than five per cent; and the mean of their mean heights is to that of the self-fertilized plants as 100 to 78. The third group consists of the species, only five in number, of which the mean heights of the self-fertilized plants exceed that of the crossed by more than five per cent; and here the mean of the mean heights of the crossed plants is to that of the self-fertilized as 100 to 109. There-fore if we exclude the species which are approximately equal, there are / thirty-seven species in which the mean of the mean heights of the crossed plants exceeds that of the self-fertilized by twenty-two per cent; whereas there are only five species in which the mean of the mean heights of the self-fertilized plants exceeds that of the crossed, and this only by nine per cent.

The truth of the conclusion – that the good effects of a cross depend on the plants having been subjected to different conditions or to their belonging to different varieties, in both of which cases they would almost certainly differ somewhat in constitution – is supported by a comparison of the Tables A and C. The latter table gives the results of crossing plants with a fresh stock or with a distinct variety; and the superiority of the crossed offspring over the self-fertilized is here much more general and much more strongly marked than in Table A, in which plants of the same stock were crossed. We have just seen that the mean of the mean heights of the crossed plants of the whole fifty-four species in Table A is to that of the self-fertilized plants as 100 to 87; whereas the mean of the mean heights of the plants crossed by a fresh stock is to that of the self-fertilized in Table C as 100 to 74. So that the crossed plants beat the self-fertilized plants by thirteen per cent in Table A, and by twenty-six per cent, or double as much, in Table C, which includes the results of a cross by a fresh stock.

TABLE B

A few words must be added on the weights of the crossed plants of the same stock, in comparison with the self-fertilized. Eleven cases are given in Table B, relating to eight species. The number of plants which were weighed is shown in the two left columns, and their relative weights in the right / column, that of the crossed plants being taken as 100. A few other cases have already been recorded in Table C in reference to plants crossed by a fresh stock. I regret that more trials of this kind were not made, as the evidence of the superiority of the crossed over the self-fertilized plants is thus shown in a more conclusive manner than by their relative heights. But this plan was not thought of until a rather late period, and there were difficulties in the way, as the seeds had to be collected when ripe, by which time the plants had often begun to wither. In only one out of the eleven cases in Table B, that of Eschscholtzia, do the self-fertilized plants exceed the crossed in weight; and we have already seen they are likewise superior to them in height, though inferior in fertility, the whole advantage of a cross being here confined to the reproductive system. With Vandellia the crossed plants were a little heavier, as they were also a little taller than the self-fertilized; but as a greater number of more productive capsules, were produced by the cleistogamic flowers on the self-fertilized plants than by those on the crossed plants, the case must be left, as remarked under Table A, altogether doubtful. The crossed and self-fertilized offspring from a partially self-sterile plant of *Reseda odorata* were almost equal in weight, though not in height. In the remaining eight cases, the crossed plants show a wonderful superiority over the self-fertilized, being more than double their weight, except in one case, and here the ratio is as high as 100 to 67. The results thus deduced from the weights of the plants confirm in a striking manner the former evidence of the beneficial effects of a cross between two plants of the same stock; and in the few cases in which plants derived from a cross with a fresh stock were weighed, the results are similar or even more striking. /

CHAPTER VIII

DIFFERENCE BETWEEN CROSSED AND
SELF-FERTILIZED PLANTS IN CONSTITUTIONAL
VIGOUR AND IN OTHER RESPECTS

Greater constitutional vigour of crossed plants – The effects of great
crowding – Competition with other kinds of plants – Self-fertilized plants
more liable to premature death – Crossed plants generally flower before
the self-fertilized – Negative effects of intercrossing flowers on the same
plant – Cases described – Transmission of the good effects of a cross to
later generations – Effects of crossing plants of closely related parentage –
Uniform colour of the flowers on plants self-fertilized during several
generations and cultivated under similar conditions.

Greater constitutional vigour of crossed plants. As in almost all my experiments an equal number of crossed and self-fertilized seeds, or more
commonly seedlings just beginning to sprout, were planted on the
opposite sides of the same pots, they had to compete with one another;
and the greater height, weight, and fertility of the crossed plants may
be attributed to their possessing greater innate constitutional vigour.
Generally the plants of the two lots whilst very young were of equal
height; but afterwards the crossed gained insensibly on their opponents, and this shows that they possessed some inherent superiority,
though not displayed at a very early period of life. There were,
however, some conspicuous exceptions to the rule of the two lots being
at first equal in height; thus the crossed seedlings of the broom
(*Sarothamnus scoparius*) when under three inches in height were more
than twice as tall as the self-fertilized plants. /
    After the crossed or the self-fertilized plants had once grown de-
cidedly taller than their opponents, a still increasing advantage would
tend to follow from the stronger plants robbing the weaker ones of
nourishment and overshadowing them. This was evidently the case
with the crossed plants of *Viola tricolor*, which ultimately quite
overwhelmed the self-fertilized. But that the crossed plants have an
inherent superiority, independently of competition, was sometimes

259

well shown when both lots were planted separately, not far distant from one another, in good soil in the open ground. This was likewise shown in several cases, even with plants growing in close competition with one another, by one of the self-fertilized plants exceeding for a time its crossed opponent, which had been injured by some accident or was at first sickly, but being ultimately conquered by it. The plants of the eighth generation of Ipomoea were raised from small seeds produced by unhealthy parents, and the self-fertilized plants grew at first very rapidly, so that when the plants of both lots were about three feet in height, the mean height of the crossed to that of the self-fertilized was as 100 to 122; when they were about six feet high the two lots were very nearly equal, but ultimately when between eight and nine feet in height, the crossed plants asserted their usual superiority, and were to the self-fertilized in height as 100 to 85.

The constitutional superiority of the crossed over the self-fertilized plants was proved in another way in the third generation of Mimulus, by self-fertilized seeds being sown on one side of a pot, and after a certain interval of time crossed seeds on the opposite side. The self-fertilized seedlings thus had (for I ascertained that the seeds germinated simultaneously) a clear advantage / over the crossed in the start for the race. Nevertheless they were easily beaten (as may be seen under the head of Mimulus) when the crossed seeds were sown two whole days after the self-fertilized. But when the interval was four days, the two lots were nearly equal throughout life. Even in this latter case the crossed plants still possessed an inherent advantage, for after both lots had grown to their full height they were cut down, and without being disturbed were transferred to a larger pot, and when in the ensuing year they had again grown to their full height they were measured; and now the tallest crossed plants were to the tallest self-fertilized plants in height as 100 to 75, and in fertility (i.e. by weight of seeds produced by an equal number of capsules from both lots) as 100 to 34.

My usual method of proceeding, namely, to plant several pairs of crossed and self-fertilized seeds in an equal state of germination on the opposite sides of the same pots, so that the plants were subjected to moderately severe mutual competition, was I think the best that could have been followed, and was a fair test of what occurs in a state of nature. For plants sown by nature generally come up crowded, and are almost always exposed to very severe competition with one another and with other kinds of plants. This latter consideration led me to make some trials, chiefly but not exclusively with Ipomoea and Mimulus, by sowing crossed and self-fertilized seeds on the opposite sides of large pots in which other plants had long been growing, or in the midst of other plants out of doors. The seedlings were thus subjected to very severe competition with plants of other kinds; and in all such cases, the crossed seedlings exhibited a great superiority in their power of growth over the self-fertilized. /

After the germinating seedlings had been planted in pairs on the opposite sides of several pots, the remaining seeds, whether or not in a state of germination, were in most cases sown very thickly on the two sides of an additional large pot; so that the seedlings came up extremely crowded, and were subjected to extremely severe competition and unfavourable conditions. In such cases the crossed plants almost invariably showed a greater superiority over the self-fertilized, than did the plants which grew in pairs in the pots.

Sometimes crossed and self-fertilized seeds were sown in separate rows in the open ground, which was kept clear of weeds; so that the seedlings were not subjected to any competition with other kinds of plants. Those however in each row had to struggle with the adjoining ones in the same row. When fully grown, several of the tallest plants in each row were selected, measured, and compared. The result was in several cases (but not so invariably as might have been expected) that the crossed plants did not exceed in height the self-fertilized in nearly so great a degree as when grown in pairs in the pots. Thus with the plants of Digitalis, which competed together in pots, the crossed were to the self-fertilized in height as 100 to 70; whilst those which were grown separately were only as 100 to 85. Nearly the same result was observed with Brassica. With Nicotiana the crossed were to the self-fertilized plants in height, when grown extremely crowded together in pots, as 100 to 54; when grown much less crowded in pots as 100 to 66, and when grown in the open ground, so as to be subjected to but little competition, as 100 to 72. On the other hand with Zea, there was a greater difference in height between the crossed and self-fertilized plants growing out of doors, than between the pairs which / grew in pots in the hothouse; but this may be attributed to the self-fertilized plants being more tender, so that they suffered more than the crossed, when both lots were exposed to a cold and wet summer. Lastly, with one out of two series of *Reseda odorata* grown out of doors in rows, as well as with *Beta vulgaris*, the crossed plants did not at all exceed the self-fertilized in height, or exceeded them by a mere trifle.

The innate power of the crossed plants to resist unfavourable conditions far better than did the self-fertilized plants, was shown on two occasions in a curious manner, namely, with Iberis and in the third generation of Petunia, by the great superiority in height of the crossed over the self-fertilized seedlings, when both sets were grown under extremely unfavourable conditions; whereas owing to special circumstaces exactly the reverse occurred with the plants raised from the same seeds and grown in pairs in pots. A nearly analogous case was observed on two other occasions with plants of the first generation of Nicotiana.

The crossed plants always withstood the injurious effects of being suddenly removed into the open air after having been kept in the greenhouse better than did the self-fertilized. On several occasions they also resisted much better cold and intemperate weather. This was manifestly the case with some crossed and self-fertilized plants of Ipomoea, which were suddenly moved from the hothouse to the coldest part of a cool greenhouse. The offspring of plants of the eighth self-fertilized generation of Mimulus crossed by a fresh stock, survived a frost which killed every single self-fertilized and intercrossed plant of the same old stock. Nearly the same result followed with some crossed and self-fertilized plants of *Viola tricolor*. Even the tips of the shoots of the crossed plants of *Sarothamnus / scoparius* were not touched by a very severe winter; whereas all the self-fertilized plants were killed halfway down to the ground, so that they were not able to flower during the next summer. Young crossed seedlings of Nicotiana withstood a cold and wet summer much better than the self-fertilized seedlings. I have met with only one exception to the rule of crossed plants being hardier than the self-fertilized: three long rows of Eschscholtzia plants, consisting of crossed seedlings from a fresh stock, of intercrossed seedlings of the same stock, and of self-fertilized ones, were left unprotected during a severe winter, and all perished except two of the self-fertilized. But this case is not so anomalous as it at first appears, for it should be remembered that the self-fertilized plants of Eschscholtzia always grow taller and are heavier than the crossed; the whole benefit of a cross with this species being confined to increased fertility.

Independently of any external cause which could be detected, the self-fertilized plants were more liable to premature death than were the crossed; and this seems to me a curious fact. Whilst the seedlings were very young, if one died its antagonist was pulled up and thrown away, and I believe that many more of the self-fertilized died at this early age than of the crossed; but I neglected to keep any record. With *Beta vulgaris*, however, it is certain that a large number of the self-fertilized seeds perished after germinating beneath the ground, whereas the crossed seeds sown at the same time did not thus suffer. When a plant died at a somewhat more advanced age the fact was recorded; and I find in my notes that out of several hundred plants, only seven of the crossed died, whilst of the self-fertilized at least twenty-nine were thus lost, that is more than four times as many. Mr Galton, / after examining some of my tables, remarks: 'It is very evident that the columns with the self-fertilized plants include the larger number of exceptionally small plants'; and the frequent presence of such puny plants no doubt stands in close relation with their liability to premature death. The self-fertilized plants of Petunia completed their growth and began to wither sooner than did the intercrossed plants; and these latter considerably before the offspring from a cross with a fresh stock.

Period of flowering. In some cases, as with Digitalis, Dianthus, and Reseda, a larger number of the crossed than of the self-fertilized plants threw up flower-stems; but this probably was merely the result of their greater power of growth; for in the first generation of *Lobelia fulgens*, in which the self-fertilized plants greatly exceeded in height the crossed plants, some of the latter failed to throw up flower-stems. With a large number of species, the crossed plants exhibited a well-marked tendency to flower before the self-fertilized ones growing in the same pots. It should however be remarked that no record was kept of the flowering of many of the species; and when a record was kept, the flowering of the first plant in each pot was alone observed, although two or more pairs grew in the same pot. I will now give three lists – one of the species in which the first plant that flowered was a crossed one – a second in which the first that flowered was a self-fertilized plant – and a third of those which flowered at the same time.

Species, of which the first plants that flowered were of crossed parentage

Ipomoea purpurea. I record in my notes that in all ten generations many of the crossed plants flowered before the self-fertilized; but no details were kept. /

Mimulus luteus (first generation). Ten flowers on the crossed plants were fully expanded before one on the self-fertilized.

Mimulus luteus (second and third generation). In both these generations a crossed plant flowered before one of the self-fertilized in all three pots.

Mimulus luteus (fifth generation). In all three pots a crossed plant flowered first; yet the self-fertilized plants, which belonged to the new tall variety, were in height to the crossed as 126 to 100.

Mimulus luteus. Plants derived from a cross with a fresh stock, as well as the intercrossed plants of the old stock, flowered before the self-fertilized plants in nine out of the ten pots.

Salvia coccinea. A crossed plant flowered before any one of the self-fertilized in all three pots.

Origanum vulgare. During two successive seasons several crossed plants flowered before the self-fertilized.

Brassica oleracea (first generation). All the crossed plants growing in pots and in the open ground flowered first.

Brassica oleracea (second generation). A crossed plant in three out of the four pots flowered before any one of the self-fertilized.

Iberis umbellata. In both pots a crossed plant flowered first.

Eschscholtzia californica. Plants derived from the Brazilian stock crossed by the English stock flowered in five out of the nine pots first; in four of them a self-fertilized plant flowered first; and not in one pot did an intercrossed plant of the old stock flower first.

Viola tricolor. A crossed plant in five out of the six pots flowered before any one of the self-fertilized.

Dianthus caryophyllus (first generation). In two large beds of plants, four of the crossed plants flowered before any one of the self-fertilized.

Dianthus caryophyllus (second generation). In both pots a crossed plant flowered first.

Dianthus caryophyllus (third generation). In three out of the four pots a crossed plant flowered first; yet the crossed were to the self-fertilized in height only as 100 to 99, but in weight as 100 to 49.

Dianthus caryophyllus. Plants derived from a cross with a fresh stock, and the intercrossed plants of the old stock, both flowered before the self-fertilized in nine out of ten pots.

Hibiscus africanus. In three out of the four pots a crossed / plant flowered before any one of the self-fertilized; yet the latter were to the crossed in height as 109 to 100.

Tropaeolum minus. A crossed plant flowered before any one of the self-fertilized in three out of the four pots, and simultaneously in the fourth pot.

Limnanthes douglasii. A crossed plant flowered before any one of the self-fertilized in four out of the five pots.

Phaseolus multiflorus. In both pots a crossed plant flowered first.

Specularia speculum. In all four pots a crossed plant flowered first.

Lobelia ramosa (first generation). In all four pots a crossed plant flowered before any one of the self-fertilized.

Lobelia ramosa (second generation). In all four pots a crossed plant flowered some days before any one of the self-fertilized.

Nemophila insignis. In four out of the five pots a crossed plant flowered first.

Borago officinalis. In both pots a crossed plant flowered first.

Petunia violacea (second generation). In all three pots a crossed plant flowered first.

Nicotiana tabacum. A plant derived from a cross with a fresh stock flowered before any one of the self-fertilized plants of the fourth generation, in fifteen out of the sixteen pots.

Cyclamen persicum. During two successive seasons a crossed plant flowered some weeks before any one of the self-fertilized in all four pots.

Primula veris (equal-styld var.). in all three pots a crossed plant flowered first.

Primula sinensis. In all four pots plants derived from an illegitimate cross between distinct plants flowered before any one of the self-fertilized plants.

Primula sinensis. A legitimately crossed plant flowered before any one of the self-fertilized plants in seven out of the eight pots.

Fagopyrum esculentum. A legitimately crossed plant flowered from one to two days before any one of the self-fertilized plants in all three pots.

Zea mays. In all four pots a crossed plant flowered first.

Phalaris canariensis. The crossed plants flowered before the self-fertilized in the open ground, but simultaneously in the pots. /

Species, of which the first plants that flowered were of self-fertilized parentage

Eschscholtzia californica (first generation). The crossed plants were at first taller than the self-fertilized, but on their second growth during the following year the self-fertilized exceeded the crossed in height, and now they flowered first in three out of the four pots.

Lupinus luteus. Although the crossed plants were to the self-fertilized in height as 100 to 82; yet in all three pots the self-fertilized plants flowered first.

Clarkia elegans. Although the crossed plants were, as in the last case, to the self-fertilized in height as 100 to 82, yet in the two pots the self-fertilized flowered first.

Lobelia fulgens (first generation). The crossed plants were to the self-fertilized in height only as 100 to 127, and the latter flowered much before the crossed.

Petunia violacea (third generation). The crossed plants were to the self-fertilized in height as 100 to 131, and in three out of the four pots a self-fertilized plant flowered first; in the fourth pot simultaneously.

Petunia violacea (fourth generation). Although the crossed plants were to the self-fertilized in height as 100 to 69, yet in three out of the five pots a self-fertilized plant flowered first; in the fourth pot simultaneously, and only in the fifth did a crossed plant flower first.

Nicotiana tabacum (first generation). The crossed plants were to the self-fertilized in height only as 100 to 178, and a self-fertilized plant flowered first in all four pots.

Nicotiana tabacum (third generation). The crossed plants were to the self-fertilized in height as 100 to 101, and in four out of the five pots a self-fertilized plant flowered first.

Canna warscewiczi. In the three generations taken together the crossed were to the self-fertilized in height as 100 to 101; in the first generation the self-fertilized plants showed some tendency to flower first, and in the third generation they flowered first in nine out of the twelve pots.

Species in which the crossed and self-fertilized plants flowered
almost simultaneously

Mimulus luteus (second generation). The crossed plants were inferior in height
  and vigour to the self-fertilized plants / which all belonged to the new white-
  flowered tall variety, yet in only half the pots did the self-fertilized plants
  flower first, and in the other half the crossed plants.
Viscaria oculata. The crossed plants were only a little taller than the self-
  fertilized (viz., as 100 to 97), but considerably more fertile, yet both lots
  flowered almost simultaneously.
Lathyrus odoratus (second generation). Although the crossed plants were to the
  self-fertilized in height as 100 to 88, yet there was no marked difference in
  their period of flowering.
Lobelia fulgens (second generation). Although the crossed plants were to the
  self-fertilized in height as 100 to 91, yet they flowered simultaneously.
Nicotiana tabacum (third generation). Although the crossed plants were to the
  self-fertilized in height as 100 to 83, yet in half the pots a self-fertilized plant
  flowered first, and in the other half a crossed plant.

These three lists include fifty-eight cases, in which the period of flowering of the crossed and self-fertilized plants was recorded. In forty-four
of them a crossed plant flowered first either in a majority of the pots or
in all; in nine instances a self-fertilized plant flowered first, and in five
the two lots flowered simultaneously. One of the most striking cases is
that of Cyclamen, in which the crossed plants flowered some weeks
before the self-fertilized in all four pots during two seasons. In the
second generation of *Lobelia ramosa*, a crossed plant flowered in all four
pots some days before any one of the self-fertilized. Plants derived
from a cross with a fresh stock generally showed a very strongly
marked tendency to flower before the self-fertilized and the intercrossed plants of the old stock; all three lots growing in the same pots.
Thus with Mimulus and Dianthus, in only one pot out of ten, and in
Nicotiana in only one pot out of sixteen, did a self-fertilized plant
flower before the plants of the two crossed kinds – these latter
flowering almost simultaneously. /
  A consideration of the two first lists, especially of the second one,
shows that a tendency to flower first is generally connected with greater
power of growth, that is, with greater height. But there are some
remarkable exceptions to this rule, proving that some other cause
comes into play. Thus the crossed plants both of *Lupinus luteus* and
Clarkia elegans were to the self-fertilized plants in height as 100 to 82,
and yet the latter flowered first. In the third generation of Nicotiana,

and in all three generations of Canna, the crossed and self-fertilized plants were of nearly equal height, yet the self-fertilized tended to flower first. On the other hand, with *Primula sinensis*, plants raised from a cross between two distinct individuals, whether these were legitimately or illegitimately crossed, flowered before the illegitimately self-fertilized plants, although all the plants were of nearly equal height in both cases. So it was with respect to height and flowering with Phaseolus, Specularia, and Borago. The crossed plants of Hibiscus were inferior in height to the self-fertilized, in the ratio of 100 to 109, and yet they flowered before the self-fertilized in three out of the four pots. On the whole, there can be no doubt that the crossed plants exhibit a tendency to flower before the self-fertilized, almost though not quite so strongly marked as to grow to a greater height, to weigh more, and to be more fertile.

A few other cases not included in the above three lists deserve notice. In all three pots of *Viola tricolor*, naturally crossed plants the offspring of crossed plants flowered before naturally crossed plants the offspring of self-fertilized plants. Flowers on two plants, both of self-fertilized parentage, of the sixth generation of *Mimulus luteus* were intercrossed, and other flowers on the same plants were self-fertilized with their own pollen; / intercrossed seedlings and seedlings of the seventh self-fertilized generation were thus raised, and the latter flowered before the intercrossed in three out of the five pots. Flowers on a plant both of *Mimulus luteus* and of *Ipomoea purpurea* were crossed with pollen from other flowers on the same plant, and other flowers were fertilized with their own pollen; intercrossed seedlings of this peculiar kind, and others strictly self-fertilized being thus raised. In the case of the Mimulus the self-fertilized plants flowered first in seven out of the eight pots, and in the case of the Ipomoea in eight out of the ten pots; so that an intercross between the flowers on the same plant was very far from giving to the offspring thus raised, any advantage over the strictly self-fertilized plants in their period of flowering.

The effects of crossing flowers on the same plant

In the discussion on the results of a cross with a fresh stock, given under Table C in the last chapter, it was shown that the mere act of crossing by itself does no good; but that the advantages thus derived depend on the plants which are crossed, either consisting of distinct varieties which will almost certainly differ somewhat in constitution, or on the progenitors of the plants which are crossed, though identical in every external character, having been subjected to somewhat different conditions and having thus acquired some slight difference in constitution. All the flowers produced by the same plant have been developed from the same seed; those which expand at the same time have been exposed to exactly the same climatic influences; and the stems have all been nourished by the same roots. Therefore in accordance with the conclusions just referred to, no good ought to result from / crossing flowers on the same plant.[1] In opposition to this conclusion is the fact that a bud is in one sense a distinct individual, and is capable of occasionally or even not rarely assuming new external characters, as well as new constitutional peculiarities. Plants raised from buds which have thus varied may be propagated for a great length of time by grafts, cuttings, etc., and sometimes even by seminal generation.[2] There exist also numerous species in which the flowers on the same plant differ from one another – as in the sexual organs of monoecious and polygamous plants – in the structure of the circumferential flowers in many Compositae, Ulbelliferae, etc. – in the structure of the central flower in some plants – in the two kinds of flowers produced by cleistogamic species – and in several other such cases. These instances clearly prove that the flowers on the same plant have often varied independently of one another in many important respects, such variations having been fixed, / like those on distinct plants during the development of species.

It was therefore necessary to ascertain by experiment what would be

[1] It is, however, possible that the stamens which differ in length or construction in the same flower may produce pollen differing in nature, and in this manner a cross might be made effective between the several flowers on the same plant. Mr Macnab states (in a communication to M. Verlot, *La Production des Variétés*, 1865, p. 42) that seedlings raised from the shorter and longer stamens of rhododendron differ in character; but the shorter stamens apparently are becoming rudimentary, and the seedlings are dwarfs, so that the result may be simply due to a want of fertilizing power in the pollen, as in the case of the dwarfed plants of Mirabilis raised by Naudin by the use of too few pollen-grains. Analogous statements have been made with respect to the stamens of Pelargonium. With some of the Melastomaceae, seedlings raised by me from flowers fertilized by pollen from the shorter stamens certainly differed in appearance from those raised from the longer stamens, with differently coloured anthers; but here, again, there is some reason for believing that the shorter stamens are tending towards abortion. In the very different case of trimorphic heterostyled plants, the two sets of stamens in the same flower have widely different fertilizing powers.

[2] I have given numerous cases of such bud-variations in my *Variation of Animals and Plants under Domestication*, chap. xi, 2nd edit., vol. i, p. 448.

the effect of intercrossing flowers on the same plant, in comparison with fertilizing them with their own pollen or crossing them with pollen from a distinct plant. Trials were carefully made on five genera belonging to four families; and in only one case, namely, Digitalis, did the offspring from a cross between the flowers on the same plant receive any benefit, and the benefit here was small compared with that derived from a cross between distinct plants. In the chapter on fertility, when we consider the effects of cross-fertilization and self-fertilization on the productiveness of the parent-plants we shall arrive at nearly the same result, namely, that a cross between the flowers on the same plant does not at all increase the numbers of the seeds, or only occasionally and to a slight degree. I will now give an abstract of the results of the five trials which were made.

(1.) *Digitalis purpurea.* Seedlings raised from intercrossed flowers on the same plant, and others from flowers fertilized with their own pollen, were grown in the usual manner in competition with one another on the opposite sides of ten pots. In this and the four following cases, the details may be found under the head of each species. In eight pots, in which the plants did not grow much crowded, the flower-stems on sixteen intercrossed plants were in height to those on sixteen self-fertilized plants, as 100 to 94. In the two other pots in which the plants grew much crowded, the flower-stems on nine intercrossed plants were in height to those on nine self-fertilized plants, as 100 to 90. That the intercrossed plants in these two latter pots had a real advantage over their self-fertilized / opponents, was well shown by their relative weights when cut down, which was as 100 to 78. The mean height of the flower-stems on the twenty-five intercrossed plants in the ten pots taken together, was to that of the flower-stems on the twenty-five self-fertilized plants, as 100 to 92. Thus the intercrossed plants were certainly superior to the self-fertilized in some degree; but their superiority was small compared with that of the offspring from a cross between distinct plants over the self-fertilized, this being in the ratio of 100 to 70 in height. Nor does this latter ratio show at all fairly the great superiority of the plants derived from a cross between distinct individuals over the self-fertilized, as the former produced more than twice as many flower-stems as the latter, and were much less liable to premature death.

(2.) *Ipomoea purpurea.* Thirty-one intercrossed plants raised from a cross between flowers on the same plants were grown in ten pots in competition with the same number of self-fertilized plants, and the former were to the latter in height as 100 to 105. So that the self-fertilized plants were a little taller than the intercrossed; and in eight out of the ten pots a self-fertilized plant flowered before any one of the crossed plants in the same pots. The plants which were not greatly crowded in nine of the pots (and these offer the fairest standard of comparison) were cut down and weighed; and the weight of the twenty-seven intercrossed plants was to that of the twenty-seven self-fertilized as 100 to 124; so that by this test the superiority of the self-fertilized was strongly marked. To this subject of the superiority of the self-fertilized plants in certain cases, I shall have to recur in a future chapter. If we now turn to the offspring from a cross between distinct plants when put into competition with self-fertilized / plants, we find that the mean height of seventy-three such crossed plants, in the course of ten generations, was to that of the same number of self-fertilized plants as 100 to 77; and in the case of the plants of the tenth generation in weight as 100 to 44. Thus the contrast between the effects of crossing flowers on the same plant, and of crossing flowers on distinct plants, is wonderfully great.

(3.) *Mimulus luteus.* Twenty-two plants raised by crossing flowers on the same plant were grown in competition with the same number of self-fertilized plants; and the former were to the latter in height as 100 to 95, or if four dwarfed plants are excluded as 100 to 101; and in weight as 100 to 103. In seven out of the eight pots a self-fertilized plant flowered before any of the intercrossed. So that here again the self-fertilized exhibit a trifling superiority over the intercrossed plants. For the sake of comparison, I may add that seedlings raised during three generations from a cross between distinct plants were to the self-fertilized plants in height as 100 to 65.

(4.) *Pelargonium zonale.* Two plants growing in separate pots, which had been propagated by cuttings from the same plant, and therefore formed in fact parts of the same individual, were intercrossed, and other flowers on one of these plants were self-fertilized; but the seedlings obtained by the two processes did not differ in height. When, on the other hand, flowers on one of the above plants were crossed with pollen taken from a distinct seedling, and other flowers were self-fertilized, the crossed offspring thus obtained were to the self-fertilized in height as 100 to 74.

(5.) *Origanum vulgare.* A plant which had been long cultivated in my kitchen garden, had spread by stolons so as to form a large bed or clump. Seedlings / raised by intercrossing flowers on these plants, which strictly consisted of the same plant, and other seedlings raised from self-fertilized flowers, were carefully compared from their earliest youth to maturity; and they did not differ at all in height or in constitutional vigour. Some flowers on these seedlings were then crossed with pollen taken from a distinct seedling, and other flowers were self-fertilized; two fresh lots of seedlings being thus raised, which were the grandchildren of the plant that had spread by stolons and formed a large clump in my garden. These differed much in height, the crossed plants being to the self-fertilized as 100 to 86. They differed, also, to a wonderful degree in constitutional vigour. The crossed plants flowered first, and produced exactly twice as many flower-stems; and they afterwards increased by stolons to such an extent as almost to overwhelm the self-fertilized plants.

Reviewing these five cases, we see that in four of them, the effect of a cross between flowers on the same plant (even on offsets of the same plant growing on separate roots, as with the Pelargonium and Origanum) does not differ from that of the strictest self-fertilization. Indeed, in two of the cases the self-fertilized plants were slightly superior to such intercrossed plants. With Digitalis a cross between the flowers on the same plant certainly did do some good, yet very slight compared with that from a cross between distinct plants. On the whole the results here arrived at, if we bear in mind that the flower-buds are to a certain extent distinct individuals and occasionally vary independently of one another, agree well with our general conclusion, that the advantages of a cross depend on the progenitors of the crossed plants possessing somewhat different constitutions, either from having been / exposed to different conditions, or to their having varied from unknown causes in a manner which we in our ignorance are forced to speak of as spontaneous. Hereafter I shall have to recur to this subject of the inefficiency of a cross between the flowers on the same plant, when we consider the part which insects play in the cross-fertilization of flowers.

On the transmission of the good effects from a cross and of the evil effects from self-fertilization. We have seen that seedlings from a cross between distinct plants almost always exceed their self-fertilized opponents in height, weight, and constitutional vigour, and, as will hereafter be shown, often in fertility. To ascertain whether this superiority would be transmitted beyond the first generation, seedlings were raised on three occasions from crossed and self-fertilized plants, both sets being fertilized in the same manner, and therefore not as in the many cases given in Table A, B, and C, in which the crossed plants were again crossed and the self-fertilized again self-fertilized.

First, seedlings were raised from self-fertilized seeds produced under a net by crossed and self-fertilized plants of *Nemophila insignis*; and the latter were to the former in height as 133 to 100. But these seedlings became very unhealthy early in life, and grew so unequally that in both lots some were five times as tall as the others. Therefore this experiment was quite worthless; but I have felt bound to give it, as opposed to my general conclusion. I should state that in this and the two following trials, both sets of plants were grown on the opposite sides of the same pots, and treated in all respects alike. The details of the experiments may be found under the head of each species.

Secondly, a crossed and a self-fertilized plant of Heartsease (*Viola tricolor*) grew near together in the / open ground and near to other plants of heartsease; and as both produced an abundance of very fine capsules, the flowers on both were certainly cross-fertilized by insects. Seeds were collected from both plants, and seedlings raised from them. Those from the crossed plants flowered in all three pots before those from the self-fertilized plants; and when fully grown the former were to the latter in height as 100 to 82. As both sets of plants were the product of cross-fertilization, the difference in their growth and period of flowering was clearly due to their parents having been of crossed and self-fertilized parentage; and it is equally clear that they transmitted different constitutional powers to their offspring, the grandchildren of the plants which were originally crossed and self-fertilized.

Thirdly, the Sweet Pea (*Lathyrus odoratus*) habitually fertilizes itself in this country. As I possessed plants, the parents and grandparents of which had been artificially crossed and other plants descended from the same parents which had been self-fertilized for many previous generations, these two lots of plants were allowed to fertilize themselves under a net, and their self-fertilized seeds saved. The seedlings thus raised were grown in competition with each other in the usual manner, and differed in their powers of growth. Those from the self-fertilized plants which had been crossed during the two previous generations were to those from the plants self-fertilized during many previous generations in height as 100 to 90. These two lots of seeds were likewise tried by being sown under very unfavourable conditions in poor exhausted soil, and the plants whose grandparents and great-grandparents had been crossed showed in an unmistakable manner their superior constitutional vigour. In this case, as in that of the heartsease, there could be no doubt that / the advantage derived from a cross between two plants was not confined to the offspring of the first generation. That constitutional vigour due to cross-parentage is transmitted for many generations may also be inferred as highly probable, from some of Andrew Knight's varieties of the common pea, which were raised by crossing distinct varieties, after which time they no doubt fertilized themselves in each succeeding generation. These varieties lasted for upwards of sixty years, 'but their glory is now departed'.[3] On the other hand, most of the varieties of the common pea, which there is no reason to suppose owe their origin to a cross, have had a much shorter existence. Some also of Mr Laxton's varieties produced by artificial crosses have retained their astonishing vigour and luxuriance for a considerable number of generations; but as Mr Laxton informs me, his experience does not extend beyond twelve generations, within which period he has never perceived any diminution of vigour in his plants.

An allied point may be here noticed. As the force of inheritance is strong with plants (of which abundant evidence could be given), it is almost certain that seedlings from the same capsule or from the same plant would tend to inherit nearly the same constitution; and as the advantage from a cross depends on the plants which are crossed differing somewhat in constitution, it may be inferred as probable that under similar conditions a cross between the nearest relations would not benefit the offspring so much as one between non-related plants. In support of this conclusion we have some evidence, as Fritz Müller has shown by his / valuable experiments on hybrid Abutilons, that the union of brothers and sisters, parents and children, and of other near relations is highly injurious to the fertility of the offspring. In one case, moreover, seedlings from such near relations possessed very weak constitutions.[4] This same observer also found[5] three plants of a Bignonia growing near together. He fertilized twenty-nine flowers on one of them with their own pollen, and they did not set a single capsule. Thirty flowers were then fertilized with pollen from a distinct

[3] See the evidence on this head in my *Variation under Domestication*, chap. ix, vol. i, 2nd edit., p. 397.

[4] *Jenaische Zeitschrift für Naturw.*; vol. vii, pp. 22 and 45, 1872; and 1873, pp. 441–50.

[5] *Bot. Zeitung*, 1868, p. 626.

plant, one of the three growing together, and they yielded only two capsules. Lastly, five flowers were fertilized with pollen from a fourth plant growing at a distance, and all five produced capsules. It seems therefore probable, as Fritz Müller suggests, that the three plants growing near together were seedlings from the same parent, and that from being closely related they had little power of fertilizing one another.[6]

Lastly, the fact of the intercrossed plants in Table A not exceeding in height the self-fertilized plants in a greater and greater degree in the later generations, is probably due to their having become more and more closely inter-related.

Uniform colour of the flowers on plants, self-fertilized and grown under similar conditions for several generations. At the commencement of my experiments, the parent-plants of *Mimulus luteus, Ipomoea purpurea, Dianthus caryophyllus,* and *Petunia violacea,* raised from purchased seeds, varied greatly in the colour / of their flowers. This occurs with many plants which have been long cultivated as an ornament for the flower-garden, and which have been propagated by seeds. The colour of the flowers was a point to which I did not at first in the least attend, and no selection whatever was practised. Nevertheless, the flowers produced by the self-fertilized plants of the above four species became absolutely uniform in tint, or very nearly so, after they had been grown for some generations under closely similar conditions. The intercrossed plants, which were more or less closely inter-related in the later generations, and which had been likewise cultivated all the time under similar conditons, became more uniform in the colour of their flowers than were the original parent-plants, but much less so than the self-fertilized plants. When self-fertilized plants of one of the later generations were crossed with a fresh stock, and seedlings thus raised, these presented a wonderful contrast in the diversified tints of their flowers compared with those of the self-fertilized seedlings. As such cases of flowers becoming uniformly coloured without any aid from selection seem to me curious, I will give a full abstract of my observations.

Mimulus luteus. A tall variety, bearing large, almost white flowers blotched with crimson, appeared among the intercrossed and self-fertilized plants of the third and fourth generations. This variety increased so rapidly, that in the sixth generation of self-fertilized

---

[6] Some remarkable cases are given in my *Variation under Domestication* (chap. xvii, 2nd edit., vol. 2, p. 121) of hybrids of Gladiolus and Cistus, any one of which could be fertilized by pollen from any other, but not by its own pollen.

plants every single one consisted of it. So it was with all the many plants which were raised, up to the last or ninth self-fertilized generation. Although this variety first appeared among the intercrossed plants, yet from their offspring being intercrossed in each succeeding generation, it never prevailed among / them; and the flowers on the several intercrossed plants of the ninth generation differed considerably in colour. On the other hand, the uniformity in colour of the flowers on the plants of all the later self-fertilized generations was quite surprising; on a casual inspection they might have been said to be quite alike, but the crimson blotches were not of exactly the same shape, or in exactly the same position. Both my gardener and myself believe that this variety did not appear among the parent-plants, raised from purchased seeds, but from its appearance among both the crossed and self-fertilized plants of the third and fourth generations; and from what I have seen of the variation of this species on other occasions, it is probable that it would occasionally appear under any circumstances. We learn, however, from the present case that under the peculiar conditions to which my plants were subjected, this particular variety, remarkable for its colouring, largeness of the corolla, and increased height of the whole plant, prevailed in the sixth and all the succeeding self-fertilized generations to the complete exclusion of every other variety.

Ipomoea purpurea. My attention was first drawn to the present subject by observing that the flowers on all the plants of the seventh self-fertilized generation were of a uniform, remarkably rich, dark purple tint. The many plants which were raised during the three succeeding generations, up to the last or tenth, all produced flowers coloured in the same manner. They were absolutely uniform in tint, like those of a constant species living in a state of nature; and the self-fertilized plants might have been distinguished with certainty, as my gardener remarked, without the aid of labels, from the intercrossed plants of the later generations. These, however, had more uniformly coloured flowers / than those which were first raised from the purchased seeds. This dark purple variety did not appear, as far as my gardener and myself could recollect, before the fifth or sixth self-fertilized generation. However this may have been, it became, through continued self-fertilization and the cultivation of the plants under uniform conditions, perfectly constant, to the exclusion of every other variety.

Dianthus caryophyllus. The self-fertilized plants of the third generation all bore flowers of exactly the same pale rose-colour; and in this respect they differed quite remarkably from the plants growing in a large bed close by and raised from seeds purchased from the same nursery garden. In this case it is not improbable that some of the parent-plants which were first self-fertilized may have borne flowers thus coloured; but as several plants were self-fertilized in the first generation, it is extremely improbable that all bore flowers of exactly the same tint as those of the self-fertilized plants of the third generation. The intercrossed plants of the third generation likewise produced flowers almost, though not quite so uniform in tint as those of the self-fertilized plants.

Petunia violacea. In this case I happened to record in my notes that the flowers on the parent-plant which was first self-fertilized were of a 'dingy purple colour'. In the fifth self-fertilized generation, every one of the twenty-one self-fertilized plants growing in pots, and all the many plants in a long row out of doors, produced flowers of absolutely the same tint, namely, of a dull, rather peculiar and ugly flesh colour; therefore, considerably unlike those on the parent-plant. I believe that this change of colour supervened quite gradually; but I kept no record, as the point did not interest me until I was struck with the uniform tint / of the flowers on the self-fertilized plants of the fifth generation. The flowers on the intercrossed plants of the corresponding generation were mostly of the same dull flesh colour, but not nearly so uniform as those on the self-fertilized plants, some few being very pale, almost white. The self-fertilized plants which grew in a long row in the open ground were also remarkable for their uniformity in height, as were the intercrossed plants in a less degree, both lots being compared with a large number of plants raised at the same time under similar conditions from the self-fertilized plants of the fourth generation crossed by a fresh stock. I regret that I did not attend to the uniformity in height of the self-fertilized seedlings in the later generations of the other species.

These few cases seem to me to possess much interest. We learn from them that new and slight shades of colour may be quickly and firmly fixed, independently of any selection, if the conditions are kept as nearly uniform as is possible, and no intercrossing be permitted. With Mimulus, not only a grotesque style of colouring, but a larger corolla and increased height of the whole plant were thus fixed; whereas with most plants which have been long cultivated for the flower-garden, no character is more variable than that of colour, excepting perhaps that of height. From the consideration of these cases we may infer that the variability of cultivated plants in the above respects is due, first, to their being subjected to somewhat diversified conditions, and, secondly, to their being often inter-crossed, as would follow from the free access of insects. I do not see how this interference can be avoided, as when the above plants were cultivated for several generations under closely similar conditions, and were intercrossed in each generation, the colour / of their flowers tended in some degree to change and to become uniform. When no intercrossing with other plants of the same stock was allowed – that is, when the flowers were fertilized with their own pollen in each generation – their colour in the later generations became as uniform as that of plants growing in a state of nature, accompanied at least in one instance by much uniformity in the height of the plants. But in saying that the diversified tints of the flowers on cultivated plants treated in the ordinary manner are due to differences in the soil, climate, etc., to which they are exposed, I do not wish to imply that such variations are caused by these agencies in any more direct manner than that in which the most diversified illnesses, as colds, inflammation of the lungs or pleura, rheumatism, etc., may be said to be caused by exposure to cold. In both cases the constitution of the being which is acted on is of preponderant importance. /

CHAPTER IX

THE EFFECTS OF CROSS-FERTILIZATION AND SELF-FERTILIZATION ON THE PRODUCTION OF SEEDS

Fertility of plants of crossed and self-fertilized parentage, both lots being fertilized in the same manner – Fertility of the parent-plants when first crossed and self-fertilized, and of their crossed and self-fertilized offspring when again crossed and self-fertilized – Comparison of the fertility of flowers fertilized with their own pollen and with that from other flowres on the same plant – Self-sterile plants – Causes of self-sterility – The appearance of highly self-fertile varieties – Self-fertilization apparently in some respects beneficial, independently of the assured production of seeds – Relative weights and rates of germination of seeds from crossed and self-fertilized flowers.

The present chapter is devoted to the Fertility of plants, as influenced by cross-fertilization and self-fertilization. The subject consists of two distinct branches; first, the relative productiveness or fertility of flowers crossed with pollen from a distinct plant and with their own pollen, as shown by the proportional number of capsules which they produce, together with the number of the contained seeds. Secondly, the degree of innate fertility or sterility of the seedlings raised from crossed and self-fertilized seeds; such seedlings being of the same age, grown under the same conditions, and fertilized in the same manner. These two branches of the subject correspond with the two which have to be considered by any one treating of hybrid plants; namely, in the first place the comparative productiveness of a species when fertilized with pollen from a distinct species and with its own pollen; and / in the second place, the fertility of its hybrid offspring. These two classes of cases do not always run parallel; thus some plants, as Gärtner has shown, can be crossed with great ease, but yield excessively sterile hybrids; while others are crossed with extreme difficulty, but yield fairly fertile hybrids.

The natural order to follow in this chapter would have been first to

278

consider the effects on the fertility of the parent-plants of crossing them, and of fertilizing them with their own pollen; but as we have discussed in the two last chapters the relative height, weight, and constitutional vigour of crossed and self-fertilized plants – that is, of plants raised from crossed and self-fertilized seeds – it will be convenient here first to consider their relative fertility. The cases observed by me are given in the following table, D, in which plants of crossed and self-fertilized parentage were left to fertilize themselves, being either crossed by insects or spontaneously self-fertilized. It should be observed that the results cannot be considered as fully trustworthy, for the fertility of a plant is a most variable element, depending on its age, health, nature of the soil, amount of water given, and temperature to which it is exposed. The number of the capsules produced and the number of the contained seeds, ought to have been ascertained on a large number of crossed and self-fertilized plants of the same age and treated in every respect alike. In these two latter respects my observations may be trusted, but a sufficient number of capsules were counted only in a few instances. The fertility, or as it may perhaps better be called the productiveness, of a plant depends on the number of capsules produced, and on the number of seeds which these contain. But from various causes, chiefly from the want of time, I was often compelled to rely on the / number of the capsules alone. Nevertheless, in the more interesting cases, the seeds were also counted or weighed. The average number of seeds per capsule is a more variable criterion of fertility than the number of capsules produced. This latter circumstance depends partly on the size of the plant; and we know that crossed plants are generally taller and heavier than the self-fertilized; but the difference in this respect is rarely sufficient to account for the difference in the number of the capsules produced. It need hardly be added that in the following table the same number of crossed and self-fertilized plants are always compared. Subject to the foregoing sources of doubt I will now give the table, in which the parentage of the plants experimented on, and the manner of determining their fertility are explained. Fuller details may be found in the previous part of this work, under the head of each species. /

This table includes thirty-three cases relating to twenty-three species and shows the degree of innate fertility of plants of crossed parentage in comparison with those of self-fertilized parentage; both lots being fertilized in the same manner. With several of the species, as with Eschscholtzia, Reseda, Viola, Dianthus, Petunia, and Primula, both lots

TABLE D

Relative fertility of plants of crossed and self-fertilized parentage, both sets being fertilized in the same manner. Fertility judged of by various standards. That of the crossed plants taken as 100

Ipomoea purpurea – first generation: seeds per capsule on crossed and self-fertilized plants, not growing much crowded, spontaneously self-fertilized under a net, in number ... as 100 to 99

Ipomoea purpurea – seeds per capsule on crossed and self-fertilized plants from the same parents as in the last case, but growing much crowded, spontaneously self-fertilized under a net, in number ... „ „ 93

Ipomoea purpurea – productiveness of the same plants, as judged by the number of capsules produced, and average number of seeds per capsule ... „ „ 45

Ipomoea purpurea – third generation: seeds per capsule on crossed and self-fertilized plants, spontaneously self-fertilized under a net, in number ... „ „ 94

Ipomoea purpurea – productiveness of the same plants, as judged by number of capsules produced, and average number of seeds per capsule ... „ „ 35/

Ipomoea purpurea – fifth generation: seeds per capsule on crossed and self-fertilized plants, left uncovered in the hothouse, and spontaneously fertilized ... „ „ 89

Ipomoea purpurea – ninth generation: number of capsules on crossed plants to those on self-fertilized plants, spontaneously self-fertilized under a net ... „ „ 26

Mimulus luteus – an equal number of capsules on plants descended from self-fertilized plants of the 8th generation crossed by a fresh stock, and on plants of the 9th self-fertilized generation, both sets having been left uncovered and spontaneously fertilized, contained seeds, by weight ... „ „ 30

Mimulus luteus – productiveness of the same plants, as judged by number of capsules produced, and average weight of seeds per capsule ... „ „ 3

Vandellia nummularifolia – seeds per capsule from cleistogamic flowers on the crossed and self-fertilized plants, in number ... „ „ 106

Salvia coccinea – crossed plants, compared with the self-fertilized plants, produced flowers, in number ... „ „ 57

Iberis umbellata – plants left uncovered in greenhouse; intercrossed plants of the 3rd generation, compared with self-fertilized plants of the 3rd generation, yielded seeds, in number ... „ „ 75

Iberis umbellata – plants from a cross between two varieties, compared with self-fertilized plants of the 3rd generation, yielded seeds, by weight ... „ „ 75

Papaver vagum – crossed and self-fertilized plants, left uncovered, produced capsules, in number ... „ „ 99

TABLE D — *continued*

Eschscholtzia californica – Brazilian stock; plants left uncovered and cross-fertilized by bees; capsules on intercrossed plants of 2nd generation, compared with capsules on self-fertilized plants of 2nd generation, contained seeds, in number   as 100 to   78

Eschscholtzia californica – productiveness of the same plants, as judged by number of capsules produced, and average number of seeds per capsule   „   „   89

Eschscholtzia californica – plants left uncovered and cross-fertilized by bees; capsules on plants derived from intercrossed plants of 2nd generation of the Brazilian stock crossed by English stock, compared with capsules on self-fertilized plants of 2nd generation, contained seeds, in number   „   „   63

Eschscholtzia californica – productiveness of the same plants, as judged by number of capsules produced, and average number of seeds per capsule   „   „   40 /

Reseda odorata – crossed and self-fertilized plants, left uncovered and cross-fertilized by bees, produced capsules in number (about)   „   „   100

Viola tricolor – crossed and self-fertilized plants, left uncovered and cross-fertilized by bees, produced capsules in number   „   „   10

Delphinium consolida – crossed and self-fertilized plants, left uncovered in the greenhouse, produced capsules in number   „   „   56

Viscaria oculata – crossed and self-fertilized plants, left uncovered in the greenhouse, produced capsules in number   „   „   77

Dianthus caryophyllus – plants spontaneously self-fertilized under a net; capsules on intercrossed and self-fertilized plants of the 3rd generation contained seeds in number   „   „   125

Dianthus caryophyllus – plants left uncovered and cross-fertilized by insects: offspring from plants self-fertilized for three generations and then crossed by an intercrossed plant of the same stock, compared with plants of the 4th self-fertilized generation, produced seeds by weight   „   „   73

Dianthus caryophyllus – plants left uncovered and cross-fertilized by insects: offspring from plants self-fertilized for three generations and then crossed by a fresh stock, compared with plants of the 4th self-fertilized generation, produced seeds by weight   „   „   33

Tropaeolum minus – crossed and self-fertilized plants, left uncovered in the greenhouse, produced seeds in number   „   „   64

Limnanthes douglasii – crossed and self-fertilized plants, left uncovered in greenhouse, produced capsules in number (about)   „   „   100

Lupinus luteus – crossed and self-fertilized plants of the 2nd generation, left uncovered in the greenhouse, produced seeds in number (judged from only a few pods)   „   „   88

Phaseolus multiflorus – crossed and self-fertilized plants, left uncovered in the greenhouse, produced seeds in number (about)   „   „   100

281

TABLE D – *continued*

Lathyrus odoratus – crossed and self-fertilized plants of the 2nd generation, left uncovered in the greenhouse, but certainly self-fertilized, produced pods in number   100 as to 91

Clarkia elegans – crossed and self-fertilized plants, left uncovered in the greenhouse, produced capsules in number   „ „ 60

Nemophila insignis – crossed and self-fertilized plants, covered by a net and spontaneously self-fertilized in the greenhouse, produced capsules in number   „ „ 29

Petunia violacea – left uncovered and cross-fertilized by insects: plants of the 5th intercrossed and self-fertilized generations produced seeds, as judged by the weight of an equal number of capsules   „ „ 86 /

Petunia violacea – left uncovered as above: offspring of plants self-fertilized for four generations and then crossed by a fresh stock, compared with plants of the 5th self-fertilized generation, produced seeds, as judged by the weight of an equal number of capsules   „ „ 46

Cyclamen persicum – crossed and self-fertilized plants, left uncovered in the greenhouse, produced capsules in number   „ „ 12

Anagallis collina – crossed and self-fertilized plants, left uncovered in the greenhouse, produced capsules in number   „ „ 8

Primula veris – left uncovered in open ground and cross-fertilized by insects: offspring from plants of the 3rd illegitimate generation crossed by a fresh stock, compared with plants of the 4th illegitimate and self-fertilized generation, produced capsules in number   „ „ 5

Same plants in the following year   „ „ 3·5

Primula veris – (equal-styled variety): left uncovered in open ground and cross-fertilized by insects: offspring from plants self-fertilized for two generations and then crossed by another variety, compared with plants of the 3rd self-fertilized generation, produced by capsules in number   „ „ 15

Primula veris – (equal-styled var.) same plants; average number of seeds per capsule   „ „ 71

Primula veris – (equal-styled var.) productiveness of the same plants, as judged by number of capsules produced and average number of seeds per capsule   „ „ 11

were certainly cross-fertilized by insects, and so it probably was with several of the others; but in some of the species, as with Nemophila, and in some of the trials with Ipomoea and Dianthus, the plants were covered up, and both lots were spontaneously self-fertilized. This also was necessarily the case with the capsules produced by the cleistogamic flowers of Vandellia. /

The fertility of the crossed plants is represented in the table by 100, and that of the self-fertilized by the other figures. There are five cases in which the fertility of the self-fertilized plants is approximately equal to that of the crossed; nevertheless, in four of these cases the crossed plants were plainly taller, and in the fifth somewhat taller than the self-fertilized. But I should state that in some of these five cases the fertility of the two lots was not strictly ascertained, as the capsules were not actually counted, from appearing equal in number and from all apparently containing a full complement of seeds. In only two instances in the table, viz., with Vandellia and in the third generation of Dianthus, the capsules on the self-fertilized plants contained more seed than those on the crossed plants. With Dianthus the ratio between the number of seeds contained in the self-fertilized and crossed capsules was as 125 to 100; both sets of plants were left to fertilize themselves under a net; and it is almost certain that the greater fertility of the self-fertilized plants was here due merely to their having varied and become less strictly dichogamous, so as to mature their anthers and stigmas more nearly at the same time than is proper to the species. Excluding the seven cases now referred to, there remain twenty-six in which the crossed plants were manifestly much more fertile, sometimes to an extraordinary degree, than the self-fertilized with which they grew in competition. The most striking instances are those in which plants derived from a cross with a fresh stock are compared with plants of one of the later self-fertilized generations; yet there are some striking cases, as that of Viola, between the intercrossed plants of the same stock and the self-fertilized, even in the first generation. The results most to be trusted are those / in which the productiveness of the plants was ascertained by the number of capsules produced by an equal number of plants, together with the actual or average number of seeds in each capsule. Of such cases there are twelve in the table, and the mean of their mean fertility is as 100 for the crossed plants, to 59 for the self-fertilized plants. The Primulaceae seem eminently liable to suffer in fertility from self-fertilization.

The following short table, E, includes four cases which have already been partly given in the last table. / These cases show us how greatly superior in innate fertility the seedlings from plants self-fertilized or intercrossed for several generations and then crossed by a fresh stock are, in comparison with the seedlings from plants of the old stock, either intercrossed or self-fertilized for the same number of generations. The three lots of plants in each case were left freely exposed to the the visits of insects, and their flowers without doubt were cross-fertilized by them.

TABLE E

Innate fertility of plants from a cross with a fresh stock, compared with that of intercrossed plants of the same stock, and with that of self-fertilized plants, all of the corresponding generation; all these sets being fertilized in the same manner. Fertility judged of by the number or weight of seeds produced by an equal number of plants.

	Plants from a cross with a fresh stock	Intercrossed plants of the same stock	Self-fertilized plants
Mimulus luteus – the intercrossed plants are derived from a cross between two plants of the 8th self-fertilized generation. The self-fertilized plants belong to the 9th generation	100	4	3
Eschscholtzia californica – the intercrossed and self-fertilized plants belong to the 2nd generation	100	45	40
Dianthus caryophyllus – the intercrossed plants are derived from self-fertilized of the 3rd generation, crossed by intercrossed plants of the 3rd generation. The self-fertilized plants belong to the 4th generation	100	45	33
Petunia violacea – the intercrossed and self-fertilized plants belong to the 5th generation	100	54	46

N.B. In the above cases, excepting in that of Eschscholtzia, the plants derived from a stock with a fresh stock belong on the mother-side to the same stock with the intercrossed and self-fertilized plants, and to the corresponding generation.

This table further shows us that in all four cases the intercrossed plants of the same stock still have a decided though small advantage in fertility over the self-fertilized plants.

With respect to the state of the reproductive organs in the self-fertilized plants of the two last tables, only a few observations were made. In the seventh and eighth generation of Ipomoea, the anthers in the flowers of the self-fertilized plants were plainly smaller than those in the flowers of the intercrossed plants. The tendency to sterility in these same plants was also shown by the first-formed flowers, after they had been carefully fertilized, often dropping off, in the same manner as frequently occurs with hybrids. The flowers likewise tended to be monstrous. In the fourth generation of Petunia, the pollen produced by the self-fertilized and intercrossed plants was compared, and there were far more empty and shrivelled grains in the former.

Relative fertility of flowers crossed with pollen from a distinct plant and with their own pollen. This heading includes flowers on the parent-plants, and on the crossed and self-fertilized seedlings of the first or a succeeding generation. I will first treat of the parent-plants, which / were raised from seeds purchased from nursery-gardens, or taken from plants growing in my garden, or growing wild, and surrounded in every case by many individuals of the same species. Plants thus circumstanced will commonly have been intercrossed by insects; so that the seedlings which were first experimented on will generally have been the product of a cross. Consequently any difference in the fertility of their flowers, when crossed and self-fertilized, will have been caused by the nature of the pollen employed; that is, whether it was taken from a distinct plant or from the same flower. The degrees of fertility shown in the following table, F, were determined in each case by the average number of seeds per capsule, ascertained either by counting or weighing.

Another element ought properly to have been taken into account, namely, the proportion of flowers which yielded capsules when they were crossed and self-fertilized; and as crossed flowers generally produce a larger proportion of capsules, their superiority in fertility, if this element had been taken into account, would have been much more strongly marked than appears in Table F. But had I thus acted, there would have been greater liability to error, as pollen applied to the stigma at the wrong time fails to produce any effect, independently of its greater or less potency. A good illustration of the great difference in the results which sometimes follows, if the number of capsules produced relatively to the number of flowers fertilized be included in the calculation, was afforded by *Nolana prostrata*. Thirty flowers on some plants of this species were crossed and produced twenty-seven capsules, each containing five seeds; thirty-two flowers on the same plants were self-fertilized and produced / only six capsules, each containing five seeds. As the number of the seeds per capsule is here the same, the fertility of the crossed and self-fertilized flowers is given in Table F as equal, or as 100 to 100. But if the flowers which failed to produce capsules be included, the crossed flowers yielded on an average 4·50 seeds, whilst the self-fertilized flowers yielded only 0·94 seeds, so that their relative fertility would have been as 100 to 21. I should here state that it has been found convenient to reserve for separate discussion the cases of flowers which are usually quite sterile with their own pollen. /

TABLE F

Relative fertility of the flowers on the parent-plants used in my experiments, when fertilized with pollen from a distinct plant and with their own pollen. Fertility judged of by the average number of seeds per capsule. Fertility of crossed flowers taken as 100.

Ipomoea purpurea – crossed and self-fertilized flowers yielded
seeds as (about) .... 100 to 100
Mimulus luteus – crossed and self-fertilized flowers yielded seeds
as (by weight) .... „   „   79
Linaria vulgaris – crossed and self-fertilized flowers yielded seeds
as .... „   „   14
Vandellia nummularifolia – crossed and self-fertilized flowers
yielded seeds as .... „   „   67?
Gesneria pendulina – crossed and self-fertilized flowers yielded
seeds as (by weight) .... „   „   100
Salvia coccinea – crossed and self-fertilized flowers yielded seeds
as (about) .... „   „   100
Brassica oleracea – crossed and self-fertilized flowers yielded seeds
as .... „   „   25
Eschscholtzia californica – (English stock) crossed and self-
fertilized flowers yielded seeds as (by weight) .... „   „   71
Eschscholtzia californica – (Brazilian stock grown in England)
crossed and self-fertilized flowers yielded seeds (by weight) as
(about) .... „   „   15
Delphinium consolida – crossed and self-fertilized flowers (self-
fertilized capsules spontaneously produced, but result supported by other evidence) yielded seeds as .... „   „   59
Viscaria oculata – crossed and self-fertilized flowers yielded seeds
as (by weight) .... „   „   38
Viscaria oculata – crossed and self-fertilized flowers (crossed
capsules compared on following year with spontaneously self-
fertilized capsules) yielded seeds as .... „   „   58
Dianthus caryophyllus – crossed and self-fertilized flowers yielded
seeds as .... „   „   92
Tropeaolum minus – crossed and self-fertilized flowers yielded
seeds as .... „   „   92
Tropaeolum tricolorum[1] – crossed and self-fertilized flowers
yielded seeds as .... „   „   115

[1] *Tropaeolum tricolorum* and *Cuphea purpurea* have been introduced into this table, although seedlings were not raised from them; but of the Cuphea only six crossed

Limnanthes douglasii – crossed and self-fertilized flowers yielded seeds as (about)		100 to 100
Sarothamnus scoparius – crossed and self-fertilized flowers yielded seeds as	„ „	41
Ononis minutissima – crossed and self-fertilized flowers yielded seeds as	„ „	65
Cuphea purpurea – crossed and self-fertilized flowers yielded seeds as	„ „	113
Passiflora gracilis – crossed and self-fertilized flowers yielded seeds as	„ „	85
Specularia speculum – crossed and self-fertilized flowers yielded seeds as	„ „	72
Lobelia fulgens – crossed and self-fertilized flowers yielded seeds as (about)	„ „	100
Nemophila insignis – crossed and self-fertilized flowers yielded seeds as (by weight)	„ „	69
Borago officinalis – crossed and self-fertilized flowers yielded seeds as	„ „	60
Nolana prostrata – crossed and self-fertilized flowers yielded seeds as	„ „	100
Petunia violacea – crossed and self-fertilized flowers yielded seeds as (by weight)	„ „	67
Nicotiana tabacum – crossed and self-fertilized flowers yielded seeds as (by weight)	„ „	150
Cyclamen persicum – crossed and self-fertilized flowers yielded seeds as	„ „	38
Anagallis collina – crossed and self-fertilized flowers yielded seeds as	„ „	96
Canna warscewiczi – crossed and self-fertilized flowers (on three generations of crossed and self-fertilized plants taken all together) yielded seeds as	„ „	85 /

A second table, G, gives the relative fertility of flowers on crossed plants again cross-fertilized, and of flowers on self-fertilized plants again self-fertilized, either in the first or in a later generation. Here two causes combine tò diminish the fertility of the self-fertilized flowers; namely, the lesser efficacy of pollen from the same flower, and the innate lessened fertility of plants derived from self-fertilized seeds, which as we have seen in the previous Table D is strongly marked. The fertility was determined in the same manner as in Table F, that is, by the average

---

and six self-fertilized capsules, and of the Tropaeolum only six crossed and eleven self-fertilized capsules, were compared. A larger proportion of the self-fertilized than of the crossed flowers of the Tropaeolum produced fruit.

number of seeds per capsule; and the same remarks as before, with respect to the different proportion of flowers which set capsules when they are cross-fertilized and self-fertilized, are here likewise applicable.

TABLE G
Relative fertility of flowers on crossed and self-fertilized plants of the first or some succeeding generation; the former being again fertilized with pollen from a distinct plant, and the latter again with their own pollen. Fertility judged of by the average. Number of seeds per capsule. Fertility of crossed flowers taken as 100.

Ipomoea purpurea – crossed and self-fertilized flowers on the crossed and self-fertilized plants of the 1st generation yielded seeds as 100 to 93

Ipomoea purpurea – crossed and self-fertilized flowers on the crossed and self-fertilized plants of the 3rd generation yielded seeds as „ „ 94

Ipomoea purpurea – crossed and self-fertilized flowers on the crossed and self-fertilized plants of the 4th generation yielded seeds as „ „ 94

Ipomoea purpurea – crossed and self-fertilized flowers on the crossed and self-fertilized plants of the 5th generation yielded seeds as „ „ 107

Mimulus luteus – crossed and self-fertilized flowers on the crossed and self-fertilized plants of the 3rd generation yielded seeds as (by weight) „ „ 65

Mimulus luteus – same plants treated in the same manner on following year yielded seeds as (by weight) „ „ 34

Mimulus luteus – crossed and self-fertilized flowers on the crossed and self-fertilized plants of the 4th generation yielded seeds as (by weight) „ „ 40/

Viola tricolor – crossed and self-fertilized flowers on the crossed and self-fertilized plants of the 1st generation yielded seeds as „ „ 69

Dianthus caryophyllus – crossed and self-fertilized flowers on the crossed and self-fertilized plants of the 1st generation yielded seeds as „ „ 65

Dianthus caryophyllus – flowers on self-fertilized plants of the 3rd generation crossed by intercrossed plants, and other flowers again self-fertilized yielded seeds as „ „ 97

Dianthus caryophyllus – flowers on self-fertilized plants of the 3rd generation crossed by a fresh stock, and other flowers again self-fertilized yielded seeds as „ „ 127

Lathyrus odoratus – crossed and self-fertilized flowers on the crossed and self-fertilized plants of the 1st generation yielded seeds as „ „ 65

Lobelia ramosa – crossed and self-fertilized flowers on the crossed and self-fertilized plants of the 1st generation yielded seeds as (by weight) „ „ 60

288

TABLE G – *continued*

Petunia violacea – crossed and self-fertilized flowers on the crossed and self-fertilized plants of the 1st generation yielded seeds as (by weight)	100 to	68
Petunia violacea – crossed and self-fertilized flowers on the crossed and self-fertilized plants of the 4th generation yielded seeds as (by weight)	„ „	72
Petunia violacea – flowers on self-fertilized plants of the 4th generation crossed by a fresh stock, and other flowers again self-fertilized yielded seeds as (by weight)	„ „	48
Nicotiana tabacum – crossed and self-fertilized flowers on the crossed and self-fertilized plants of the 1st generation yielded seeds as (by weight)	„ „	97
Nicotiana tabacum – flowers on self-fertilized plants of the 2nd generation crossed by intercrossed plants, and other flowers again self-fertilized yielded seeds as (by estimation)	„ „	110
Nicotiana tabacum – flowers on self-fertilized plants of the 3rd generation crossed by a fresh stock, and other flowers again self-fertilized yielded seeds as (by estimation)	„ „	110
Anagallis collina – flowers on a red variety crossed by a blue variety, and other flowers on the red variety self-fertilized yielded seeds as	„ „	48
Canna warscewiczi – crossed and self-fertilized flowers on the crossed and self-fertilized plants of three generations taken together yielded seeds as	„ „	85

As both these tables relate to the fertility of flowers fertilized by pollen from another plant and by their own pollen, they may be considered together. The difference between them consists in the self-fertilized / flowers in the second table, G, being produced by self-fertilized parents, and the crossed flowers by crossed parents, which in the later generations had become somewhat closely inter-related, and had been subjected all the time to nearly the same conditions. These two tables include fifty cases relating to thirty-two species. The flowers on many other species were crossed and self-fertilized, but as only a few were thus treated, the results cannot be trusted, as far as fertility is concerned, and are not here given. Some other cases have been rejected, as the plants were in an unhealthy condition. If we look to the figures in the two tables expressing the ratios between the mean relative fertility of the crossed and self-fertilized flowers, we see that in a majority of the cases (i.e., in thirty-five out of fifty) flowers fertilized by pollen from a distinct plant yield more, sometimes many more, seeds than flowers fertilized with their own pollen; and they commonly set a larger proportion of capsules. The degree of infertility of the self-fertilized flowers differs extremely in the different species, and even, as we shall see in the section on self-sterile plants, in the individuals of the same species, as well as under slightly changed conditions of life. Their fertility ranges from zero to fertility equalling that of the crossed flowers; and of this fact no explanation can be offered. There are fifteen cases in the two tables in which the number of seeds per capsule produced by the self-fertilized flowers equals or even exceeds that yielded by the crossed flowers. Some few of these cases are, I believe, accidental; that is, would not recur on a second trial. This was apparently the case with the plants of the fifth generation of Ipomoea, and in one of the experiments with Dianthus. Nicotiana offers the most anomalous case of any, / as the self-fertilized flowers on the parent-plants, and on their descendants of the second and third generations, produced more seeds than did the crossed flowers; but we shall recur to this case when we treat of highly self-fertile varieties.

It might have been expected that the difference in fertility between the crossed and self-fertilized flowers would have been more strongly marked in Table G, in which the plants of one set were derived from self-fertilized parents, than in Table F, in which flowers on the parent-plants were self-fertilized for the first time. But this is not the case, as far as my scanty materials allow of any judgement. There is therefore no evidence at present, that the fertility of plants goes on diminishing in successive self-fertilized generations, although there is some rather weak evidence that this does occur with respect to their height or growth. But we should bear in mind that in the later generations the crossed plants had become more or less closely inter-related, and had been subjected all the time to nearly uniform conditions.

It is remarkable that there is no close correspondence, either in the parent-plants or in the successive generations, between the relative number of seeds produced by the crossed and self-fertilized flowers, and the relative powers of growth of the seedlings raised from such seeds. Thus, the crossed and self-fertilized flowers on the parent-plants of Ipomoea, Gesneria, Salvia, Limnanthes, *Lobelia fulgens*, and Nolana produced a nearly equal number of seeds, yet the plants raised from the crossed seeds exceeded considerably in height those raised from the self-fertilized seeds. The crossed flowers of Linaria and Viscaria yielded far more seeds than the self-fertilized flowers; and although the plants raised from the former were taller / than those from the latter, they were not so in any corresponding degree. With

Nicotiana the flowers fertilized with their own pollen were more productive than those crossed with pollen from a slightly different variety; yet the plants raised from the latter seeds were much taller, heavier, and more hardy than those raised from the self-fertilized seeds. On the other hand the crossed seedlings of Eschscholtzia were neither taller nor heavier than the self-fertilized, although the crossed flowers were far more productive than the self-fertilized. But the best evidence of a want of correspondence between the number of seeds produced by crossed and self-fertilized flowers, and the vigour of the offspring raised from them, is afforded by the plants of the Brazilian and European stocks of Eschscholtzia, and likewise by certain individual plants of *Reseda odorata*; for it might have been expected that the seedlings from plants, the flowers of which were excessively self-sterile, would have profited in a greater degree by a cross, than the seedlings from plants which were moderately or fully self-fertile, and therefore apparently had no need to be crossed. But no such result followed in either case: for instance, the crossed and self-fertilized offspring from a highly self-fertile plant of *Reseda odorata* were in average height to each other as 100 to 82; whereas the similar offspring from an excessively self-sterile plant were as 100 to 92 in average height.

With respect to the innate fertility of the plants of crossed and self-fertilized parentage, given in the previous Table D – that is, the number of seeds produced by both lots when their flowers were fertilized in the same manner – nearly the same remarks are applicable, in reference to the absence of any close correspondence between their fertility and powers of / growth, as in the case of the plants in the Tables F and G, just considered. Thus the crossed and self-fertilized plants of Ipomoea, Papaver, *Reseda odorata*, and Limnanthes were almost equally fertile, yet the former exceeded considerably in height the self-fertilized plants. On the other hand, the crossed and self-fertilized plants of Mimulus and Primula differed to an extreme degree in innate fertility, but by no means to a corresponding degree in height or vigour.

In all the cases of self-fertilized flowers included in Tables E, F, and G, these were fertilized with their own pollen; but there is another form of self-fertilization, viz., by pollen from other flowers on the same plant; but this latter method made no difference in comparison with the former in the number of seeds produced, or only a slight difference. Neither with Digitalis nor Dianthus were more seeds produced by the one method than by the other, to any trustworthy degree. With Ipomoea rather more seeds, in the proportion of 100 to 91, were produced from a cross between flowers on the same plant than from strictly self-fertilized flowers; but I have reason to suspect that the result was accidental. With *Origanum vulgare*, however, a cross between flowers on plants propagated by stolons from the same stock certainly increased slightly their fertility. This likewise occurred, as we shall see in the next section, with Eschscholtzia, perhaps with *Corydalis cava* and Oncidium; but not so with Bignonia, Abutilon, Tabernaemontana, Senecio, and apparently *Reseda odorata*.

Self-sterile plants

The cases here to be described might have been introduced in Table F, which gives the relative fertility of flowers fertilized with their own pollen, and / with that from a distinct plant; but it has been found more convenient to keep them for separate discussion. The present cases must not be confounded with those to be given in the next chapter relatively to flowers which are sterile when insects are excluded; for such sterility depends not merely on the flowers being incapable of fertilization with their own pollen, but on mechanical causes, by which their pollen is prevented from reaching the stigma, or on the pollen and stigma of the same flower being matured at different periods.

In the seventeenth chapter of my *Variation of Animals and Plants under Domestication* I had occasion to enter fully on the present subject; and I will therefore here give only a brief abstract of the cases there described, but others must be added, as they have an important bearing on the present work. Kölreuter long ago described plants of *Verbascum phoeniceum* which during two years were sterile with their own pollen, but were easily fertilized by that of four other species; these plants however afterwards became more or less self-fertile in a strangely fluctuating manner. Mr Scott also found that this species, as well as two of its varieties, were self-sterile, as did Gärtner in the case of *Verbascum nigrum*. So it was, according to this latter author, with two plants of *Lobelia fulgens*, though the pollen and ovules of both were in an efficient state in relation to other species. Five species of Passiflora and certain individuals of a sixth species have been found sterile with their own pollen; but slight changes in their conditions, such as being grafted on another stock or a change of temperature, rendered them self-fertile. Flowers on a completely self-impotent plant of *Passiflora alata* fertilized with pollen from its own self-impotent seedlings were quite fertile. Mr Scott, and afterwards Mr Munro, found that some / species of Oncidium and of Maxillaria cultivated in a hothouse in Edinburgh were quite sterile with their own pollen; and Fritz Müller found this to be the case with a large number of Orchidaceous genera growing in their native home of South Brazil.[2] He also discovered that the pollen-masses of some orchids acted on their own stigmas like a poison; and it appears that Gärtner formerly observed indications of this extraordinary fact in the case of some other plants.

Fritz Müller also states that a species of Bignonia and *Tabernaemontana echinata* are both sterile with their own pollen in their native country of Brazil.[3] Several Amaryllidaceous and Liliaceous plants are in the same predicament. Hildebrand observed with care *Corydalis cava*, found it completely self-sterile;[4] but according to Caspary a few self-fertilized seeds are occasionally produced: *Corydalis halleri* is only slightly self-sterile, and *C. intermedia* not at all so.[5] In another Fumariaceous genus, Hypecoum, Hildebrand observed[6] that *H. grandiflorum* was highly self-sterile, whilst *H. procumbens* was fairly self-fertile. *Thunbergia alata* kept by me in a warm greenhouse was self-sterile early in the season, but at a later period produced many spontaneously self-fertilized fruits. So it was with *Papaver vagum*: another species, *P. alpinum*, was found by Professor H. Hoffmann to be quite self-sterile excepting on one occasion;[7] whilst *P. somniferum* has been with me always completely self-fertile.

Eschscholtzia californica. This species deserves a fuller consideration. A plant cultivated by Fritz / Müller in South Brazil happened to flower a month before any of the others, and it did not produce a single capsule. This led him to make further observations during the next six generations, and he found that all his plants were complately sterile, unless they were crossed by insects or were artificially fertilized with pollen from a distinct plant, in which case they were completely fertile.[8] I was much surprised at this fact, as I had found that English plants, when covered by a net, set a considerable number of capsules; and that these contained seeds by weight, compared with those on

---

[2] *Bot. Zeitung*, 1868, p. 114.
[3] Ibid., 1868, p. 626, and 1870, p. 274.
[4] *Report of the International Hort. Congress*, 1866.
[5] *Bot. Zeitung*, 27 June, 1873.
[6] *Jahrb. für wiss. Botanik*, vol. vii, p. 464.
[7] *Zur Speciesfrage*, 1875, p. 47.
[8] *Bot. Zeitung*, 1868, p. 115, and 1869, p. 223.

plants intercrossed by the bees, as 71 to 100. Professor Hildebrand, however, found this species much more self-sterile in Germany than it was with me in England, for the capsules produced by self-fertilized flowers, compared with those from intercrossed flowers, contained seeds in the ratio of only 11 to 100. At my request Fritz Müller sent me from Brazil seeds of his self-sterile plants, from which I raised seedlings. Two of these were covered with a net, and one produced spontaneously only a single capsule containing no good seeds, but yet, when artificially fertilized with its own pollen, produced a few capsules. The other plant produced spontaneously under the net eight capsules, one of which contained no less than thirty seeds, and on an average about ten seeds per capsule. Eight flowers on these two plants were artificially self-fertilized, and produced seven capsules, containing on an average twelve seeds; eight other flowers were fertilized with pollen from a distinct plant of the Brazilian stock, and produced eight capsules, containing on an average about eighty seeds: this gives a ratio of 15 seeds for the self-fertilized capsules to 100 for the crossed / capsules. Later in the season twelve other flowers on these two plants were artificially self-fertilized; but they yielded only two capsules, containing three and six seeds. It appears therefore that a lower temperature than that of Brazil favours the self-fertility of this plant, whilst a still lower temperature lessens it. As soon as the two plants which had been covered by the net were uncovered, they were visited by many bees, and it was interesting to observe how quickly they became, even the more sterile plant of the two, covered with young capsules. On the following year eight flowers on plants of the Brazilian stock of self-fertilized parentage (i.e., grandchildren of the plants which grew in Brazil) were again self-fertilized, and produced five capsules, containing on an average 27·4 seeds, with a maximum in one of forty-two seeds; so that their self-fertility had evidently increased greatly by being reared for two generations in England. On the whole we may conclude that plants of the Brazilian stock are much more self-fertile in this country than in Brazil, and less so than plants of the English stock in England; so that the plants of Brazilian parentage retained by inheritance some of their former sexual constitution. Conversely, seeds from English plants sent by me to Fritz Müller and grown in Brazil, were much more self-fertile than his plants which had been cultivated there for several generations; but he informs me that one of the plants of English parentage which did not flower the first year, and was thus exposed for two seasons to the climate of Brazil, proved quite self-sterile, like a Brazilian plant, showing how quickly the climate had acted on its sexual constitution.

Abutilon darwinii. Seeds of this plant were sent me by Fritz Müller, who found it, as well as some other species of the same genus, quite sterile in its / native home of South Brazil, unless fertilized with pollen from a distinct plant, either artificially or naturally by humming-birds.[9] Several plants were raised from these seeds and kept in the hothouse. They produced flowers very early in the spring, and twenty of them were fertilized, some with pollen from the same flower, and some with pollen from other flowers on the same plants; but not a single capsule was thus produced, yet the stigmas twenty-seven hours after the application of the pollen were penetrated by the pollen-tubes. At the same time nineteen flowers were crossed with pollen from a distinct plant, and these produced thirteen capsules, all abounding with fine seeds. A greater number of capsules would have been pro-duced by the cross, had not some of the nineteen flowers been on a plant which was afterwards proved to be from some unknown cause completely sterile with pollen of any kind. Thus far these plants behaved exactly like those in Brazil; but later in the season, in the latter part of May and in June, they began to produce under a net a few spontaneously self-fertilized capsules. As soon as this occurred, sixteen flowers were fertilized with their own pollen, and these produced five capsules, containing on an average 3·4 seeds. At the same time I selected by chance four capsules from the uncovered plants growing close by, the flowers of which I had seen visited by humble-bees, and these contained on an average 21·5 seeds; so that the seeds in the naturally intercrossed capsules to those in the self-fertilized capsules were as 100 to 16. The interesting point in this case is that these plants, which were unnaturally treated by being grown in pots in a hothouse, under another / hemisphere, with a complete reversal of the seasons, were thus rendered slightly self-fertile, whereas they seem always to be completely self-sterile in their native home.

Senecio cruentus (greenhouse varieties, commonly called cinerarias, probably derived from several fruticose or herbaceous species much intercrossed.)[10] Two

---

[9] *Jenaische Zeitschr. für Naturwiss*, vol. vii, 1872, p. 22, and 1873, p. 441.

[10] I am much obliged to Mr Moore and to Mr Thiselton Dyer for giving me information with respect to the varieties on which I experimented. Mr Moore believes that *Senecio cruentus, tussilaginis,* and perhaps *heritieri, maderensis* and *populifolius* have all been more or less blended together in our cinerarias.

purple-flowered varieties were placed under a net in the greenhouse, and four corymbs on each were repeatedly brushed with flowers from the other plant, so that their stigmas were well covered with each other's pollen. Two of the eight corymbs thus treated produced very few seeds, but the other six produced on an average 41·3 seeds per corymb, and these germinated well. The stigmas on four other corymbs on both plants were well smeared with pollen from the flowers on their own corymbs; these eight corymbs produced altogether ten extremely poor seeds, which proved incapable of germinating. I examined many flowers on both plants, and found the stigmas spontaneously covered with pollen; but they produced not a single seed. These plants were afterwards left uncovered in the same house where many other Cinerarias were in flower; and the flowers were frequently visited by bees. They then produced plenty of seed, but one of the two plants less than the other, as this species shows some tendency to be dioecious.

The trial was repeated on another variety with white petals tipped with red. Many stigmas on two corymbs were covered with pollen from the foregoing purple variety, and these produced eleven and twenty-two / seeds, which germinated well. A large number of the stigmas on several of the other corymbs were repeatedly smeared with pollen from their own corymb; but they yielded only five very poor seeds, which were incapable of germination. Therefore the above three plants belonging to two varieties, though growing vigorously and fertile with pollen from either of the other two plants, were utterly sterile with pollen from other flowers on the same plant.

Reseda odorata. Having observed that certain individuals were self-fertile, I covered during the summer of 1868 seven plants under separate nets, and will call these plants A, B, C, D, E, F, G. They all appeared to be quite sterile with their own pollen, but fertile with that of any other plant.

Fourteen flowers on A were crossed with pollen from B or C, and produced thirteen fine capsules. Sixteen flowers were fertilized with pollen from other flowers on the same plant, but yielded not a single capsule.

Fourteen flowers on B were crossed with pollen from A, C, or D, and all produced capsules; some of these were not very fine, yet they contained plenty of seeds. Eighteen flowers were fertilized with pollen from other flowers on the same plant, and produced not one capsule.

Ten flowers on C were crossed with pollen from A, B, D, or E, and produced nine fine capsules. Nineteen flowers were fertilized with pollen from other flowers on the same plant, and produced no capsules.

Ten flowers on D were crossed with pollen from A, B, C, or E, and produced nine fine capsules. Eighteen flowers were fertilized with pollen from other flowers on the same plant, and produced no capsules.

Seven flowers on E were crossed with pollen from / A, C, or D, and all produced fine capsules. Eight flowers were fertilized with pollen from other flowers on the same plant, and produced no capsules.

On the plants F and G no flowers were crossed, but very many (number not recorded) were fertilized with pollen from other flowers on the same plants, and these did not produce a single capsule.

We thus see that fifty-five flowers on five of the above plants were reciprocally crossed in various ways; several flowers on each of these plants being fertilized with pollen from several of the other plants. These fifty-five flowers produced fifty-two capsules, almost all of which were of full size and contained an abundance of seeds. On the other hand, seventy-nine flowers (besides many others not recorded) were fertilized with pollen from other flowers on the same plants, and these did not produce a single capsule. In one case in which I examined the stigmas of the flowers fertilized with their own pollen, these were penetrated by the pollen-tubes, although such penetration produced no effect. Pollen falls generally, and I believe always, from the anthers on the stigmas of the same flower; yet only three out of the above seven protected plants produced spontaneously any capsules, and these it might have been thought must have been self-fertilized. There were altogether seven such capsules; but as they were all seated close to the artificially crossed flowers, I can hardly doubt that a few grains of foreign pollen had accidentally fallen on their stigmas. Besides the above seven plants, four others were kept covered under the *same* large net; and some of these produced here and there in the most capricious manner little groups of capsules; and this makes me believe that a bee, many of which settled on the outside of the net, being / attracted by the odour, had on some one occasion found an entrance, and had intercrossed a few of the flowers.

In the spring of 1869 four plants raised from fresh seeds were carefully protected under separate nets; and now the result was widely different to what it was before. Three of these protected plants became actually loaded with capsules, especially during the early part of the summer; and this fact indicates that temperature produces some effect, but the experiment given in the following paragraph shows that the innate constitution of the plant is a far more important element. The fourth plant produced only a few capsules, many of them of small size; yet it was far more self-fertile than any of the seven plants tried during the previous year. The flowers on four small branches of this semi-self-sterile plant were smeared with pollen from one of the other plants, and they all produced fine capsules.

As I was much surprised at the difference in the results of the trials made during the two previous years, six fresh plants were protected by separate nets in the year 1870. Two of these proved almost completely self-sterile, for on carefully searching them I found only three small capsules, each containing either one or two seeds of small size, which; however, germinated. A few flowers on both these plants were reciprocally fertilized with each other's pollen, and a few with pollen from one of the following self-fertile plants, and all these flowers produced fine capsules. The four other plants whilst still remaining protected beneath the nets presented a wonderful contrast (though one of them in a somewhat less degree than the others), for they became actually covered with spontaneously self-fertilized capsules, as / numerous as, or very nearly so, and as fine as those on the unprotected plants growing near.

The above three spontaneously self-fertilized capsules produced by the two almost completely self-sterile plants, contained altogether five seeds; and from these I raised in the following year (1871) five plants, which were kept under separate nets. They grew to an extraordinarily large size, and on 29 August were examined. At first sight they appeared entirely destitute of capsules; but on carefully searching their many branches, two or three capsules were found on three of the plants, half-a-dozen on the fourth, and about eighteen on the fifth plant. But all these capsules were small, some being empty; the greater number contained only a single seed, and very rarely more than one. After this examination the nets were taken off, and the bees immediately carried pollen from one of these almost self-sterile plants to the other, for no other plants grew near. After a few weeks the ends of the branches on all five plants became covered with capsules, presenting a curious contrast with the lower and naked parts of the same long branches. These five plants therefore inherited almost exactly the same sexual constitution as their parents; and without doubt a self-sterile race of Mignonette could have been easily established.

Reseda lutea. Plants of this species were raised from seeds gathered from a group of wild plants growing at no great distance from my garden. After casually observing that some of these plants were self-sterile, two plants taken by hazard were protected under separate nets. One of these soon became covered with spontaneously self-fertilized capsules, as numerous as those on the surrounding unprotected plants; so that it was evidently quite self-fertile. / The other plant was partially self-sterile, producing very few capsules, many of which were of small size. When, however, this plant had grown tall, the uppermost branches became pressed against the net and grew crooked, and in this position the bees were able to suck the flowers through the meshes, and brought pollen to them from the neighbouring plants. These branches then became loaded with capsules; the other and lower branches remaining almost bare. The sexual constitution of this species is therefore similar to that of *Reseda odorata.*

Concluding remarks on self-sterile plants

In order to favour as far as possible the self-fertilization of some of the foregoing plants, all the flowers on *Reseda odorata* and some of those on the Abutilon were fertilized with pollen from other flowers on the same plant, instead of with their own pollen, and in the case of the Senecio with pollen from other flowers on the same corymb; but this made no difference in the result. Fritz Müller tried both kinds of self-fertilization in the case of Bignonia, Tabernaemontana and Abutilon, likewise with no difference in the result. With Eschscholtzia, however, he found that pollen from other flowers on the same plant was a little more effective than pollen from the same flower. So did Hildebrand[11] in Germany; as thirteen out of fourteen flowers of Eschscholtzia thus fertilized set capsules, these containing on an average 9·5 seeds; whereas only fourteen flowers out of twenty-one fertilized with their own pollen set capsules, these containing on an average 9·0 seeds. Hildebrand / found a trace of a similar difference with *Corydalis cava,* as did Fritz Müller with an Oncidium.[12]

In considering the several cases above given of complete or almost complete self-sterility, we are first struck with their wide distribution throughout the vegetable kingdom. Their number is not at present large, for they can be discovered only by protecting plants from insects

[11] *Pringsheim's Jahrbuch. für wiss. Botanik,* vii, p. 467.
[12] *Var. under Dom.,* chap. xvii, 2nd edit., vol. ii, pp. 113–15.

and then fertilizing them with pollen from another plant of the same species and with their own pollen; and the latter must be proved to be in an efficient state by other trials. Unless all this be done, it is impossible to know whether their self-sterility may not be due to the male or female reproductive organs, or to both, having been affected by changed conditions of life. As in the course of my experiments I have found three new cases, and as Fritz Müller has observed indications of several others, it is probable that they will hereafter be proved to be far from rare.[13]

As with plants of the same species and parentage, some individuals are self-sterile and others self-fertile, of which fact *Reseda odorata* offers the most striking instances, it is not at all surprising that species of the same genus differ in this same manner. Thus *Verbascum phoeniceum* and *nigrum* are self-sterile, whilst *V. thapsus* and *lychnitis* are quite self-fertile, as I know by trial. There is the same difference between some of the species of Papaver, Corydalis, and of other genera. Nevertheless, the tendency to self-sterility certainly runs to a certain extent in groups, as we see / in the genus Passiflora, and with the Vandeae among Orchids.

Self-sterility differs much in degree in different plants. In those extraordinary cases in which pollen from the same flower acts on the stigma like a poison, it is almost certain that the plants would never yield a single self-fertilized seed. Other plants, like *Corydalis cava*, occasionally, though very rarely, produce a few self-fertilized seeds. A large number of species, as may be seen in Table F, are less fertile with their own pollen than with that from another plant; and lastly, some species are perfectly self-fertile. Even with the individuals of the same species, as just remarked, some are utterly self-sterile, others moderately so, and some perfectly self-fertile. The cause, whatever it may be, which renders many plants more or less sterile with their own pollen, that is, when they are self-fertilized, must be different, at least to a certain extent, from that which determines the difference in height, vigour, and fertility of the seedlings raised from self-fertilized and crossed seeds; for we have already seen that the two classes of cases do not by any means run parallel. This want of parallelism would be intelligible, if it could be shown that self-sterility depended solely on the incapacity of the pollen-tubes to penetrate the stigma of the same flower deeply enough to reach the ovules; whilst the greater or less vigorous growth of the seedlings no

[13] Mr Wilder, the editor of a horticultural journal in the U. States (quoted in *Gard. Chron.* 1868, p. 1286) states that *Lilium auratum, Impatiens pallida* and *fulva*, cannot be fertilized with their own pollen. Rimpan shows that rye is probably sterile with its own pollen.

doubt depends on the nature of the contents of the pollen-grains and ovules. Now it is certain that with some plants the stigmatic secretion does not properly excite the pollen-grains, so that the tubes are not properly developed, if the pollen is taken from the same flower. This is the case according to Fritz Müller with Eschscholtzia, for he found that the pollen-tubes did not penetrate / the stigma deeply;[14] and with the Orchidaceous genus Notylia they failed altogether to penetrate it.

With dimorphic and trimorphic species, an illegitimate union between plants of the same form presents a close analogy with self-fertilization, whilst a legitimate union closely resembles cross-fertilization; and here again the lessened fertility or complete sterility of an illegitimate union depends, at least in part, on the incapacity for interaction between the pollen-grains and stigma. Thus with *Linum grandiflorum*, as I have elsewhere shown,[15] not more than two or three out of hundreds of pollen-grains, either of the long-styled or short-styled form, when placed on the stigma of their own form, emit their tubes, and these do not penetrate deeply; nor does the stigma itself change colour, as occurs when it is legitimately fertilized.

On the other hand the difference in innate fertility, as well as in growth between plants raised from crossed and self-fertilized seeds, and the difference in fertility and growth between the legitimate and illegitimate offspring of dimorphic and trimorphic plants, must depend on some incompatibility between the sexual elements contained within the pollen-grains and ovules, as it is through their union that new organisms are developed.

If we now turn to the more immediate cause of self-sterility, we clearly see that in most cases it is determined by the conditions to which the plants have been subjected. Thus Eschscholtzia is completely self-sterile in the hot climate of Brazil, but is perfectly fertile there with the pollen of any other individual. The offspring of Brazilian plants became in England / in a single generation partially self-fertile, and still more so in the second generation. Conversely, the offspring of English plants, after growing for two seasons in Brazil, became in the first generation quite self-sterile. Again, *Abutilon darwinii*, which is self-sterile in a single generation in an English hothouse. Some other plants are self-sterile during the early part of the year, and later in the season become self-fertile. *Passiflora alata* lost its self-sterility when grafted on another species. With Reseda, however, in which some individuals of

---

[14] *Bot. Zeitung*, 1868, pp. 114, 115.   [15] *The Different Forms of Flowers*, etc., p. 87.

the same parentage are self-sterile and others are self-fertile, we are forced in our ignorance to speak of the cause as due to spontaneous variability; but we should remember that the progenitors of these plants, either on the male or female side, may have been exposed to somewhat different conditions. The power of the environment thus to affect so readily and in so peculiar a manner the reproductive organs, is a fact which has many important bearings; and I have therefore thought the foregoing details worth giving. For instance, the sterility of many animals and plants under changed conditions of life, such as confinement, evidently comes within the same general principle of the sexual system being easily affected by the environment. It has already been proved, that a cross between plants which have been self-fertilized or intercrossed during several generations, having been kept all the time under closely similar conditions, does not benefit the offspring; and on the other hand, that a cross between plants that have been subjected to different conditions benefits the offspring to an extraordinary degree. We may therefore conclude that some degree of differentiation in the sexual system is necessary for / the full fertility of the parent-plants and for the full vigour of their offspring. It seems also probable that with those plants which are capable of complete self-fertilization, the male and female elements and organs already differ to an extent sufficient to excite their mutual interaction; but that when such plants are taken to another country, and become in consequence self-sterile, their sexual elements and organs are so acted on as to be rendered too uniform for such interaction, like those of a self-fertilized plant long cultivated under the same conditions. Conversely, we may further infer that plants which are self-fertile in their native country, but become self-fertile under changed conditions, have their sexual elements so acted on, that they become sufficiently differentiated for mutual interaction.

We know that self-fertilized seedlings are inferior in many respects to those from a cross; and as with plants in a state of nature pollen from the same flower can hardly fail to be often left by insects or by the wind on the stigma, it seems at first sight highly probable that self-sterility has been gradually acquired through Natural Selection in order to prevent self-fertilization. It is no valid objection to this belief that the structure of some flowers, and the dichogamous condition of many others, suffice to prevent the pollen reaching the stigma of the same flower; for we should remember that with most species many flowers expand at the same time, and that pollen from the same plant is equally injurious or nearly so as that from the same flower. Nevertheless, the belief that self-sterility is a quality which has been gradually acquired for the special purpose of preventing self-fertilization must, I believe, be rejected. In the first place, there is no close correspondence in degree / between the sterility of the parent-plants when self-fertilized, and the extent to which their offspring suffer in vigour by this process; and some such correspondence might have been expected if self-sterility had been acquired on account of the injury caused by self-fertilization. The fact of individuals of the same parentage differing greatly in their degree of self-sterility is likewise opposed to such a belief; unless, indeed, we suppose that certain individuals have been rendered self-sterile to favour intercrossing, whilst other individuals have been rendered self-fertile to ensure the propagation of the species. The fact of self-sterile individuals appearing only occasionally, as in the case of Lobelia, does not countenance this latter view. But the strongest argument against the belief that self-sterility has been acquired to prevent self-fertilization, is the immediate and powerful effect of changed conditions in either causing or in removing self-sterility. We are not therefore justified in admitting that this peculiar state of the reproductive system has been gradually acquired through Natural Selection; but we must look at it as an incidental result, dependent on the conditions to which the plants have been subjected, like the ordinary sterility caused in the case of animals by confinement, and in the case of plants by too much manure, heat, etc., I do not, however, wish to maintain that self-sterility may not sometimes be of service to a plant in preventing self-fertilization; but there are so many other means by which this result might be prevented or rendered difficult, including as we shall see in the next chapter the prepotency of pollen from a distinct individual over a plant's own pollen, that self-sterility seems an almost superfluous acquirement for this purpose.

Finally, the most interesting point in regard to self-sterile / plants is the evidence which they afford of the advantage, or rather of the necessity, of some degree or kind of differentiation in the sexual elements, in order that they should unite and give birth to a new being. It was ascertained that the five plants of *Reseda odorata* which were selected by chance, could be perfectly fertilized by pollen taken from any one of them, but not by their own pollen; and a few additional trials were made with some other individuals, which I have not thought worth recording. So again, Hildebrand and Fritz Müller frequently speak of self-sterile plants being fertile with the pollen of any other individual; and if there had been any exceptions to the rule, these could hardly have escaped their observation and my own. We may therefore confidently assert that a self-sterile plant can be fertilized by the pollen of any one out of a thousand or ten thousand individuals of the same species, but not by its own. Now it is obviously impossible that the sexual organs and elements of every individual can have been specialized with respect to every other individual. But there is no difficulty in believing that the sexual elements of each differ slightly in the same diversified manner as do their external characters; and it has often been remarked that no two individuals are absolutely alike. Therefore we can hardly avoid the conclusion, that differences of an analogous and indefinite nature in the reproductive system are sufficient to excite the mutual action of the sexual elements, and that unless there be such differentiation fertility fails.

The appearance of highly self-fertile varieties. We have just seen that the degree to which flowers are capable of being fertilized with their own pollen differs much, both with the species of the same genus, and / sometimes with the individuals of the same species. Some allied cases of the appearance of varieties which, when self-fertilized, yield more seed and produce offspring growing taller than their self-fertilized parents, or than the intercrossed plants of the corresponding generation, will now be considered.

First, in the third and fourth generations of *Mimulus luteus*, a tall variety, often alluded to, having large white flowers blotched with crimson, appeared among both the intercrossed and self-fertilized plants. It prevailed in all the later self-fertilized generations to the exclusion of every other variety, and transmitted its characters faithfully, but disappeared from the intercrossed plants, owing no doubt to their characters being repeatedly blended by crossing. The self-fertilized plants belonging to this variety were not only taller, but more fertile than the intercrossed plants; though these latter in the earlier generations were much taller and more fertile than the self-fertilized plants. Thus in the fifth generation the self-fertilized plants were to the intercrossed in height as 126 to 100. In the sixth generation they were likewise much taller and finer plants, but were not actually measured; they produced capsules compared with those on the intercrossed plants, in number, as 147 to 100; and the self-fertilized capsules contained a greater number of seeds. In the seventh generation the self-fertilized plants were to the crossed in height as 137 to 100; and twenty flowers on these self-fertilized plants fertilized with their own pollen yielded nineteen very fine capsules – a degree of self-fertility which I have not seen equalled in any other case. This variety seems to have become specially adapted to profit in every way by self-fertilization, although this process was so injurious to the parent-plants during the first / four generations. It should however be remembered that seedlings raised from this variety, when crossed by a fresh stock, were wonderfully superior in height and fertility to the self-fertilized plants of the corresponding generation.

Secondly, in the sixth self-fertilized generation of Ipomoea, a single plant named the Hero appeared, which exceeded by a little in height its intercrossed opponent – a case which had not occurred in any previous generation. Hero transmitted the peculiar colour of its flowers, as well as its increased tallness and a high degree of self-fertility, to its children, grandchildren, and great-grandchildren. The self-fertilized children of Hero were in height to other self-fertilized plants of the same stock as 100 to 85. Ten self-fertilized capsules produced by the grandchildren contained on an average 5·2 seeds; and this is a higher average than was yielded in any other generation by the capsules of self-fertilized flowers. The great-grandchildren of Hero derived from a cross with a fresh stock were so unhealthy, from having been grown at an unfavourable season, that their average height in comparison with that of the self-fertilized plants cannot be judged of with any safety; but it did not appear that they had profited even by a cross of this kind.

Thirdly, the plants of Nicotiana on which I experimented appear to come under the present class of cases; for they varied in their sexual constitution and were more or less highly self-fertile. They were probably the offspring of plants which had been spontaneously self-fertilized under glass for several generations in this country. The flowers on the parent-plants which were first fertilized by me with their own pollen yielded half again as many seeds as did / those which were crossed; and the seedlings raised from these self-fertilized seeds exceeded in height those raised from the crossed seeds to an extra-ordinary degree. In the second and third generations, although the self-fertilized plants did not exceed the crossed in height, yet their self-fertilized flowers yielded on two occasions considerably more seeds than the crossed flowers, even than those which were crossed with pollen from a distinct stock or variety.

Lastly, as certain individual plants of *Reseda odorata* and *lutea* are incomparably more self-fertile than other individuals, the former might be included under the present heading of the appearance of new and highly self-fertile varieties. But in this case we should have to look at these two species as normally self-sterile; and this, judging by my experience, appears to be the correct view.

We may therefore conclude from the facts now given, that varieties sometimes arise which when self-fertilized possess an increased power of producing seeds and of growing to a greater height, than the intercrossed or self-fertilized plants of the corresponding generation – all the plants being of course subjected to the same conditions. The appearance of such varieties is interesting, as it bears on the existence under nature of plants which regularly fertilize themselves, such as *Ophrys apifera* and some other orchids, or as *Leersia oryzoides*, which produces an abundance of cleistogamic flowers, but most rarely flowers capable of cross-fertilization.[16]

Some observations made on other plants lead me to suspect that self-fertilization is in some respects beneficial; although the benefit thus derived is as a / rule very small compared with that from a cross with a distinct plant. Thus we have seen in the last chapter that seedlings of Ipomoea and Mimulus raised from flowers fertilized with their own pollen, which is the strictest possible form of self-fertilization, were superior in height, weight, and in early flowering to the seedlings raised from flowers crossed with pollen from other flowers on the same plant; and this superiority apparently was too strongly marked to be accidental. Again, the cultivated varieties of the common pea are highly self-fertile, although they have been self-fertilized for many generations; and they exceeded in height seedlings from a cross between two plants belonging to the same variety in the ratio of 115 to 100; but then only four pairs of plants were measured and compared. The self-fertility of *Primula veris* increased after several generations of illegitimate fertilization, which is a process analogous to self-fertilization, but only as long as the plants were cultivated under the same favourable conditions. I have also elsewhere shown[17] that with several species of Primula equal-styled varieties occasionally appear which possess the sexual organs of the two forms combined in the same flower. Consequently they fertilize themselves in a legitimate

---

[16] On Leersia, see *Different Forms of Flowers*, etc., p. 335.

[17] *Different Forms of Flowers*, etc., p. 272.

manner and are highly self-fertile; but the remarkable fact is that they are rather more fertile than ordinary plants of the same species legitimately fertilized by pollen from a distinct individual. Formerly it appeared to me probable, that the increased fertility of these hetero-styled plants might be accounted for by the stigma lying so close to the anthers that it was impregnated at the most favourable age and time of the day; but this explanation is not applicable to the / above given cases, in which the flowers were artificially fertilized with their own pollen.

Considering the facts now adduced, including the appearance of those varieties which are more fertile and taller than their parents and than the intercrossed plants of the corresponding generation, it is difficult to avoid the suspicion that self-fertilization is in some respects advantageous; though if this be really the case,[18] any such advantage is as a rule quite insignificant compared with that from a cross with a distinct plant, and especially with one of a fresh stock. Should this suspicion be hereafter verified, it would throw light, as we shall see in the next chapter, on the existence of plants bearing small and incon-spicuous flowers which are rarely visited by insects, and therefore are rarely intercrossed.

Relative weight and period of germination of seeds from crossed and self-fertilized flowers. An equal number of seeds from flowers fertilized with pollen from another plant, and from flowers fertilized with their own pollen, were weighed, but only in sixteen cases. Their relative weights are given in the following list; that of the seeds from the crossed flowers being taken as 100.

Ipomoea purpurea (parent plants)	as 100 to 127
„ „ (third generation)	„ „ 87
Salvia coccinea	„ „ 100
Brassica oleracea	„ „ 103
Iberis umbellata (second generation)	„ „ 136
Delphinium consolida	„ „ 45
Hibiscus africanus	„ „ 105
Tropaeolum minus	„ „ 115
Lathyrus odoratus (about)	„ „ 100
Sarothamnus scoparius	„ „ 88 /

[18] M. Errara, who intends publishing on the present subject, has been so kind as to send me his MS. to read. He is convinced that self-fertilization is never more beneficial than a cross with another flower. I hope that his view may hereafter be proved correct, as the subject of cross and self-fertilization would be thus much simplified.

Specularia speculum	as 100 to 86
Nemophila insignis	„ „ 105
Borago officinalis	„ „ 111
Cyclamen persicum (about)	„ „ 50
Fagopyrum esculentum	„ „ 82
Canna warscewiczi (three generations)	„ „ 102

It is remarkable that in ten out of these sixteen cases the self-fertilized seeds were either superior or equal to the crossed in weight; nevertheless, in six out of the ten cases (viz., with Ipomoea, Salvia, Brassica, Tropaeolum, Lathyrus, and Nemophila) the plants raised from these self-fertilized seeds were very inferior in height and in other respects to those raised from the crossed seeds. The superiority in weight of the self-fertilized seeds in at least six out of the ten cases, namely, with Brassica, Hibiscus, Tropaeolum, Nemophila, Borago, and Canna, may be accounted for in part by the self-fertilized capsules containing fewer seeds; for when a capsule contains only a few seeds, these will be apt to be better nourished, so as to be heavier, than when many are contained in the same capsule. It should, however, be observed that in some of the above cases, in which the crossed seeds were the heaviest, as with Sarothamnus and Cyclamen, the crossed capsules contained a larger number of seeds. Whatever may be the explanation of the self-fertilized seeds being often the heaviest, it is remarkable in the case of Brassica Tropaeolum, Nemophila, and of the first generation of Ipomoea, that the seedlings raised from them were inferior in height and in other respects to the seedlings raised from the crossed seeds. This fact shows how superior in constitutional vigour the crossed seedlings must have been, for it cannot be doubted that heavy and fine seeds tend to yield the finest plants. Mr Galton has shown that this holds good with *Lathyrus odoratus*; as has Mr A. J. Wilson with the swedish / turnip, *Brassica campestri ruta baga*. Mr Wilson separated the largest and smallest seeds of this latter plant, the ratio between the weights of the two lots being as 100 to 59, and he found that the seedlings 'from the larger seeds took the lead and maintained their superiority to the last, both in height and thickness of stem'.[19]

---

[19] *Gardeners' Chronicle*, 1867, p. 107. Loiseleur-Deslongchamp (*Les Céréales*, 1842, pp. 208–19) was led by his observations to the extraordinary conclusion that the smaller grains of cereals produce as fine plants as the large. This conclusion is, however, contradicted by Major Hallet's great success in improving wheat by the selection of the finest grains. It is possible, however, that man, by long-continued selection, may have given to the grains of the cereals a greater amount of starch or other matter, than the seedlings can utilize for their growth. There can be little

Nor can this difference in the growth of the seedling turnips be attributed to the heavier seeds having been of crossed, and the lighter of self-fertilized origin, for it is known that plants belonging to this genus are habitually intercrossed by insects.

With respect to the relative period of germination of crossed and self-fertilized seeds, a record was kept in only twenty-one cases; and the results are very perplexing. Neglecting one case in which the two lots germinated simultaneously, in ten cases or exactly one-half many of the self-fertilized seeds germinated before the crossed, and in the other half many of the crossed before the self-fertilized. In four out of these twenty cases, seeds derived from a cross with a fresh stock were compared with self-fertilized seeds from one of the later self-fertilized generations; and here again in half the cases the crossed seeds, and in the other half the self-fertilized seeds, germinated first. Yet the plants of Mimulus raised from such self-fertilized seeds were inferior in all respects to the crossed plants, and in / the case of Eschscholtzia they were inferior in fertility. Unfortunately the relative weight of the two lots of seeds was ascertained in only a few instances in which their germination was observed; but with Ipomoea and I believe with some of the other species, the relative lightness of the self-fertilized seeds apparently determined their early germination, probably owing to the smaller mass being favourable to the more rapid completion of the chemical and morphological changes necessary for germination.[20] On the other hand, Mr Galton gave me seeds (no doubt all self-fertilized) of *Lathyrus odoratus*, which were divided into two lots of heavier and lighter seeds; and several of the former germinated first. It is evident that many more observations are necessary before anything can be decided with respect to the relative period of germination of crossed and self-fertilized seeds. /

---

doubt, as Humboldt long ago remarked, that the grains of cereals have been rendered attractive to birds in a degree which is highly injurious to the species.

[20] Mr J. Scott remarks (*Manual of Opium Husbandry*, 1877, p. 131) that the smaller seeds of *Papaver somniferum* germinate first. He also states that the larger seeds yield the finer crop of plants. With respect to this latter subject see an abstract in Burbidge's *Cultivated Plants*, 1877, p. 33, on the important experiments showing the same results, by Dr Marck and Professor Lehmann.

CHAPTER X

MEANS OF FERTILIZATION

Sterility and fertility of plants when insects are excluded – The means by which flowers are cross-fertilized – Structures favourable to self-fertilization – Relation between the structure and conspicuousness of flowers, the visits of insects, and the advantages of cross-fertilization – The means by which flowers are fertilized with pollen from a distinct plant – Greater fertilizing power of such pollen – Anemophilous species – Conversion of anemophilous species into entomophilous – Origin of nectar – Anemophilous plants generally have their sexes separated – Conversion of diclinous into hermaphrodite flowers – Trees often have their sexes separated.

In the introductory chapter I briefly specified the various means by which cross-fertilization is favoured or ensured, namely, the separation of the sexes – the maturity of the male and female sexual elements at different periods – the heterostyled or dimorphic and trimorphic condition of certain plants – many mechanical contrivances – the more or less complete inefficiency of a flower's own pollen on the stigma – and the prepotency of pollen from any other individual over that from the same plant. Some of these points require further consideration; but for full details I must refer the reader to the several excellent works mentioned in the introduction. I will in the first place give two lists: the first, of plants which are either quite sterile or produce less than about half the full complement of seeds, when insects are excluded; and a second list of plants which, when thus treated, are fully fertile or produce at least half the full complement / of seeds. These lists have been compiled from the several previous tables, with some additional cases from my own observations and those of others. The species are arranged nearly in the order followed by Lindley in his *Vegetable Kingdom*. The reader should observe that the sterility or fertility of the plants in these two lists depends on two wholly distinct causes; namely, the absence or presence of the proper means by which pollen is applied to the stigma, and its less or greater efficiency when thus applied. As it is obvious that with plants in which the sexes are separate, pollen must be carried by some means from flower to flower, such species are excluded from the lists; as are likewise heterostyled plants, in which the same necessity occurs to a limited extent. Experience has proved to me that, independently of the exclusion of insects, the seed-bearing power of a plant is not lessened by covering it while in flower under a thin net supported on a frame; and this might indeed have been inferred from the consideration of the two following lists, as they include a considerable number of species belonging to the same genera, some of which are quite sterile and others quite fertile when protected by a net from the access of insects.

List of plants which, when insects are excluded, are either quite sterile, or produce, as far as I could judge, less than half the number of seeds produced by unprotected plants

Passiflora alata, racemosa, coerulea, edulis, laurifolia, and some individuals of *P. quadrangularis* (Passifloraceae), are quite sterile under these conditions: see *Variation of Animals and Plants under Domestication,* chap. xvii, 2nd edit., vol. ii, p. 118.

Viola canina (Violaceae). Perfect flowers quite sterile unless fertilized by bees, or artificially fertilized. /

V. tricolor. Sets very few and poor capsules.

Reseda odorata (Resedaceae). Some individuals quite sterile.

R. lutea. Some individuals produce very few and poor capsules.

Abutilon darwinii (Malvaceae). – Quite sterile in Brazil: see previous discussion on self-sterile plants.

Nymphaea (Nymphaeaceae). Professor Caspary informs me that some of the species are quite sterile if insects are excluded.

Euryale amazonica (Nymphaeaceae). Mr J. Smith, of Kew, informs me that capsules from flowers left to themselves, and probably not visited by insects, contained from eight to fifteen seeds; those from flowers artificially fertilized with pollen from other flowers on the same plant contained from fifteen to thirty seeds; and that two flowers fertilized with pollen brought from another plant at Chatsworth contained respectively sixty and seventy-five seeds. I have given these statements because Professor Caspary advances this plant as a case opposed to the doctrine of the necessity or advantage of cross-fertilization: see *Sitzungsberichte der Phys.-ökon. Gesell. zu Königsberg,* vol. vi, p. 20.

Delphinium consolida (Ranunculaceae). Produces many capsules, but these contain only about half the number of seeds compared with capsules from flowers naturally fertilized by bees.

Eschscholtzia californica (Papaveraceae). Brazilian plants quite sterile: English plants produce a few capsules.

Papaver vagum (Papaveraceae). In the early part of the summer produced very few capsules, and these contained very few seeds.

P. alpinum. H. Hoffmann (*Speciesfrage*, 1875, p. 47) states that this species produced seeds capable of germination only on one occasion.

Corydalis cava (Fumariaceae). Sterile: see the previous discussion on self-sterile plants.

C. solida. I had a single plant in my garden (1863), and saw many hive-bees sucking the flowers, but not a single seed was produced. I was much surprised at this fact, as Professor Hildebrand's discovery that *C. cava* is sterile with its own pollen had not then been made. He likewise concludes from the few experiments which he made on the present species that it is self-sterile. The two foregoing cases are interesting, because botanists formerly thought (see, for / instance, Lecoq, *De la Fécondation et de l'Hybridation*, 1845, p. 61, and Lindley, *Vegetable Kingdom*, 1853, p. 436) that all the species of the Fumariaceae were specially adapted for self-fertilization.

C. lutea. A covered-up plant produced (1861) exactly half as many capsules as an exposed plant of the same size growing close alongside. When humble-bees visit the flowers (and I repeatedly saw them thus acting) the lower petals suddenly spring downwards and the pistil upwards; this is due to the elasticity of the parts, which takes effect, as soon as the coherent edges of the hood are separated by the entrance of an insect. Unless insects visit the flowers the parts do not move. Nevertheless, many of the flowers on the plants which I had protected produced capsules, notwithstanding that their petals and pistils still retained their original position; and I found to my surprise that these capsules contained more seeds than those from flowers, the petals of which had been artificially separated and allowed to spring apart. Thus, nine capsules produced by undisturbed flowers contained fifty-three seeds; whilst nine capsules from flowers, the petals of which had been artificially separated, contained only thirty-two seeds. But we should remember that if bees had been permitted to visit these flowers, they would have visited them at the best time for fertilization. The flowers, the petals of which had been artificially separated, set their capsules before those which were left undisturbed under the net. To show with what certainty the flowers are visited by bees, I may add that on one occasion all the flowers on some unprotected plants were examined, and every single one had its petals separated; and, on a second occasion, forty-one out of forty-three flowers were in this state. Hildebrand states (*Pring. Jahr. f. wiss. Botanik*, vol. vii, p. 450) that the mechanism of the parts in this species is nearly the same as in *C. ochroleuca*, which he has fully described.

Hypercoum grandiflorum (Fumariaceae). Highly self-fertile (Hildebrand, ibid.).

Kalmia latifolia (Ericaceae). Mr W. J. Beal says (*American Naturalist*, 1867) that flowers protected from insects wither and drop off, with 'most of the anthers still remaining in the pockets'.

Pelargonium zonale (Geraniaceae). Almost sterile; one plant / produced two fruits. It is probable that different varieties would differ in this respect, as some are only feebly dichogamous.

Dianthus caryophyllus (Caryophyllaceae). Produces very few capsules, which contain any good seeds.

Phaseolus multiflorus (Leguminosae). Plants protected from insects produced on two occasions about one-third and one-eighth of the full number of seeds:

see my article in *Gardeners' Chronicle*, 1857, p. 225, and 1858, p. 828; also *Annals and Mag. of Natural History*, 3rd series, vol. ii, 1858, p. 462. Dr Ogle (*Pop. Science Review*, 1870, p. 168) found that a plant was quite sterile when covered up. The flowers are not visited by insects in Nicaragua, and, according to Mr Belt, the species is there quite sterile: *The Naturalist in Nicaragua*, p. 70.

Vicia faba (Leguminosae). Seventeen covered-up plants yielded 40 beans, whilst seventeen plants left unprotected and growing close alongside produced 135 beans; these latter plants were, therefore, between three and four times more fertile than the protected plants: see *Gardeners' Chronicle* for fuller details, 1858, p. 828.

Erythrina (*sp.* ?) (Leguminosae). Sir W. MacArthur informed me that in New South Wales the flowers do not set, unless the petals are moved in the same manner as is done by insects.

Lathyrus grandiflorus (Leguminosae). Is in this country more or less sterile. It never sets pods unless the flowers are visited by humble-bees (and this happens only rarely), or unless they are artificially fertilized: see my article in *Gardeners' Chronicle*, 1858, p. 828.

Sarothamnus scoparius (Leguminosae). Extremely sterile when the flowers are neither visited by bees, nor disturbed by being beaten by the wind against the surrounding net.

Melilotus officinalis (Leguminosaw). An unprotected plant visited by bees produced at least thirty times more seeds than a protected one. On this latter plant many scores of racemes did not produce a single pod; several racemes produced each one or two pods; five produced three; six produced four; and one produced six pods. On the unprotected plant each of several racemes produced fifteen pods; nine produced between sixteen and twenty-two pods, and one produced thirty pods. /

Lotus corniculatus (Leguminosae). Several covered-up plants produced only two empty pods, and not a single good seed.

Trifolium repens (Leguminosae). Several plants were protected from insects, and the seeds from ten flower-heads on these plants, and from ten heads on other plants growing outside the net (which I saw visited by bees), were counted; and the seeds from the latter plants were very nearly ten times as numerous as those from the protected plants. The experiment was repeated on the following year; and twenty protected heads now yielded only a single aborted seed, whilst twenty heads on the plants outside the net (which I saw visited by bees) yielded 2,290 seeds, as calculated by weighing all the seed, and counting the number in a weight of two grains.

T. pratense. One hundred flower-heads on plants protected by a net did not produce a single seed, whilst 100 heads on plants growing outside, which were visited by bees, yielded 68 grains weight of seeds; and as eighty seeds weighed two grains, the 100 heads must have yielded 2,720 seeds. I have often watched this plant, and have never seen hive-bees sucking the flowers, except from the outside through holes bitten by humble-bees, or deep down between the flowers, as if in search of some secretion from the calyx, almost in the same manner as described by Mr Farrer, in the case of Coronilla (*Nature*, 2 July, 1874, p. 169). I must, however, except one occasion, when an adjoining field of sainfoin (*Hedysarum onobrychis*) had just been cut down, and when the bees seemed driven to desperation. On this occasion most of the flowers of the clover were somewhat withered, and contained an extraordinary quantity of nectar, which the bees were able to suck. An experienced apiarian, Mr Miner, says that in the United States hive-bees never suck the red clover; and Mr R. Colgate informs me that he has observed the same fact in New Zealand after the introduction of the hive-bee into that island. On the other hand, H. Müller (*Befruchtung*, p. 224) has often seen hive-bees visiting this plant in Germany, for the sake both of pollen and nectar, which latter they obtained by breaking apart the petals. It is at least certain that humble-bees are the chief fertilizers of the common red clover.

T. incarnatum. The flower-heads containing ripe seeds, on some covered and uncovered plants, appeared equally fine, but / this was a false appearance; 60 heads on the latter yielded 349 grains weight of seeds, whereas 60 on the covered-up plants yielded only 63 grains, and many of the seeds in the latter lot were poor and aborted. Therefore the flowers which were visited by bees produced between five and six times as many seeds as those which were protected. The covered-up plants not having been much exhausted by seed-bearing, bore a second considerable crop of flower-stems, whilst the exposed plants did not do so.

Cytisus laburnum (Leguminosae). Seven flower-racemes ready to expand were enclosed in a large bag made of net, and they did not seem in the least injured by this treatment. Only three of them produced any pods, each a single one; and these three pods contained one, four, and five seeds. So that only a single pod from the seven racemes included a fair complement of seeds.

Cuphea purpurea (Lythraceae). Produced no seeds. Other flowers on the same plant artificially fertilized under the net yielded seeds.

Vinca major (Apocynaceae). Is generally quite sterile, but sometimes sets seeds when artificially cross-fertilized: see my notice, *Gardeners' Chronicle*, 1861, p. 552.

V. rosea. Behaves in the same manner as the last species: *Gardeners' Chronicle*, 1861, pp. 699, 736, 831.

Tabernaemontana echinata (Apocynaceae). Quite sterile.

Petunia violacea (Solanaceae). Quite sterile, as far as I have observed.

Solanum tuberosum (Solanaceae). Tinzmann says (*Gardeners' Chronicle*, 1846, p. 183) that some varieties are quite sterile unless fertilized by pollen from another variety.

Primula scotica (Primulaceae). A non-dimorphic species, which is fertile with its own pollen, but is extremely sterile if insects are excluded. J. Scott in *Journal Linn. Soc. Bot.*, vol. viii, 1864, p. 119.

Cortusa matthioli (Primulaceae). Protected plants completely sterile; artificially self-fertilized flowers perfectly fertile. J. Scott, ibid., p. 84.

Cyclamen persicum (Primulaceae). During one season several covered-up plants did not produce a single seed.

Borago officinalis (Boraginaceae). Protected plants produced about half as many seeds as the unprotected.

Salvia tenori (Labiate). Quite sterile; but two or three flowers / on the summits of the three of the spikes, which touched the net when the wind blew, produced a few seeds. This sterility was not due to the injurious effects of the net, for I fertilized five flowers with pollen from an adjoining plant, and these all yielded fine seeds. I removed the net, whilst one little branch still bore a few not completely faded flowers, and these were visited by bees and yielded seeds.

S. coccinea. Some covered-up plants produced a good many fruits, but not, I think, half as many as did the uncovered plants; twenty-eight of the fruits spontaneously produced by the protected plant contained on an average only 1·45 seeds, whilst some artificially self-fertilized fruits on the same plant contained more than twice as many, viz., 3·3 seeds.

Bignonia (unnamed species) (Bignonianceae). Quite sterile: see my account of self-sterile plants.

Digitalis purpurea (Scrophulariaceae). Extremely sterile, only a few poor capsules being produced.

Linaria vulgaris (Scrophulariaceae). Extremely sterile.

Antirrhinum majus, red var. (Scrophulariaceae). Fifty pods gathered from a large plant under a net contained 9·8 grains weight of seeds; but many (unfortunately not counted) of the fifty pods contained no seeds. Fifty pods on a plant fully exposed to the visits of humble-bees contained 23·1 grains weight of seed, that is, more than twice the weight; but in this case again, several of the fifty pods contained no seeds.

A. majus (white var., with a pink mouth to the corolla). Fifty pods, of which only a very few were empty, on a covered-up plant contained 20 grains weight of seed; so that this variety seems to be much more self-fertile than the previous one. With Dr W. Ogle (*Pop. Science Review*, January, 1870, p. 52) a plant of this species was much more sterile when protected from insects than with me, for it produced only two small capsules. As showing the efficiency of bees, I may add that Mr Crocker castrated some young flowers and left them uncovered; and these produced as many seeds as the unmutilated flowers.

A. majus (peloric var.). This variety is quite fertile when artificially fertilized with its own pollen, but is utterly sterile when left to itself and uncovered, as humble-bees cannot crawl into the narrow tubular flowers. /

Verbascum phoeniceum (Scrophulariaceae). Quite sterile⎱ See my account of
V. nigrum. Quite sterile   ⎰   self-sterile plants.

Campanula carpathica (Lobeliaceae). Quite sterile.

Lobelia ramosa (Lobeliaceae). Quite sterile.

L. fulgens. This plant is never visited in my garden by bees, and is quite sterile; but in a nursery-garden at a few miles distance I saw humble-bees visiting the flowers, and they produced some capsules.

Isotoma (a white-flowered var.) (Lobeliaceae). Five plants left unprotected in my greenhouse produced twenty-four fine capsules, containing altogether 12·2 grains weight of seed, and thirteen other very poor capsules, which were rejected. Five plants protected from insects, but otherwise exposed to the same conditions as the above plants, produced sixteen fine capsules, and twenty other very poor and rejected ones. The sixteen fine capsules contained seeds by weight in such proportion that twenty-four would have yielded 4·66 grains. So that the unprotected plants produced nearly thrice as many seeds by weight as the protected plants.

Leschenaultia formosa (Goodeniaceae). Quite sterile. My experiments on this plant, showing the necessity of insect aid, are given in the *Gardeners' Chronicle*, 1871, p. 1166.

Senecio cruentus (Compositae). Quite sterile: see my account of self-sterile plants.

Heterocentron mexicanum (Melastomaceae). Quite sterile; but this species and the following members of the group produce plenty of seed when artificially self-fertilized.

Rhexia glandulosa (Melastomaceae). Set spontaneously only two or three capsules.

Centradenia floribunda (Melastomaceae). During some years produced spontaneously two or three capsules, sometimes none.

Pleroma (unnamed species from Kew) (Melastomaceae). During some years produced spontaneously two or three capsules, sometimes none.

Monochaetum ensiferum (Melastomaceae). During some years produced spontaneously two or three capsules, sometimes none.

Hecychium (unnamed species) (Marantaceae). Almost self-sterile without aid.

Orchideae. An immense proportion of the species sterile, if insects are excluded. /

List of plants, which when protected from insects are either quite fertile, or yield more than half the number of seeds produced by unprotected plants

Passiflora gracilis (Passifloraceae). Produces many fruits, but these contain fewer seeds than fruits from intercrossed flowers.

Brassica oleracea (Cruciferae). Produces many capsules, but these generally not so rich in seed as those on uncovered plants.

Raphanus sativus (Cruciferae). Half of a large branching plant was covered by a net, and was as thickly covered with capsules as the other and unprotected half; but twenty of the capsules on the latter contained on an average 3·5 seeds, whilst twenty of the protected capsules contained only 1·85 seeds, that is, only a little more than half the number. This plant might perhaps have been more properly included in the former list.

Iberis umbellata (Cruciferae). Highly fertile.

I. amara. Highly fertile.

Reseda odorata and *lutea* (Resedaceae). Certain individuals completely self-fertile.

Euryale ferox (Nymphaeaceae). Professor Caspary informs me that this plant is highly self-fertile when insects are excluded. He remarks in the paper before referred to, that his plants (as well as those of the *Victoria regia*) produce only one flower at a time; and that as this species is an annual, and was introduced in 1809, it must have been self-fertilized for the last fifty-six generations; but Dr Hooker assures me that to his knowledge it has been repeatedly introduced, and that at Kew the same plant both of the Euryale and of the Victoria produce several flowers at the same time.

Nymphaea (Nymphaeaceae). Some species, as I am informed by Professor Caspary, are quite self-fertile when insects are excluded.

Adonis aestivalis (Ranunculaceae). Produces, according to Professor H. Hoffmann (*Speciesfrage*, p. 11), plenty of seeds when protected from insects.

Ranunculus acris (Ranunculaceae). Produces plenty of seeds under a net.

Papaver somniferum (Papaveraceae). Thirty capsules from uncovered plants yielded 15·6 grains weight of seed, and thirty capsules from covered-up plants, growing in the same bed, / yielded 16·5 grains weight; so that the latter plants were more productive than the uncovered. Professor H. Hoffmann (*Speciesfrage*, 1875, p. 53) also found this species self-fertile when protected from insects.

P. vagum. Produced late in the summer plenty of seeds, which germinated well.

P. argemonoides } According to Hildebrand (*Jahrbuch für w.*
Glaucium luteum (Papaveraceae) } *Bot.*, vol. vii, p. 466), spontaneously self-
Argemone ochroleuca (Papaveraceae) } fertilized flowers are by no means sterile.

Adlumia cirrhosa (Fumariaceae). Sets an abundance of capsules.

Hypecoum procumbens (Fumariaceae). Hildebrand says (idem), with respect to protected flowers, that 'eine gute Fruchtbildung eintrete'.

Fumaria officinalis (Fumariaceae). Covered-up and unprotected plants apparently produced an equal number of capsules, and the seeds of the former seemed to the eye equally good. I have often watched this plant, and so has Hildebrand, and we have never seen an insect visit the flowers. H. Müller has likewise been struck with the rarity of the visits of insects to it, though he has sometimes seen hive-bees at work. The flowers may perhaps be visited by small moths, as is probably the case with the following species.

F. capreolata. Several large beds of this plant growing wild were watched by me during many days, but the flowers were never visited by any insects, though a humble-bee was once seen closely to inspect them. Nevertheless, as the nectary contains much nectar, especially in the evening, I felt convinced that they were visited, probably by moths. The petals do not naturally separate or open in the least; but they had been opened by some means in a certain proportion of the flowers, in the same manner as follows when a thick bristle is pushed into the nectary; so that in this respect they resemble the flowers of *Corydalis lutea.* Thirty-four heads, each including many flowers, were examined, and twenty of them had from one to four flowers, whilst fourteen had not a single flower thus opened. It is therefore clear that some of the flowers had been visited by insects, while the majority had not; yet almost all produced capsules.

Linum usitatissimum (Linaceae). Appears to be quite fertile. H. Hoffmann, *Bot. Zeitung*, 1876, p. 566.

Impatiens barbigera (Balsaminaceae). The flowers, though excellently / adapted for cross-fertilization by the bees which freely visit them, set abundantly under a net.

I. noli-me-tangere (Balsaminaceae). This species produces cleistogamic and perfect flowers. A plant was covered with a net, and some perfect flowers, marked with threads, produced eleven spontaneously self-fertilized capsules, which contained on an average 3·45 seeds. I neglected to ascertain the number of seeds produced by perfect flowers exposed to the visits of insects, but I believe it is not greatly in excess of the above average. Mr A. W. Bennett has carefully described the structure of the flowers of *I. fulva* in *Journal Linn. Soc.*, vol. xiii, *Bot.*, 1872, p. 147. This latter species is said to be sterile with its own pollen (*Gard. Chronicle*, 1868, p. 1286), and if so, it presents a remarkable contrast with *I. barbigerum* and *noli-me-tangere*.

Limnanthes douglasii (Geraniaceae). Highly fertile.

Viscaria oculata (Caryophyllaceae). Produces plenty of capsules with good seeds.

Stellaria media (Caryophyllaceae). Covered-up and uncovered plants produced an equal number of capsules, and the seeds in both appeared equally numerous and good.

Beta vulgaris (Chenopodiaceae). Highly self-fertile.

Vicia sativa (Leguminosae). Protected and unprotected plants produced an equal number of pods and equally fine seeds. If there was any difference between the two lots, the covered-up plants were the most productive.

V. hirsuta. This species bears the smallest flowers of any British leguminous plant. The result of covering up plants was exactly the same as in the last species.

Pisum sativum (Leguminosae). Fully fertile.

Lathyrus odoratus (Leguminosae). Fully fertile.

L. nissolia. Fully fertile.

Lupinus luteus (Leguminosae). Fairly productive.

L. pilosus. Produced plenty of pods.

Ononis minutissima (Leguminosae). Twelve perfect flowers on a plant under a net were marked by threads, and produced eight pods, containing on an average 2·38 seeds. Pods produced by flowers visited by insects would probably have contained on an average 3·66 seeds, judging from the effects of artificial cross-fertilization.

Phaseolus vulgaris (Leguminosae). Quite fertile.

Trifolium arvense (Leguminosae). The excessively small flowers / are incessantly visited by hive and humble-bees. When insects were excluded the flower-heads seemed to produce as many and as fine seeds as the exposed heads.

T. procumbens. On one occasion covered-up plants seemed to yield as many seeds as the uncovered. On a second occasion sixty uncovered flower-heads yielded 9·1 grains weight of seeds, whilst sixty heads on protected plants yielded no less than 17·7 grains; so that these latter plants were much more productive; but this result I suppose was accidental. I have often watched this plant, and have never seen the flowers visited by insects; but I suspect that the flowers of this species, and more especially of *Trifolium minus*, are frequented by small nocturnal moths which, as I hear from Mr Bond, haunt the smaller clovers.

Medicago lupulina (Leguminosae). On account of the danger of losing the seeds, I was forced to gather the pods before they were quite ripe; 150 flower-heads on plants visited by bees yielded pods weighing 101 grains; whilst 150 heads on protected plants yielded pods weighing 77 grains. The inequality would probably have been greater if the mature seeds could have been all safely collected and compared. Ig. Urban (Keimung, Bluthen, etc., bei Medicago, 1873) has described the means of fertilization in this genus, as has the Rev. G. Henslow in the *Journal of Linn. Soc. Bot.*, vol. ix, 1866, pp. 327 and 355.

Nicotiana tabacum (Solanaceae). Fully self-fertile.

Ipomoea purpurea (Convolvulaceae). Highly self-fertile.

Leptosiphon androsaceus (Polemoniaceae). Plants under a net produced a good many capsules.

Primula mollis (Primulaceae). A homomorphic species, self-fertile: J. Scott, in *Journal Linn. Soc. Bot.*, vol. viii, 1864, p. 120.

Nolana prostrata (Nolanaceae). Plants covered up in the greenhouse, yielded seeds by weight compared with uncovered plants, the flowers of which were visited by many bees, in the ratio of 100 to 61.

Ajuga reptans (Labiatae). Set a good many seeds; but none of the stems under a net produced so many as several uncovered stems growing closely by.

Euphrasia officinalis (Scrophulariaceae). Covered-up plants produced plenty of seed; whether less than the exposed plants I cannot say. I saw two small Dipterous insects (*Dolichopos nigripennis* and *Empis chioptera*) repeatedly sucking the / flowers; as they crawled into them, they rubbed against the bristles which project from the anthers, and became dusted with pollen.

Veronica agrestis (Scrophulariaceae). Covered-up plants produced an abundance of seeds. I do not know whether any insects visit the flowers; but I have observed Syrphidae repeatedly covered with pollen visiting the flowers of *V. hederaefolia* and *chamaedrys*.

Mimulus luteus (Scrophulariaceae). Highly self-fertile.

Calceolaria (greenhouse variety) (Scrophulariaceae). Highly self-fertile.

Verbascum thapsus (Scrophulariaceae). Highly self-fertile.

V. lychnitis. Highly self-fertile.

Vandellia nummularifolia (Scrophulariaceae). Perfect flowers produce a good many capsules.

Bartsia odontites (Scrophulariaceae). Covered-up plants produced a good many seeds; but several of these were shrivelled, nor were they so numerous as those produced by unprotected plants, which were incessantly visited by hive and humble-bees.

Specularia speculum (Lobeliaceae). Covered plants produced almost as many capsules as the uncovered.

Lactuca sativa (Compositae). Covered plants produced some seeds, but the summer was wet and unfavourable.

Galium aparine (Rubiaceae). Covered plants produced quite as·many seeds as the uncovered.

Apium petroselinum (Umbelliferae). Covered plants apparently were as productive as the uncovered.

Zea mays (Gramineae). A single plant in the greenhouse produced a good many grains.

Canna warscewiczi (Marantaceae). Highly self-fertile.

Orchidaeceae. In Europe *Ophrys apifera* is as regularly self-fertilized as is any cleistogamic flower. In the United States, South Africa, and Australia there are a few species which are perfectly self-fertile. These several cases are given in the 2nd edit. of my work on the *Fertilisation of Orchids.*

Allium cepa (blood red var.) (Liliaceae). Four flower-heads were covered with a net, and they produced somewhat fewer and smaller capsules than those on the uncovered heads. The capsules were counted on one uncovered head, and were 289 in number; whilst those on a fine head from under the net were only 199. /

Each of these lists contains by a mere accident the same number of genera, viz., forty-nine.[1] The genera in the first list include sixty-five species, and those in the second sixty; the Orchideae in both being excluded. If the genera in this latter order, as well as in the Asclepiadae and Apocynaceae, had been included, the number of species which are sterile if insects are excluded would have been greatly increased; but the lists are confined to species which were actually experimented on. The results can be considered as only approximately accurate, for fertility is so variable a character, that each species ought to have been tried many times. The above number of species, namely, 125, as as nothing to the host of living plants; but the mere fact of more than half of them being sterile within the specified degree, when insects are excluded, is a striking one; for whenever pollen has to be carried from the anthers to the stigma in order to ensure full fertility, there is at least a good chance of cross-fertilization. I do not, however, believe that if all known plants were tried in the same manner, half would be found to be sterile within the specified limits; for / many flowers were selected for experiment which presented some remarkable structure; and such flowers often require insect-aid. Thus out of the forty-nine genera in the first list, about thirty-two have flowers which are asymmetrical or present some remarkable peculiarity; whilst in the second list, including species which are fully or moderately fertile when insects were excluded, only about

---

[1] The plants in these two lists are entomophilous, or adapted for fertilization by insects, with the exception of Zea and Beta, which are anemophilous or fertilized by the wind. I may therefore here repeat that, according to Rimpan (*Landwirth. Jarbuch*, vol. vi, 1877, pp. 192–233, and p. 1073), Rye is sterile if the access of pollen from other plants is prevented; whereas wheat and barley are quite fertile under these conditions. Rimpan states (p. 199) that the different varieties of wheat behave differently with respect to self and cross-fertilization. He removed at an early age all the anthers from the florets of one variety of wheat, which nevertheless produced a considerable number of grains, being fertilized by the surrounding plants. I state this fact, because Mr A. S. Wilson concludes from his excellent experiments (*Gardeners' Chronicle*, 21 March, 1874, p. 375) that wheat is invariably self-fertilized, and no doubt it is so generally. Mr Wilson believes that all the pollen shed by the exserted anthers is absolutely useless. This is a conclusion which it would require very rigid proof to make me to admit.

twenty-one out of the forty-nine are asymmetrical or present any remarkable peculiarity.

Means of cross-fertilization. The most important of all the means by
which pollen is carried from the anthers to the stigma of the same
flower, or from flower to flower, are insects, belonging to the orders of
Hymenoptera, Lepidoptera, and Diptera; and in some parts of the
world, birds.[2] Next in importance, but / in a quite subordinate degree,
is the wind; and with some aquatic plants, according to Delpino,[3]
currents of water. The simple fact of the necessity in many cases of
extraneous aid for the transport of the pollen, and the many contrivances for this purpose, render it highly probable that some great
benefit is thus gained; and this conclusion has now been firmly
established by the proved superiority in growth, vigour, and fertility of
plants of crossed parentage over those of self-fertilized parentage. But
we should always keep in mind that two somewhat opposed ends have
to be gained; the first and more important one being the production of
seeds by any means, and the second, cross-fertilization.

The advantages derived from cross-fertilization throw a flood of
light on most of the chief characters of flowers. We can thus understand their large size and bright colours, and in some cases the
bright tints of the adjoining parts, such as the peduncles, bracteae,

[2] I will here give all the cases known to me of birds fertilizing flowers. In South
Brazil, humming-birds certainly fertilize various plants which are sterile without
their aid: (Fritz Müller, *Bot. Zeit.*, 1870, pp. 274–5, and *Jen. Zeit. f. Naturwiss.*, vol. vii,
1872, 24). Long-beaked humming-birds visit the flowers of Brugmansia, whilst
some of the short-beaked species often penetrate its large corolla in order to obtain
the nectar in an illegitimate manner, in the same manner as do bees in all parts of
the world. It appears, indeed, that the beaks of humming-birds are specially
adapted to the various kinds of flowers which they visit: on the Cordillera they suck
the Salviae, and lacerate the flowers of the Tacsoniae; in Nicaragua, Mr Belt saw
them sucking the flowers of Marcgravia and Erythrina, and thus they carried pollen
from flower to flower. In North America they are said to frequent the flowers of
Impatiens: (Gould, *Introduction to the Trochilidae*, 1861, pp. 15, 120; *Gard. Chronicle*,
1869, p. 389; *The Naturalist in Nicaragua*, p. 129; *Journal of Linn. Soc. Bot.*, vol. xiii,
1872, p. 151). I may add that I often saw in Chile a Mimus with its head yellow with
pollen from, as I believe, a Cassia. I have been assured that at the Cape of Good
Hope, Strelitzia is fertilized by the Nectarinidae. There can hardly be a doubt that
many Australian flowers are fertilized by the many honey-sucking birds of that
country. Mr Wallace remarks (Address to the Biological Section, Brit. Assoc. 1876)
that he has often observed the beaks and faces of the brush-tongued lories of the
Moluccas covered with pollen'. In New Zealand many specimens of the *Anthornis
melanura* had their heads coloured with pollen from the flowers of an endemic
species of Fuchsia: (Potts, *Transact. New Zealand Institute*, vol. iii, 1870, p. 72).
[3] See also Dr Ascherson's interesting essay in *Bot. Zeitung*, 1871, p. 444.

even true leaves, as with Poinsettia, etc. By this means they are rendered conspicuous to insects, on the same principle that almost every fruit which is devoured by birds presents a strong contrast in colour with the green foliage, in order that it may be seen and its seeds freely disseminated. With some flowers conspicuousness is gained at the expense even of the reproductive organs, as with the ray-florets of many Compositae, the exterior flowers of Hydrangea, and the terminal flowers of the Feather-hyacinth or Muscari. There is also reason to believe, and this was the opinion of Sprengel, that flowers differ in colour in accordance with the kinds of insects which frequent them. /

Not only do the bright colours of flowers serve to attract insects, but dark-coloured streaks and marks are often present, which Sprengel long ago maintained served as guides to the nectary. These marks follow the veins in the petals, or lie between them. They may occur on only one, or on all excepting one or more of the upper or lower petals; or they may form a dark ring round the tubular part of the corolla, or be confined to the lips of an irregular flower. In the white varieties of many flowers, such as of *Digitalis purpurea, Antirrhinum majus*, several species of Dianthus, Phlox, Myosotis, Rhododendron, Pelargonium, Primula, and Petunia, the marks generally persist, whilst the rest of the corolla has become of a pure white; but this may be due merely to their colour being more intense and thus less readily obliterated. Sprengel's notion of the use of these marks as guides appeared to me for a long time fanciful; for insects, without such aid, readily discover the nectary and bite holes through it from the outside. They also discover the minute nectar-secreting glands on the stipules and leaves of certain plants. Moreover, some few plants, such as certain poppies, which are not nectariferous, have guiding marks; but we might perhaps expect that some few plants would retain traces of a former nectariferous condition. On the other hand, these marks are much more common on asymmetrical flowers, the entrance into which would be apt to puzzle insects, than on regular flowers. Sir J. Lubbock has also proved that bees readily distinguish colours, and that they lose much time if the position of honey which they have once visited be in the least changed.[4] The following case affords, I think, the best evidence / that these marks have really been developed in correlation with the nectary. The two upper petals of the common Pelargonium are thus

[4] *British Wild Flowers in relation to Insects*, 1875, p. 44.

marked near their bases; and I have repeatedly observed that when the flowers vary so as to become peloric or regular, they lose their nectaries and at the same time the dark marks. When the nectary is only partially aborted, only one of the upper petals loses its mark. Therefore the nectary and these marks clearly stand in some sort of close relation to one another; and the simplest view is that they were developed together for a special purpose; the only conceivable one being that the marks serve as a guide to the nectary. It is, however, evident from what has been already said, that insects could discover the nectar without the aid of guiding marks. They are of service to the plant, only by aiding insects to visit and suck a great number of flowers within a given time than would otherwise be possible; and thus there will be a better chance of fertilization by pollen brought from a distinct plant, and this we know is of paramount importance.

The odours emitted by flowers attract insects, as I have observed in the case of plants covered by a muslin net. Nägeli affixed artificial flowers to branches, scenting some with essential oils and leaving others unscented; and insects were attracted to the former in an unmistakable manner.[5] It would appear that they must be guided by the simultaneous action of sight and smell, for M. Plateau[6] found that excellently made, but not scented, artificial flowers never deceived them. It will be shown in the next chapter that the flowers of certain plants remain fully expanded for / days or weeks and do not attract any insects; and it is probable that they are neglected from not having as yet secreted any nectar or become odoriferous. Nature may be said occasionally to try on a large scale the same experiment as that by M. Plateau. Not a few flowers are both conspicuous and odoriferous. Of all colours, white is the prevailing one; and of white flowers a considerably larger proportion smell sweetly than of any other colour, namely, 14·6 per cent; of red, only 8·2 per cent are odoriferous.[7] The fact of a larger proportion of white flowers smelling sweetly may depend in part on those which are fertilized by moths requiring the double aid of conspicuousness in the dusk and of odour. Most flowers which are fertilized by crepuscular or nocturnal insects emit their odour chiefly or exclusively in the evening, and they are thus less likely to be

[5] *Enstehung, etc., der Naturhist. Art.*, 1865, p. 23.

[6] *Proceedings of the French Assoc. for the Advancement of Science*, 1876.

[7] The colours and odours of the flowers of 4,200 species have been tabulated by Landgrabe, and by Schübler and Köhler. I have not seen their original works, but a very full abstract is given in Loudon's *Gardeners' Mag.*, vol. xiii, 1837, p. 367.

visited and have their nectar stolen by ill-adapted diurnal insects. Some flowers, however, which are highly odoriferous depend solely on this quality for their fertilization, such as the night-flowering stock (Hesperis) and some species of Daphne; and these present the rare case of flowers which are fertilized by insects being obscurely coloured.

The storage of a supply of nectar in a protected place is manifestly connected with the visits of insects. So is the position which the stamens and pistils occupy, either permanently or at the proper period through their own movements; for when mature they invariably stand in the pathway leading to the nectary. The shape of the nectary and of the adjoining parts are likewise related to the particular kinds of insects which / habitually visit the flowers; this has been well shown by H. Müller by his comparison of lowland species which are chiefly visited by bees, with alpine species belonging to the same genera which are visited by butterflies.[8] Flowers may also be adapted to certain kinds of insects, by secreting nectar particularly attractive to them, and unattractive to other kinds; of which fact *Epipactis latifolia* offers the most striking instance known to me, as it is visited exclusively by wasps. Structures also exist, such as the hairs within the corolla of the foxglove (Digitalis), which apparently serve to exclude insects that are not well fitted to bring pollen from one flower to another.[9] I need say nothing here of the endless contrivances, such as the viscid glands attached to the pollen-masses of the Orchideae and Asclepiadae, or the viscid or roughened state of the pollen-grains of many plants, or the irritability of their stamens which move when touched by insects, etc. – as all these contrivances evidently favour or ensure cross-fertilization.

All ordinary flowers are so far open that insects can force an entrance into them, notwithstanding that some, like the Snapdragon (Antirrhinum), various Papilionaceous and Fumariaceous flowers, are in appearance closed. It cannot be maintained that their openness is necessary for fertility, as cleistogamic flowers which are permanently closed yield a full complement of seeds. Pollen contains much nitrogen and phosphorus / – the two most precious of all the elements for the growth of plants – but in the case of most open flowers, a large

[8] *Nature*, 1874, p. 110; 1875, p. 190; 1876, pp. 210, 289.

[9] Belt, *The Naturalist in Nicaragua*, 1874, p. 132. Kerner has shown in his admirable essay, 'Die Schutzmittel der Blüthen gegen unberufene Gäste, 1826', that many structures – hairs, viscid glands, the position of the parts, etc. – protect the flowers from the access of crawling or wingless insects, which would steal the nectar, and yet, as they do not commonly carry pollen from one plant to another, but only from flower to flower on the same plant, would confer no benefit to the species.

quantity of pollen is consumed by pollen-devouring insects, and a large quantity is destroyed during long-continued rain. With many plants this latter evil is guarded against, as far as is possible, by the anthers opening only during dry weather[10] – by the position and form of some or all of the petals – by the presence of hairs, etc., and as Kerner has shown in his interesting essay,[11] by the movements of the petals or of the whole flower during cold and wet weather. In order to compensate the loss of pollen in so many ways, the anthers produce a far larger amount than is necessary for the fertilization of the same flower. I know this from my own experiments on Ipomoea, given in the Introduction; and it is still more plainly shown by the astonishingly small quantity produced by cleistogamic flowers, which lose none of their pollen, in comparison with that produced by the open flowers borne by the same plants; and yet this small quantity suffices for the fertilization of all their numerous seeds. Mr Hassall took pains in estimating the number of pollen-grains produced by a flower of the Dandelion (Leontodon), and found the number to be 243,600, and in a Paeony 3,654,000.[12] A single plant of Typha produced 144 grains by weight of pollen, and as this plant is anemophilous with very small pollen-grains, / their number in the above weight must have been prodigious. We may judge of this from the following facts: Dr Blackley ascertained[13] by an ingenious method, that in the three following anemophilous plants, a single grain-weight of the pollen of *Lolium perenne* contained 6,032,000 grains; the same weight of the pollen of *Plantago lanceolata* contained 10,124,000 grains; and that of *Scirpus lacustris*, 27,302,050 grains. Again Mr A. S. Wilson estimated by micromeasurement[14] that a single floret of rye yielded 60,000 pollen-grains, whilst one of spring wheat yielded only 6,864 grains. The editor of the *Botanical Register* counted the ovules in the flowers of *Wistaria sinensis*, and carefully estimated the number of pollen-grains, and he found that for each ovule there were 7,000 grains.[15] With Mirabilis, three or

[10] Mr Blackley observed that the ripe anthers of rye did not dehisce whilst kept under a bell-glass in a damp atmosphere, whilst other anthers exposed to the same temperature in the open air dehisced freely. He also found much more pollen adhering to the sticky slides, which were attached to kites and sent high up in the atmosphere, during the first fine and dry days after wet weather, than at other times: *Experimental Researches on Hay Fever*, 1873, p. 127.

[11] *Die Schutzmittel des Pollens*, 1873.

[12] *Annals and Mag. of Nat. Hist.*, vol. viii, 1842, p. 108.

[13] *New Observations on Hay Fever*, 1877, p. 14.

[14] *Gardeners' Chronicle*, March, 1874, p. 376.

[15] Quoted in *Gard. Chron.*, 1846, p. 771.

four of the very large pollen-grains are sufficient to fertilize an ovule; but I do not know how many grains a flower produces. With Hibiscus, Kölreuter found that sixty grains were necessary to fertilize all the ovules of a flower, and he calculated that 4,863 grains were produced by a single flower, or eighty-one times too many. With *Geum urbanum*, however, according to Gärtner, the pollen is only ten times too much.[16] As we thus see that the open state of all ordinary flowers, and the consequent loss of much pollen, necessitate the development of so prodigious an excess of this precious substance, why, it may be asked, are flowers always left open? As many plants exist throughout the vegetable kingdom which bear cleistogamic flowers, there can hardly be a doubt that all / open flowers might easily have been converted into closed ones. The graduated steps by which this process could have been effected may be seen at the present time in *Lathyrus nissolia*, *Biophytum sensitivum*, and several other plants. The answer to the above question obviously is, that with permanently closed flowers there could be no cross-fertilization.

The frequency, almost regularity, with which pollen is transported by insects from flower to flower, often from a considerable distance, well deserves attention.[17] This is best shown by the impossibility in many cases of raising two varieties of the same species pure, if they grow at all near together; but to this subject I shall presently return; also by the many cases of hybrids which have appeared spontaneously both in gardens and a state of nature. With respect to the distance from which pollen is often brought, no one who has had any experience would expect to obtain pure cabbage-seed, for instance, if a plant of another variety grew within two or three hundred yards. An accurate observer, the late Mr Masters of Canterbury, assured me that he once had his whole stock of seeds 'seriously affected with purple bastards', by some plants of purple kale which flowered in a cottager's

[16] Kölreuter, *Vorläufige Nachricht*, 1761, p. 9. Gärtner, *Beitrage zur Kenntniss*, etc., p. 346.
[17] An experiment made by Kölreuter (*Fortsetzung*, etc. 1763, p. 69) affords good evidence on this head. *Hibiscus vesicarius* is strongly dichogamous, its pollen being shed before the stigmas are mature. Kölreuter marked 310 flowers, and put pollen from other flowers on their stigmas every day, so that they were thoroughly fertilized; and he left the same number of other flowers to the agency of insects. Afterwards he counted the seeds of both lots: the flowers which he had fertilized with such astonishing care produced 11,237 seeds, whilst those left to the insects produced 10,886; that is, a less number by only 351; and this small inferiority is fully accounted for by the insects not having worked during some days, when the weather was cold with continued rain.

garden at the distance of half a mile; no other plant of this variety growing any / nearer.[18] But the most striking case which has been recorded is that by M. Godron,[19] who shows by the nature of the hybrids produced that *Primula grandiflora* must have been crossed with pollen brought by bees from *P. officinalis*, growing at the distance of above two kilometres, or of about one English mile and a quarter.

All those who have long attended to hybridization, insist in the strongest terms on the liability of castrated flowers to be fertilized by pollen brought from distant plants of the same species.[20] The following case shows this in the clearest manner: Gärtner, before he had gained much experience, castrated and fertilized 520 flowers on various species with pollen of other genera or other species, but left them unprotected; for, as he says, he thought it a laughable idea that pollen should be brought from flowers of the same species, none of which grew nearer than between 500 and 600 yards.[21] The result was that 289 of these 520 flowers yielded no seed, or none that germinated; the seed of 29 flowers produced hybrids, such as might have been expected from the nature of the pollen employed; and lastly, the seed of the remaining 202 flowers produced perfectly / pure plants, so that these flowers must have been fertilized by pollen brought by insects from a distance of between 500 and 600 yards.[22] It is of course possible that some of these 202 flowers might have been fertilized by pollen left accidentally in them when they were castrated; but to show how improbable this is, I may add that Gärtner, during the next eighteen years, castrated no less than 8,042 flowers and hybridized

[18] Mr W. C. Marshall caught no less than seven specimens of a moth (*Cucullia umbratica*) with the pollinia of the butterfly-orchis (*Habenaria chlorantha*) sticking to their eyes, and, therefore, in the proper position for fertilizing the flowers of this species, on an island in Derwentwater, at the distance of half a mile from any place where this plant grew: *Nature*, 1872, p. 393.

[19] *Revue des Sc. Nat.*, 1875, p. 331.

[20] See, for instance, the remarks by Herbert, *Amaryllidaceae*, 1837, p. 349. Also Gärtner's strong expressions on this subject in his *Bastarderzeugung*, 1849, p. 670; and *Kenntniss der Befruchtung*, 1844, pp. 510, 573. Also Lecoq, *De la Fécondation*, etc., 1845, p. 27. Some statements have been published during late years of the extraordinary tendency of hybrid plants to revert to their parent forms; but as it is not said how the flowers were protected from insects, it may be suspected that they were often fertilized with pollen brought from a distance from the parent-species.

[21] *Kenntniss der Befruchtung*, pp. 539, 550, 575, 576.

[22] Henschel's experiments (quoted by Gärtner, *Kenntniss*, etc., p. 574), which are worthless in all other respects, likewise show how largely flowers are intercrossed by insects. He castrated many flowers on thirty-seven species, belonging to twenty-two genera, and put on their stigmas either no pollen, or pollen from distinct genera, yet they all seeded, and all the seedlings raised from them were of course pure.

them in a closed room; and the seeds from only seventy of these, that is considerably less than 1 per cent, produced pure or unhybridized offspring.[23]

From the various facts now given, it is evident that most flowers are adapted in an admirable manner for cross-fertilization. Nevertheless, the greater number likewise present structures which are manifestly adapted, though not in so striking a manner, for self-fertilization. The chief of these is their hermaphrodite condition; that is, their including within the same corolla both the male and female reproductive organs. These often stand close together and are mature at the same time; so that pollen from the same flower cannot fail to be deposited at the proper period on the stigma. There are also various details of structure adapted for self-fertilization.[24] Such structures are best shown in those curious cases discovered by H. Müller, in which a species exists under two forms – one bearing conspicuous flowers fitted for cross-fertilization, and the other smaller flowers fitted for self-fertilization, / with many parts in the latter slightly modified for this special purpose.[25]

As two objects in most respects opposed, namely cross-fertilization and self-fertilization, have in many cases to be gained, we can understand the co-existence in so many flowers of structures which appear at first sight unnecessarily complex and of an opposed nature. We can thus understand the great contrast in structure between cleistogamic flowers, which are adapted exclusively for self-fertilization, and ordinary flowers on the same plant, which are adapted so as to allow of at least occasional cross-fertilization.[26] The former are always minute, completely closed, with their petals more or less rudimentary and never brightly coloured; they never secrete nectar, never are odoriferous, have very small anthers which produce only a

---

[23] *Kenntniss*, etc., pp. 555, 576.
[24] H. Müller, *Die Befruchtung*, etc., p. 448.
[25] *Nature*, 1873, pp. 44, 433.
[26] Fritz Müller has discovered in the animal kingdom (*Jenaische Zeitschr.*, vol. iv, p. 451) a case curiously analogous to that of the plants which bear cleistogamic and perfect flowers. He finds in the nests of Termites, in Brazil, males and females with imperfect wings, which do not leave the nests and propagate the species in a cleistogamic manner, but only if a fully-developed queen after swarming does not enter the old nest. The fully-developed males and females are winged, and individuals from distinct nests can hardly fail often to intercross. In the act of swarming they are destroyed in almost infinite numbers by a host of enemies, so that a queen may often fail to enter an old nest; and then the imperfectly developed male and females propagate and keep up the stock.

few grains of pollen, and their stigmas are but little developed. Bearing in mind that some flowers are cross-fertilized by the wind (called anemophilous by Delpino), and others by insects (called ento-mophilous), we can further understand, as was pointed out by me several years ago,[27] the great contrast in appearance between these two classes of flowers. Anemophilous flowers resemble in many respects cleistogamic flowers, but differ widely in not being closed, in pro-ducing an extraordinary / amount of pollen which is always incoherent and in the stigma often being largely developed or plumose. We certainly owe the beauty and odour of our flowers and the storage of a large supply of honey to the existence of insects.

On the relation between the structure and conspicuousness of flowers, the visits of insects, and the advantages of cross-fertilization

It has already been shown that there is no close relation between the number of seeds produced by flowers when crossed and self-fertilized, and the degree to which their offspring are affected by the two processes. I have also given reasons for believing that the inefficiency of a plant's own pollen is in most cases an incidental result, or has not been specially acquired for the sake of preventing self-fertilization. On the other hand, there can hardly be a doubt that dichogamy, which prevails according to Hildebrand[28] in the greater number of species – that the heterostyled condition of certain plants – and that many mechanical structures – have all been acquired so as both to check self-fertilization and to favour cross-fertilization. The means for favouring cross-fertilization must have been acquired before those which prevent self-fertilization; as it would manifestly be injurious to a plant that its stigma should fail to receive its own pollen, unless it had already become well adapted for receiving pollen from another individual. It should be observed that many plants still possess a high power of self-fertilization, although their flowers are excellently. constructed for cross-fertilization – for instance, those of many papilionaceous species. /

It may be admitted as almost certain that some structures, such as a narrow elongated nectary, or a long tubular corolla, have been developed in order that certain kinds of insects alone should obtain the nectar. These insects would thus find a store of nectar preserved from

[27] *Journal of Linn. Soc.*, vol. vii, *Bot.*, 1863, p. 77.
[28] *Die Geschlechter Vertheilung*, etc., p. 32.

the attacks of other insects; and they would thus be led to visit frequently such flowers and to carry pollen from one to the other.[29] It
might perhaps have been expected that plants having their flowers
thus peculiarly constructed would profit in a greater degree by being
crossed, than ordinary or simple flowers; but this does not seem to
hold good. Thus *Tropaeolum minus* has a long nectary and an irregular
corolla, whilst *Limnanthes douglasii* has a regular flower and no proper
nectary, yet the crossed seedlings of both species are to the self-
fertilized in height as 100 to 79. *Salvia coccinea* has an irregular corolla,
with a curious apparatus by which insects depress the stamens, while
the flowers of Ipomoea are regular; and the crossed seedlings of the
former are in height to the self-fertilized as 100 to 76, whilst those of
the Ipomoea are as 100 to 77. Fagopyrum is heterostyled and *Anagallis
collina* is homostyled, and the crossed seedlings of both are in height to
the self-fertilized as 100 to 69.

With all European plants, excepting the comparatively rare anemophilus kinds, the possibility of distinct individuals intercrossing depends on the visits of insects; and H. Müller has proved by his
valuable observations, that large conspicuous flowers are visited much
more frequently and by many more kinds of insects, than are small
inconspicuous flowers. He further remarks that the flowers which are
rarely / visited must be capable of self-fertilization, otherwise they
would quickly become extinct.[30] There is, however, some liability to
error in forming a judgement on this head, from the extreme
difficulty of ascertaining whether flowers which are rarely or never
visited during the day (as in the above given case of *Fumaria capreolata*) are not visited by small nocturnal Lepidoptera, which are so
numerous and are known to be strongly attracted by sugar.[31] The
two lists given in the early part of this chapter support Müller's
conclusion that small and inconspicuous flowers are completely self-
fertile; for only eight or nine out of the 125 species in the two lists
come under this head, and all of these were proved to be highly
fertile when insects were excluded. The singularly inconspicuous
flowers of the Fly Ophrys (*O. muscifera*), as I have elsewhere shown,

[29] See the interesting discussion on this subject by H. Müller, *Die Befruchtung*,
etc., p. 431.
[30] *Befruchtung*, etc., p. 426. *Nature*, 1873, p. 433.
[31] In answer to a question by me, the editor of an entomological journal writes –
'The Depressariae, as is notorious to every collector of Noctuae, come very freely
to sugar, and no doubt naturally visit flowers': the *Entomologists' Weekly Intelligencer*,
1860, p. 103.

are rarely visited by insects; and it is a strange instance of imperfection, in contradiction to the above rule, that these flowers are not self-fertile, so that a large proportion of them do not produce seeds. The converse of the rule that plants bearing small and inconspicuous flowers are self-fertile, namely, that plants with large and conspicuous flowers are self-sterile, is far from true, as may be seen in our second list of spontaneously self-fertile species; for this list includes such species as *Ipomoea purpurea*, *Adonis aestivalis*, *Verbascum thapsus*, *Pisum sativum*, *Lathyrus odoratus*, some species of Papaver and of Nymphaea, and others.

The rarity of the visits of insects to small flowers, / does not depend altogether on their inconspicuousness, but likewise on the absence of some sufficient attraction; for the flowers of *Trifolium arvense* are extremely small, yet are incessantly visited by hive and humble-bees, as are the small and dingy flowers of the asparagus. The flowers of *Linaria cymbalaria* are small and not very conspicuous, yet at the proper time they are freely visited by hive-bees. I may add that, according to Mr Bennett,[32] there is another and quite distinct class of plants which cannot be much frequented by insects, as they flower either exclusively or often during the winter, and these seem adapted for self-fertilization, as they shed their pollen before the flowers expand.

That many flowers have been rendered conspicuous for the sake of guiding insects to them is highly probable or almost certain; but it may be asked, have other flowers been rendered inconspicuous so that they may not be frequently visited, or have they merely retained a former and primitive condition? If a plant were much reduced in size, so probably would be the flowers through correlated growth, and this may possibly account for some cases; but the corolla, as I have else-where shown (*Different Forms of Flowers*, 1877, p. 143), is also liable to be greatly reduced, through the direct action of unfavourable climate. Size and colour are both extremely variable characters, and it can hardly be doubted that if large and brightly-coloured flowers were advantageous to any species, these could be acquired through Natural Selection within a moderate lapse of time. Papilionaceous flowers are manifestly constructed in relation to the visits of insects, and it seems improbable, from the usual character of the / group, that the progenitors of the genera Vicia and Trifolium produced such minute and unattractive flowers as those of *V. hirsuta* and *T. procumbens*. We

[32] *Nature*, 1869, p. 11.

are thus led to infer that some plants either have not had their flowers increased in size, or have actually had them reduced and purposely rendered inconspicuous, so that they are now but little visited by insects. In either case they must also have acquired or retained a high degree of self-fertility.

If it became from any cause advantageous to a species to have its capacity for self-fertilization increased, there is little difficulty in believing that this could readily be effected; for three cases of plants varying in such a manner as to be more fertile with their own pollen than they originally were, occurred in the course of my few experiments, namely, with Mimulus, Ipomoea, and Nicotiana. Nor is there any reason to doubt that many kinds of plants are capable under favourable circumstances of propagating themselves for very many generations by self-fertilization. This is the case with the varieties of *Pisum sativum* and of *Lathyrus odoratus* which are cultivated in England, and with *Ophrys apifera* and some other plants in a state of nature. Nevertheless, most or all of these plants retain structures in an efficient state which cannot be of the least use except for cross-fertilization. We have also seen reason to suspect that self-fertilization is in some peculiar manner beneficial to certain plants; but if this be really the case, the benefit thus derived is far more than counterbalanced by a cross with a fresh stock or with a slightly different variety.

Notwithstanding the several considerations just advanced, it seems to me highly improbable that plants bearing small and inconspicuous flowers have been or should continue to be subjected to self-fertilization / for a long series of generations. I think so, not from the evil which manifestly follows from self-fertilization, in many cases even in the first generation, as with *Viola tricolor*, Sarothamnus, Nemophila, Cyclamen, etc.; nor from the probability of the evil increasing after several generations, for on this latter head I have not sufficient evidence, owing to the manner in which my experiments were conducted. But if plants bearing small and inconspicuous flowers were not occasionally intercrossed, and did not profit by the process, all their flowers would probably have been rendered cleistogamic, as they would thus have largely benefited by having to produce only a small quantity of safely-protected pollen. In coming to this conclusion, I have been guided by the frequency with which plants belonging to distinct orders have been rendered cleistogamic. But I can hear of no instance of a species with all its flowers rendered permanently cleistogamic. Leersia makes the nearest approach to this state; but as already stated, it has been known to produce perfect flowers in one part of Germany. Some other plants of the cleistogamic class, for instance Aspicarpa, have failed to produce perfect flowers during several years in a hothouse; but it does not follow that they would fail to do so in their native country, any more than with a Vandellia and Viola, which with me produced only cleistogamic flowers during certain years.[33] Plants belonging to this class commonly bear both kinds of flowers every season, and the perfect flowers of *Viola canina* yield fine capsules, but only when visited by bees. We have also seen that the seedlings of *Ononis minutissima*, raised from the perfect flowers fertilized with pollen from another plant, were finer / than those from self-fertilized flowers; and this was likewise the case to a certain extent with Vandellia. As therefore no species which at one time bore perfect though small and inconspicuous flowers has had all its flowers rendered cleistogamic, I must believe that plants now bearing small and inconspicuous flowers profit by their still remaining open, so as to be occasionally intercrossed by insects. It has been one of the greatest oversights in my work that I did not experimentize on such flowers, owing to the difficulty of fertilizing them, and to my not having seen the importance of the subject.[34]

It should be remembered that in two of the cases in which highly self-fertile varieties appeared among my experimental plants, namely, with Mimulus and Nicotiana, such varieties were greatly benefited by a cross with a fresh stock or with a slightly different variety; and this likewise was the case with the cultivated varieties of *Pisum sativum* and *Lathyrus odoratus*, which have been long propagated by self-fertilization. Therefore until the contrary is distinctly proved, I must believe that as a general rule small and inconspicuous flowers are occasionally

---

[33] These cases are given in ch. viii, of my *Different Forms of Flowers*.

[34] Some of the species of Solanum would be good ones for such experiments, for they are said by H. Müller (*Befruchtung*, p. 434) to be unattractive to insects from not secreting nectar, not producing much pollen, and not being very conspicuous. Hence probably it is that, according to Verlot (*Production des Variétiés*, 1865, p. 72), the varieties of 'les aubergines et les tomates' (Species of Solanum) do not intercross when they are cultivated near together; but it should be remembered that these are not endemic species. On the other hand, the flowers of the common potato (*S. tuberosum*, though they do not secrete nectar (Kurr, *Bedeutung der Nektarien*, 1833, p. 40), yet cannot be considered as inconspicuous, and they are sometimes visited by Diptera (Müller) and, as I have seen, by humble-bees. Tinzmann (as quoted in *Gardeners' Chronicle*, 1846, p. 183) found that some of the varieties did not bear seed when fertilized with pollen from the same variety, but were fertile with that from another variety.

intercrossed by insects; and that after long-continued self-fertilization, /
if they are crossed with pollen brought from a plant growing under
somewhat different conditions, or descended from one thus growing,
their offspring would profit greatly. It cannot be admitted, under our
present state of knowledge, that self-fertilization continued during
many successive generations is ever the most beneficial method of
reproduction.

*The means which favour or ensure flowers being fertilized with pollen
from a distinct plant*

We have seen in four cases that seedlings raised from a cross between
flowers on the same plant, even on plants appearing distinct from
having been propagated by stolons or cuttings, were not superior to
seedlings from self-fertilized flowers; and in a fifth case (Digitalis)
superior only in a slight degree. Therefore we might expect that with
plants growing in a state of nature a cross between flowers on distinct
individuals, and not merely between the flowers on the same plant,
would generally or often be effected by some means. The fact of bees
and of some Diptera visiting the flowers of the same species as long as
they can, instead of promiscuously visiting various species, favours the
intercrossing of distinct plants. On the other hand, insects usually
search a large number of flowers on the same plant before they fly to
another, and this is opposed to cross-fertilization. The extraordinary
number of flowers which bees are able to search within a very short
space of time, as will be shown in a future chapter, increases the chance
of cross-fertilization; as does the fact that they are not able to perceive
without entering a flower whether other bees have exhausted the
nectar. For instance, H. Müller found[35] that four-fifths of the / flowers
of *Lamium album* which a humble-bee visited had been already exhausted of their nectar. In order that distinct plants should be intercrossed, it is of course indispensable that two or more individuals
should grow near one another; and this is generally the case. Thus A.
de Candolle remarks that in ascending a mountain the individuals of
the same species do not commonly disappear near its upper limit quite
gradually, but rather abruptly. This fact can hardly be explained by
the nature of the conditions, as these graduate away in an insensible
manner, and it probably depends in large part on vigorous seedlings

[35] *Die Befruchtung*, etc., p. 311.

being produced only as high up the mountain as many individuals can subsist together.

With respect to dioecious plants, distinct individuals must always fertilize each other. With monoecious plants, as pollen has to be carried from flower to flower, there will always be a good chance of its being carried from plant to plant. Delpino has also observed[36] the curious fact that certain individuals of the monoecious walnut (*Juglans regia*) are proterandrous, and others proterogynous, and these will reciprocally fertilize each other. So it is with the common nut (*Corylus avellana*),[37] and, what is more surprising, with some few hermaphrodite plants, as observed by H. Müller.[38] These latter plants cannot fail to act on each other like dimorphic or trimorphic heterostyled species, in which the union of two individuals is necessary for full and normal fertility. With ordinary hermaphrodite species, the expansion of only a few flowers at the same time is one of the simplest means for favouring the intercrossing of distinct individuals; but this would / render the plants less conspicuous to insects, unless the flowers were of large size, as in the case of several bulbous plants. Kerner thinks[39] that it is for this object that the Australian *Villarsia parnassifolia* produces daily only a single flower. Mr Cheeseman also remarks,[40] that as certain Orchids in New Zealand which require insect-aid for their fertilization bear only a single flower, distinct plants cannot fail to intercross. So it is with the American species of Drosera,[41] and, as I hear from Professor Caspary, with water-lilies.

Dichogamy, which prevails so extensively throughout the vegetable kingdom, much increases the chance of distinct individuals intercrossing. With proterandrous species, which are far more common than proterogynous, the young flowers are exclusively male in function, and the older ones exclusively female; and as bees habitually alight low down on the spikes of flowers in order to crawl upwards, they get dusted with the pollen from the upper flowers, which they carry to the stigmas of the lower and older flowers on the next spike which they visit. The degree to which distinct plants will thus be intercrossed depends on the number of spikes in full flower at the same time on the

[36] *Ult. Osservazioni*, etc., part ii, fasc. ii, p. 337.
[37] *Nature*, 1875, p. 26.
[38] *Die Befruchtung*, etc., pp. 285, 339.
[39] *Die Schutzmittel*, etc., p. 23.
[40] *Transact. New Zealand Institute*, vol, v, 1873, p. 356.
[41] Asa Gray, in a review of this work, in *American Journal of Science*, vol. xiii, February, 1877, p. 135.

same plant. With proterogynous flowers and with depending racemes, the manner in which insects visit the flowers ought to be reversed in order that distinct plants should be intercrossed. But this whole subject requires further investigation, as the great importance of crosses between distinct individuals, instead of merely between distinct flowers, has hitherto been hardly recognized. /

In some few cases the special movements of certain organs almost ensure pollen being carried from plant to plant. Thus with many orchids, the pollen-masses after becoming attached to the head or proboscis of an insect do not move into the proper position for striking the stigma, until ample time has elapsed for the insect to fly to another plant. With *Spiranthes autumnalis*, the pollen-masses cannot be applied to the stigma until the labellum and rostellum have moved apart, and this movement is very slow.[42] With *Posoqueria fragrans* (one of the Rubiaceae) the same end is gained by the movement of a specially constructed stamen, as described by Fritz Müller.

We now come to a far more general and therefore more important means by which the mutual fertilization of distinct plants is effected, namely, the fertilizing power of pollen from another variety or individual being greater than that of a plant's own pollen. The simplest and best known case of prepotent action in pollen, though it does not bear directly on our present subject, is that of a plant's own pollen over that from a distinct species. If pollen from a distinct species be placed on the stigma of a castrated flower, and then after the interval of several hours, pollen from the same species be placed on the stigma, the effects of the former are wholly obliterated, excepting in some rare cases. If two varieties are treated in the same manner, the result is analogous, though of a directly opposite nature; for pollen from any other variety is often or generally prepotent over that from the same flower. I will give some instances: the pollen of *Mimulus luteus* regularly falls on the stigma of its own flower, for the / plant is highly fertile when insects are excluded. Now several flowers on a remarkably constant whitish variety were fertilized without being castrated with pollen from a yellowish variety; and of the twenty-eight seedlings thus raised, every one bore yellowish flowers, so that the pollen of the yellow variety completely overwhelmed that of the mother-plant. Again, *Iberis umbellata* is spontaneously self-fertile, and I saw an

---

[42] *The Various Contrivances by which British and Foreign Orchids are fertilised*, 1st edit., p. 128. Second edit., 1877, p. 110.

abundance of pollen from their own flowers on the stigmas; neverthe-less, of thirty seedlings raised from non-castrated flowers of a crimson variety crossed with pollen from a pink variety, twenty-bore pink flowers, like those of the male or pollen-bearing parent.

In these two cases flowers were fertilized with pollen from a distinct variety, and this was shown to be prepotent by the character of the offspring. Nearly similar results often follow when two or more self-fertile varieties are allowed to grow near one another and are visited by insects. The common cabbage produces a large number of flowers on the same stalk, and when insects are excluded these set many cap-sules, moderately rich in seeds. I planted a white Kohl-rabi, a purple Kohl-rabi, a Portsmouth broccoli, a Brussels sprout, and a Sugar-loaf cabbage near together and left them uncovered. Seeds collected from each kind were sown in separate beds; and the majority of the seed-lings in all five beds were mongrelized in the most complicated man-ner, some taking more after one variety, and some after another. The effects of the Kohl-rabi were particularly plain in the enlarged stems of many of the seedlings. Altogether 233 plants were raised, of which 155 were mongrelized in the plainest manner, and of the remaining 78 not half were absolutely pure. I repeated the experiment by planting near together two varieties of cabbage with purple-green and white-green / laciniated leaves; and of the 325 seedlings raised from the purple-green variety, 165 had white-green and 160 purple-green leaves. Of the 466 seedlings raised from the white-green variety, 220 had purple-green and 246 white-green leaves. These cases show how largely pollen from a neighbouring variety of the cabbage effaces the action of the plant's own pollen. We should bear in mind that pollen must be carried by the bees from flower to flower on the same large branching stem much more abundantly than from plant to plant; and in the case of plants the flowers of which are in some degree dichogamous, those on the same stem would be of different ages, and would thus be as ready for mutual fertilization as the flowers on distinct plants, were it not for the prepotency of pollen from another variety.[43]

Several varieties of the radish (*Raphanus sativus*), which is moderately self-fertile when insects are excluded, were in flower at the same time

[43] A writer in the *Gardeners' Chronicle* (1855, p. 730) says that he planted a bed of turnips (*Brassica rapa*) and of rape (*B. napus*) close together, and sowed the seeds of the former. The result was that scarcely one seedling was true to its kind, and several closely resembled rape.

in my garden. Seed was collected from one of them, and out of twenty-two seedlings thus raised only twelve were true to their kind.[44]

The onion produces a large number of flowers, all crowded together into a large globular head, each flower having six stamens; so that the stigmas receive plenty of pollen from their own and the adjoining anthers. Consequently the plant is fairly self-fertile when protected from insects. A blood-red, silver, globe and Spanish onion were planted near together; / and seedlings were raised from each kind in four separate beds. In all the beds mongrels of various kinds were numerous, except among the ten seedlings from the blood-red onion, which included only two. Altogether forty-six seedlings were raised, of which thirty-one had been plainly crossed. A similar result is known to follow with the varieties of many other plants, if allowed to flower near together: I refer here only to species which are capable of fertilizing themselves, for if this be not the case, they would of course be liable to be crossed by any other variety growing near. Horticulturists do not commonly distinguish between the effects of variability and intercrossing; but I have collected evidence on the natural crossing of varieties of the tulip, hyacinth, anemone, ranunculus, strawberry, *Leptosiphon androsaceus*, orange, rhododendron and rhubarb, all of which plants I believe to be self-fertile.[45] Much other indirect evidence could be given with respect to the extent to which varieties of the same species spontaneously intercross.

Gardeners who raise seed for sale are compelled by dearly bought experience to take extraordinary precautions against intercrossing. Thus Messrs Sharp / 'have land engaged in the growth of seed in no less than eight parishes'. The mere fact of a vast number of plants belonging to the same variety growing together is a considerable

[44] Duhamel, as quoted by Godron, *De l'Espèce*, vol. i, p. 50, makes an analogous statement with respect to this plant.

[45] With respect to tulips and some other flowers, see Godron, *De l'Espèce*, vol. i, p. 252. For anemones, *Gard. Chron.*, 1859, p. 98. For strawberries, see Herbert in *Transact. of Hort. Soc.*, vol. iv, p. 17. The same observer elsewhere speaks of the spontaneous crossing of rhododendrons. Gallesio makes the same statement with respect to oranges. I have myself known extensive crossing to occur with the common rhubarb. For Leptosiphon, Verlot *Des Variétés*, 1865, p. 20. I have not included in my list the Carnation, Nemophila, or Antirrhinum, the varieties of which are known to cross freely, because these plants are not always self-fertile. I know nothing about the self-fertility of Trollius (Lecoq, *De la Fécondation*, 1862, p. 93), Mahonia, and Crinum, in which genera the species intercross largely. With respect to Mahonia, it is now scarcely possible to procure in this country pure specimens of *M. aquifolium* or *repens*; and the various species of Crinum sent by Herbert (*Amaryllidacae*, p. 32) to Calcutta, crossed there so freely that pure seed could not be saved.

protection, as the chances are strong in favour of plants of the same variety intercrossing; and it is in chief part owing to this circumstance, that certain villages have become famous for pure seed of particular varieties.[46] Only two trials were made by me to ascertain after how long an interval of time, pollen from a distinct variety would obliterate more or less completely the action of a plant's own pollen. The stigmas in two lately expanded flowers on a variety of cabbage, called Ragged Jack, were well covered with pollen from the same plant. After an interval of twenty-three hours, pollen from the Early Barnes Cabbage growing at a distance was placed on both stigmas; and as the plant was left uncovered, pollen from other flowers on the Ragged Jack would certainly have been left by the bees during the next two or three days on the same two stigmas. Under these circumstances it seemed very unlikely that the pollen of the Barnes cabbage would produce any effect; but three out of the fifteen plants raised from the two capsules thus produced were plainly mongrelized: and I have no doubt that the twelve other plants were affected, for they grew much more vigorously than the self-fertilized seedlings from the Ragged Jack planted at the same time and under the same conditions. Secondly, I placed on several stigmas of a long-styled cowslip (*Primula veris*) plenty of pollen from the same plant, and after twenty-four hours added some from a short-styled dark-red Polyanthus, which is a variety of / the cowslip. From the flowers thus treated thirty seedlings were raised, and all these without exception bore reddish flowers; so that the effect of the plant's own pollen, though placed on the stigmas twenty-four hours previously was quite destroyed by that of the red variety. It should, however, be observed that these plants are heterostyled, and that the second union was a legitimate one, whilst the first was illegitimate; but flowers illegitimately fertilized with their own pollen yield a moderately fair supply of seeds.

We have hitherto considered only the prepotent fertilizing power of pollen from a distinct variety over a plant's own pollen – both kinds of pollen being placed on the same stigma. It is a much more remarkable fact that pollen from another individual of the same variety is prepotent over a plant's own pollen, as shown by the superiority of the seedlings raised from a cross of this kind over seedlings from self-fertilized flowers. Thus in Tables A, B, and C, there are at least fifteen

[46] With respect to Messrs Sharp, see *Gardeners' Chronicle*, 1856, p. 823. Lindley's *Theory of Horticulture*, p. 319.

species which are self-fertile when insects are excluded; and this implies that their stigmas must receive their own pollen; nevertheless, most of the seedlings which were raised by fertilizing the non-castrated flowers of these fifteen species with pollen from another plant were greatly superior, in height, weight, and fertility, to the self-fertilized offspring.[47] For instance, with *Ipomoea purpurea* every single inter-crossed plant exceeded in height its self-fertilized opponent until the sixth generation; and so it was with *Mimulus luteus* until the fourth generation. Out of six pairs of crossed and self-fertilized cabbages, every / one of the former was much heavier than the latter. With *Papaver vagum*, out of fifteen pairs, all but two of the crossed plants were taller than their self-fertilized opponents. Of eight pairs of *Lupinus luteus*, all but two of the crossed were taller; of eight pairs of *Beta vulgaris* all but one; and of fifteen pairs of *Zea mays* all but two were taller. Of fifteen pairs of *Limnanthes douglasii*, and of seven pairs of *Lactuca sativa*, every single crossed plant was taller than its self-fertilized opponent. It should also be observed that in these experiments no particular care was taken to cross-fertilize the flowers immediately after their expansion; it is therefore almost certain that in many of these cases some pollen from the same flower will have already fallen on and acted on the stigma.

There can hardly be a doubt that several other species of which the crossed seedlings are more vigorous than the self-fertilized, as shown in Tables A, B, and C, besides the above fifteen, must have received their own pollen and that from another plant at nearly the same time; and if so, the same remarks as those just given are applicable to them. Scarcely any result from my experiments has surprised me so much as this of the prepotency of pollen from a distinct individual over each plant's own pollen, as proved by the greater constitutional vigour of the crossed seedlings. The evidence of prepotency is here deduced from the comparative growth of the two lots of seedlings; but we have similar evidence in many cases from the much greater fertility of the non-castrated flowers on the mother-plant, when these received at the same time their own pollen and that from a distinct plant, in comparison with the flowers which received only their own pollen.

From the various facts now given on the spontaneous intercrossing

---

[47] These fifteen species consist of *Brassica oleracea*, *Reseda odorata* and *lutea*, *Limnants douglasii*, *Papaver vagum*, *Viscaria oculata*, *Beta vulgaris*, *Lupinus luteus*, *Ipomoea purpurea*, *Mimulus luteus*, *Calceolaria*, *Verbascum thapsus*, *Vandellia nummular-ifolia*, *Luctuca sativa*, and *Zea mays*.

of varieties growing near together, and on / the effects of cross-fertilizing flowers which are self-fertile and have not been castrated, we may conclude that pollen brought by insects or by the wind from a distinct plant will generally prevent the action of pollen from the same flower, even though it may have been applied some time before; and thus the intercrossing of plants in a state of nature will be greatly favoured or ensured.

The case of a great tree covered with innumerable hermaphrodite flowers seems at first sight strongly opposed to the belief in the frequency of intercrosses between distinct individuals. The flowers which grow on the opposite sides of such a tree will have been exposed to somewhat different conditions, and a cross between them may perhaps be in some degree beneficial; but it is not probable that it would be nearly so beneficial as a cross between flowers on distinct trees, as we may infer from the inefficiency of pollen taken from plant which have been propagated from the same stock, though growing on separate roots. The number of bees which frequent certain kinds of trees when in full flower is very great, and they may be seen flying from tree to tree more frequently than might have been expected. Nevertheless, if we consider how numerous are the flowers on a great tree, an incomparably larger number must be fertilized by pollen brought from other flowers on the same tree, than from flowers on a distinct tree. But we should bear in mind that with many species only a few flowers on the same peduncle produce a seed; and that these seeds are often the product of only one out of several ovules within the same ovarium. Now we know from the experiments of Herbert and others[48] that if one flower / is fertilized with pollen which is more efficient than that applied to the other flowers on the same peduncle, the latter often drop off; and it is probable that this would occur with many of the self-fertilized flowers on a large tree, if other and adjoining flowers were cross-fertilized. Of the flowers annually produced by a great tree, it is almost certain that a large number would be self-fertilized; and if we assume that the tree produced only 500 flowers, and that this number of seeds were requisite to keep up the stock, so that at least one seedling should hereafter struggle to maturity, then a large proportion of the seedlings would necessarily be derived from self-fertilized seeds. But if the tree annually produced 50,000 flowers, of which the self-fertilized dropped off without yielding seeds, then the cross-fertilized

[48] *Variation under Domestication*, ch. xvii, 2nd edit., vol. ii, p. 120.

flowers might yield seeds in sufficient number to keep up the stock, and most of the seedlings would be vigorous from being the product of a cross between distinct individuals. In this manner the production of a vast number of flowers, besides serving to entice numerous insects and to compensate for the accidental destruction of many flowers by spring-frosts or otherwise, would be a very great advantage to the species; and when we behold our orchard-trees covered with a white sheet of bloom in the spring, we should not falsely accuse nature of wasteful expenditure, though comparatively little fruit is produced in the autumn.

Anemophilous plants

The nature and relations of plants which are fertilized by the wind have been admirably discussed by Delpino[49] and H. Müller; and / I have already made some remarks on the structure of their flowers in contrast with those of entomophilous species. There is good reason to believe that the first plants which appeared on this earth were cryptogamic; and judging from what now occurs, the male fertilizing element must either have possessed the power of spontaneous movement through the water or over damp surfaces, or have been carried by currents of water to the female organs. That some of the most ancient plants, such as ferns, possessed true sexual organs there can hardly be a doubt; and this shows, as Hildebrand remarks,[50] at how early a period the sexes were separated. As soon as plants became phanerogamic and grew on the dry ground, if they were ever to intercross, it would be indispensable that the male fertilizing element should be transported by some means through the air; and the wind is the simplest means of transport. There must also have been a period when winged insects did not exist, and plants would not then have been rendered entomophilous. Even at a somewhat later period the more specialized orders of the Hymenoptera, Lepidoptera, and Diptera, which are now chiefly concerned with the transport of pollen, did not exist. Therefore the earliest terrestrial plants known to us,

---

[49] Delpino, *Ult. Osservazioni sulla Dicogamia*, part ii, fasc. i, 1870; and *Studi sopra un Lignaggio anemofilo*, etc. ,1871. H. Müller, *Die Befruchtung*, etc., pp. 412, 442. Both these authors remark that plants must have been anemophilous before they were entomophilous. H. Müller further discusses in a very interesting manner the steps by which entomophilous flowers became nectariferous and gradually acquired their present structure through successive beneficial changes.

[50] *Die Geschlechter-Vertheilung*, 1867, pp. 84–90.

namely, the Coniferae and Cycadeae, no doubt were anemophilous, like the existing species of these same groups. A vestige of this early state of things is likewise shown by some other groups of plants which are anemophilous, as these on the whole stand lower in the scale than entomophilous species. /

There is no great difficulty in understanding how an anemophilous plant might have been rendered entomophilous. Pollen is a nutritious substance, and would soon have been discovered and devoured by insects; and if any adhered to their bodies it would have been carried from the anthers to the stigma of the same flower, or from one flower to another. One of the chief characteristics of the pollen of anemophilous plants is its incoherence; but pollen in this state can adhere to the hairy bodies of insects, as we see with some Leguminosae, Ericaceae, and Melastomaceae. We have, however, better evidence of the possibility of a transition of the above kind in certain plants being now fertilized partly by the wind and partly by insects. The common rhubarb (*Rheum rhaponticum*) is so far in an intermediate condition, that I have seen many Diptera sucking the flowers, with much pollen adhering to their bodies; and yet the pollen is so incoherent, that clouds of it are emitted if the plant be gently shaken on a sunny day, some of which could hardly fail to fall on the large stigmas of the neighbouring flowers. According to Delpino and H. Müller,[51] some species of Plantago are in a similar intermediate condition.

Although it is probable that pollen was aboriginally the sole attraction to insects, and although many plants now exist whose flowers are frequented exclusively by pollen-devouring insects, yet the great majority secrete nectar as the chief attraction. Many years ago I suggested that primarily the saccharine matter in nectar was excreted[52] as a waste product of chemical changes in the sap; and that when the excretion / happened to occur within the envelopes of a flower, it was utilized for the important object of cross-fertilization, being subsequently much increased in quantity and stored in various ways. This view is rendered probable by the leaves of some trees excreting, under certain climatic conditions, without the aid of special glands, a saccharine fluid, often called honey-dew. This is the case with the leaves of the lime; for although some authors have disputed the fact, a most capable judge, Dr Maxwell Masters, informs me that, after having heard the

[51] *Die Befruchtung*, etc., p. 342.
[52] Nectar was regarded by De Candolle and Dunal as an excretion, as stated by Martinet in *Annal. des Sc. Nat.*, 1872, vol. xiv, p. 211.

discussions on this subject before the Horticultural Society, he feels no doubt on this head. Professor H. Hoffmann has lately (1876) described the case of the leaves of a young camellia secreting profusely, without the possibility of the intervention of aphides. The leaves, as well as the cut stems, of the manna ash (*Fraxinus ornus*) secrete in a like manner saccharine matter.[53] Acording to Treviranus, so do the upper surfaces of the leaves of *Carduus arctioides* during hot weather. Many analogous facts could be given.[54] There are, however, a considerable number of plants which bear small glands[55] on their leaves, petioles, phyllodia, stipules, / bracteae, or flower peduncles, or on the outside of their calyx, and these glands secrete minute drops of a sweet fluid, which is eagerly sought by sugar-loving insects, such as ants, hive-bees, and wasps. In the case of the glands on the stipules of *Vicia sativa*, the excretion manifestly depends on changes in the sap, consequent on the sun shining brightly; for I repeatedly observed that as soon as the sun was hidden behind clouds the secretion ceased, and the hive-bees left the field; but as soon as the sun broke out again, they returned to their feast.[56] I have observed an analogous fact with the secretion of true nectar in the flowers of *Lobelia erinus*.

[53] *Gard. Chron.*, 1876, p. 242.

[54] Kurr, *Untersuchungen über die Bedeutung der Nektarien*, 1833, p. 115.

[55] A large number of cases are given by Delpino in the *Bulletino Entomologico*, Anno vi, 1874. To thse may be added those given in my text, as well as the excretion of saccharine matter from the calyx of two species of Iris, and from the bracteae of certain Orchideae: see Kurr, *Bedeutung der Nektarien*, 1833, pp. 25, 28. Belt also refers (*Nicaragua*, p. 224) to a similar excretion by many epiphytal orchids and passion-flowers. Mr Rodgers has seen much nectar secreted from the bases of the flower-peduncles of Vanilla. Link says that the only example of a hypopetalous nectary known to him is externally at the base of the flowers of *Chironia decussata*: see *Reports on Botany, Ray Society*, 1846, p. 355. An important memoir bearing on this subject has lately appeared by Reinke (*Göttingen Nachrichten*, 1873, p. 825), who shows that in many plants the tips of the serrations / on the leaves in the bud bear glands which secrete only at a very early age, and which have the same morphological structure as true nectar-secreting glands. He further shows that the nectar-secreting glands on the petioles of *Prunus avium* are not developed at a very early age, yet wither away on the old leaves. They are homologous with those on the serrations of the blades of the same leaves, as shown by their structure and by transition-forms; for the lowest serrations on the blades of most of the leaves secrete nectar instead of resin (harz).

[56] I published a brief notice of this case in the *Gard. Chronicle*, 21 July, 1855, p. 487, and afterwards made further observations. Besides the hive-bee, another species of bee, a moth, ants, and two kinds of flies sucked the drops of fluid on the stipules. The larger drops tasted sweet. The hive-bees never even looked at the flowers which were open at the same time; whilst two species of humble-bees neglected the stipules and visited only the flowers.

Delpino, however, maintains that the power of secreting a sweet fluid by any extra-floral organ has been in every case specially gained, for the sake of attracting ants and wasps as defenders of the plant against their enemies; but I have never seen any reason to believe that this is so with the three species observed by me, namely, *Prunus laurocerasus*, *Vicia sativa*, and *V. faba*. No plant is so little attacked by enemies of any kind in this country as the common bracken-fern (*Pteris aquilina*); and yet, as my son / Francis has discovered, the large glands at the bases of the fronds, but only whilst young, excrete much sweetish fluid, which is eagerly sought by innumerable ants, chiefly belonging to Myrmica; and these ants certainly do not here serve as a protection against any enemy. In S. Brazil, however, ants attracted by the secretion to this plant, defend it, according to Fritz Müller,[57] against other leaf-devouring and highly destructive ants; so that, if this fern originated in tropical S. America, the capacity of secretion may have been acquired for this special purpose. Delpino argues that sugar-secreting glands ought never to be considered as merely excretory, because if they were so, they would be present in every species; but I cannot see much force in this argument, as the leaves of some plants excrete sugar only during certain states of the weather. That in some cases the secretion serves to attract insects as defenders of the plant, and may have been developed to a high degree for this special purpose, I have not the least doubt, from the observations of Delpino, and more especially from those of Mr Belt on *Acacia sphaerocephala*, and on passion-flowers. This acacia likewise produces, as an additional attraction to ants, small bodies containing much oil and protoplasm, and analogous bodies are developed by a Cecropia for the same purpose, as described by Fritz Müller.[58]

The excretion of a sweet fluid by glands seated / outside of a flower is rarely utilized as a means for cross-fertilization by the aid of insects; but this is the case with several species of Euphorbia and with the bracteae of the Marcgraviaceae, as the late Dr Crüger informed me from actual observation in the West Indies, and as Delpino inferred with much acuteness from the relative position of the several parts of

[57] See a letter in *Nature*, June, 1877, p. 100, by my son Francis, with interesting extracts from a letter by Fritz Müller.

[58] Mr Belt has given a most interesting account (*The Naturalist in Nicaragua*, 1874, p. 218) of the paramount importance of ants as defenders of the above Acacia. With respect to the Cecropia, see *Nature*, 1876, p. 304. My son Francis has described the microscopical structure and development of these wonderful food-bodies in a paper read before the Linnean Society. *Bot.* vol. xv, p. 398.

their flowers.[59] Mr Farrer has also shown[60] that the flowers of
Coronilla are curiously modified, so that bees may fertilize them whilst
sucking the fluid secreted from the outside of the calyx. With one of
Malpighiaceae, bees gnaw the glands on the calyx, and in doing so get
their abdomens dusted with pollen, which they carry to other
flowers.[61] It further appears probable from the observations of Rev.
W. A. Leighton, that the fluid so abundantly secreted by glands on the
phyllodia of the Australian *Acacia magnifica*, which stand near the
flowers, is connected with their fertilization.[62]

The amount of pollen produced by anemophilous plants, and the
distance to which it is often transported by the wind, are both
surprisingly great. Mr Hassall found, as before stated, that the weight
of pollen produced by a single plant of the bulrush / (Typha) was 144
grains. Bucketfuls of pollen, chiefly of Coniferae and Gramineae, have
been swept off the decks of vessels near the North American shore;
and Mr Riley has seen the ground near St Louis, in Missouri, covered
with pollen, as if sprinkled with sulphur; and there was good reason to
believe that this has been transported from the pine-forests at least 400
miles to the south. Kerner has seen the snow-fields on the higher Alps
similarly dusted; and Mr Blackley found numerous pollen-grains, in
one instance 1,200, adhering to sticky slides, which were sent up to
a height of from 500 to 1,000 feet by means of a kite, and then
uncovered by a special mechanism. It is remarkable that in these
experiments there were on an average nineteen times as many pollen-
grains in the atmosphere at the higher than at the lower levels.[63]

[59] *Ult. Osservaz. Dicogamia*, 1868–9, p. 188.
[60] *Nature*, 1874, p. 169.
[61] As described by Fritz Müller in *Nature*, November, 1877, p. 28.
[62] *Annals and Mag. of Nat. Hist.*, vol. xvi, 1865, p. 14. In my work on the *Fertilisation
of Orchids*, and in a paper subsequently published in the *Annals and Mag. of Nat.
History*, it has been shown that although certain kinds of orchids possess a nectary, no
nectar is actually secreted by it; but that insects penetrate the inner walls and suck the
fluid contained in the intercellular spaces. I further suggested, in the case of some
other orchids which do not secrete nectar, that insects gnawed the labellum; and this
suggestion has since been proved true. H. Müller and Delpino have now shown that
some other plants have thickened petals which are sucked or gnawed by insects, their
fertilization being thus aided. All the known facts on this head have been collected by
Delpino in his *Ult. Osserv.*, part ii, fasc. ii, 1875, pp. 59–63.
[63] for Mr Hassall's observations see *Annals and Mag. of Nat. Hist.*, vol. viii, 1842, p.
108. In the *North American Journal of Science*, January, 1842, there is an account of the
pollen swept off the decks of a vessel. Riley, *Fifth Report on the Noxious Insects of
Missouri*, 1873, p. 86. Kerner, *Die Schutzmittel des Pollens*, 1873, p. 6. This author has
also seen a lake in the Tyrol so covered with pollen, that the water no longer appeared
blue. Mr Blackley, *Experimental Researches on Hay-fever*, 1873, pp. 132, 141–52.

Considering these facts, it is not so surprising as it at first appears that all, or nearly all the stigmas of anemophilous plants should receive pollen brought to them by mere chance by the wind. During the early part of summer every object is thus dusted with pollen; for instance, I examined for another purpose the labella of a large number of flowers of the Fly Ophrys (which is rarely visited by insects), and found on all very many pollen-grains of other plants, which had been caught by their velvety surfaces.

The extraordinary quantity and lightness of the / pollen of ane-mophilous plants are no doubt both necessary, as their pollen has generally to be carried to the stigmas of other and often distant flowers; for, as we shall soon see, most anemophilous plants have their sexes separated. The fertilization of these plants is generally aided by the stigmas being of large size or plumose; and in the case of the Coniferae, by the naked ovules secreting a drop of fluid, as shown by Delpino. Although the number of anemophilous species is small, as the author just quoted remarks, the number of individuals is large in comparison with that of entomophilous species. This holds good especially in cold and temperate regions, where insects are not so numerous as under a warmer climate, and where consequently ento-mophilous plants are less favourably situated. We see this in our forests of Coniferae and other trees, such as oaks, beeches, birches, ashes, etc.; and in the Gramineae, Cyperaceae, and Juncaceae, which clothe our meadows and swamps; all these trees and plants being fertilized by the wind. As a large quantity of pollen is wasted by anemophilous plants, it is surprising that so many vigorous species of this kind abounding with individuals should still exist in any part of the world; for if they had been rendered entomophilous, their pollen would have been transported by the aid of the senses and appetites of insects with incomparably greater safety than by the wind. That such a conversation is possible can hardly be doubted, from the remarks lately made on the existence of intermediate forms; and apparently it has been effected in the group of willows, as we may infer from the nature of their nearest allies.[64]

It seems at first sight a still more surprising fact / that plants, after having been once rendered entomophilous, should ever again have become anemophilous; but this has occasionally though rarely occur-red, for instance, with the common *Poterium sanguisorba*, as may be

[64] H. Müller, *Die Befruchtung*, etc., p. 149.

inferred from its belonging to the Rosaceae. Such cases are, however, intelligible, as almost all plants require to be occasionally intercrossed; and if any entomophilous species ceased altogether to be visited by insects, it would probably perish unless it were rendered anemophilous, or acquired a full capacity for self-fertilization; but in this latter case we may suspect that it would be apt to suffer from the long-continued want of cross-fertilization. A plant would be neglected by insects if nectar failed to be secreted, unless indeed a large supply of attractive pollen was present; and from what we have seen of the excretion of saccharine fluid from leaves and glands being largely governed in several cases by climatic influences, and from some few flowers which do not now secrete nectar still retaining coloured guiding-marks, the failure of the secretion cannot be considered as a very improbable event. The same result would follow to a certainty, if winged insects ceased to exist in any district, or became very rare. Now there is only a single plant in the great order of the Cruciferae, namely, Pringlea, which is anemophilous, and this plant is an inhabitant of Kerguelen Land,[65] where there are hardly any winged insects, owing probably, as was suggested by me in the case of Madeira, to the risk which they run of being blown out to sea and destroyed.

A remarkable fact with respect to anemophilous plants is that they are often diclinous, that is, they are / either monoecious with their sexes separated on the same plant, or dioecious with their sexes on distinct plants. In the class Monoecia of Linnaeus, Delpino shows[66] that the species of twenty-eight genera are anemophilous, and of seventeen genera entomophilous. In the class Dioecia, the species of ten genera are anemophilous and of nineteen entomophilous. The larger proportion of entomophilous genera in this latter class is probably the indirect result of insects having the power of carrying pollen to another and sometimes distant plant much more securely than the wind. In the above two classes taken together there are thirty-eight anemophilous and thirty-six entomophilous genera; whereas in the great mass of hermaphrodite plants the proportion of anemophilous to entomophilous genera is extremely small. The cause of this remarkable difference may be attributed to anemophilous plants having retained in a greater degree than the entomphilous a primordial condition, in which the sexes were separated and their mutual

---

[65] The Rev. A. E. Eaton in *Proc. Royal Soc.*, vol. xxiii, 1875, p. 351.
[66] *Studi sopra un Lignaggio anemofilo delle Compositae*, 1871.

fertilization effected by means of the wind. That the earliest and lowest members of the vegetable kingdom had their sexes separated, as is still the case to a large extent, is the opinion of a high authority, Nägeli.[67] It is indeed difficult to avoid this conclusion, if we admit the view, which seems highly probable, that the conjugation of the Algae and of some of the simplest animals is the first step towards sexual reproduction; and if we further bear in mind that a greater and greater degree of differentiation between the cells which conjugate can be traced, thus leading apparently to the development of the / two sexual forms.[68] We have also seen that as plants became affixed to the ground and were more highly developed so as to be rendered phanerogamic, they would be compelled to be anemophilous in order to intercross. Therefore all plants which have not since been greatly modified, would tend still to be both diclinous and anemophilous; and we can thus understand the connection between these two states, although they appear at first sight quite disconnected. If this view is correct, plants must have been rendered hermaphrodites at a later though still very early period, and entomophilous at a yet later period, namely, after the development of winged insects. So that the relationship between hermaphroditism and fertilization by means of insects is likewise to a certain extent intelligible.

Why the descendants of plants which were originally dioecious, and which therefore profited by always intercrossing with another individual, should have been converted into hermaphrodites, may perhaps be explained by the risk which they ran, especially as long as they were anemophilous, of not being always fertilized, and consequently of not leaving offspring. This latter / evil, the greatest of all to any organism, would have been much lessened by their becoming hermaphrodites, though with the contingent disadvantage of frequent self-fertilization.

[67] *Entstehung und Begriff der naturhist. Art*, 1865. p. 22.
[68] See the interesting discussion on this whole subject by O. Bütschli in his *Studien über die ersten Entwickelungsvorgänge der Eizelle*, etc., 1876, pp. 207–19. Also, Dr A. Dodel, 'Die Kraushaar-Alge', *Pringsheims Jahrb. f. wiss. Bot.*, vol. x. Also, Engelmann, 'Ueber Entwickelung von Infusorien', *Morphol. Jahrbuch*, vol. i, p. 573. An abstract of his important memoir has appeared in *Archives de Zoolog. expérimentale*, vol. v, 1876, p. xxxiii. Engelmann concludes that the conjugation of various Infusoria, whether permanent or temporary (in this latter case called by him copulation) does not lead to the development of true ova, but to the reorganization or rejuvenescence of the individual. There seems to be a close analogy in such a result with that which follows from the union of the male and female elements of distinct plants, for the seedlings thus raised may be said to show regeneration or rejuvenescence in their greatly increased constitutional vigour.

By what graduated steps an hermaphrodite condition was acquired we do not know. But we can see that if a lowly organized form, in which the two sexes were represented by somewhat different individuals, were to increase by budding either before or after conjugation, the two incipient sexes would be capable of appearing by buds on the same stock, as occasionally occurs with various characters at the present day. The organism would then be in a monoecious condition, and this is probably the first step towards hermaphroditism; for if very simple male and female flowers on the same stock, each consisting of a single stamen or pistil, were brought close together and surrounded by a common envelope, in nearly the same manner as with the florets of the Compositae, we should have an hermaphrodite flower.[69]

There seems to be no limit to the changes which organisms undergo changing conditions of life; and some hermaphrodite plants, descended as I am led to believe from aboriginally diclinous plants, have had their sexes again separated. That this has occurred, we may infer from the presence of rudimentary stamens in the flowers of some individuals, and of rudimentary pistils in the flowers of other individuals, for example in *Lychnis dioica*. But a conversion of this kind will / not have occurred unless cross-fertilization was already assured, generally by the agency of insects; but why the production of male and female flowers on distinct plants should have been advantageous to the species, cross-fertilization having been previously assured, is far from obvious. A plant might indeed produce twice as many seeds as were necessary to keep up its numbers under new or changed conditions of life; and if it did not vary by bearing fewer flowers, and did vary in the state of its reproductive organs (as often occurs under cultivation), a wasteful expenditure of seeds and pollen would be saved by the flowers becoming diclinous.

A related point is worth notice. I remarked in my *Origin of Species* that in Britain a much larger proportion of trees and bushes than of herbaceous plants have their sexes separated; and so it is, according to

[69] Mr W. Thiselton Dyer, in a very able review of this work (*Nature*, February, 1877, p. 329), takes an exactly opposite view, and advances weighty arguments in favour of the belief that all plants were aboriginally hermaphrodites. I will only remark that I had in my mind organisms much lower in the scale than Ferns or *Selaginella*. Mr Dyer adds that my notion of very simple male and female flowers being brought together and surrounded by a common envelope, offers very considerable morphological difficulties.

Asa Gray and Hooker, in North America and New Zealand.[70] It is, however, doubtful how far this rule holds good generally, and it certainly does not do so in Australia. But I have been assured that the flowers of the prevailing Australian trees, namely, the Myrtaceae, swarm with insects, and if they are dichogamous / they would be practically diclinous.[71] As far as anemophilous plants are concerned, we know that they are apt to have their sexes separated, and we can see that it would be an unfavourable circumstance for them to bear their flowers very close to the ground, as their pollen is liable to be blown high up in the air;[72] but as the culms of grasses give sufficient elevation, we cannot thus account for so many trees and bushes being diclinous. We may infer from our previous discussion that a tree bearing numerous hermaphrodite flowers would rarely intercross with another tree, except by means of the pollen of a distinct individual being prepotent over the plant's own pollen. Now the separation of the sexes, whether the plant were anemophilous or entomophilous, would most effectually bar self-fertilization, and this may be the cause of so many trees and bushes being diclinous. Or to put the case in another way, a plant would be better fitted for development into a tree, if the sexes were separated, than if it were hermaphrodite; for in the former case its numerous flowers would be less liable to continued self-fertilization. But it should also be observed that the long life of a tree or bush permits of the separation of the sexes, with much less risk of

[70] I find in the *London Catalogue of British Plants*, that there are thirty-two indigenous trees and bushes in Great Britain, classed under nine families; but to err on the safe side, I have counted only six species of willows. Of the thirty-two trees and bushes, nineteen, or more than half, have their sexes separated; and this is an enormous proportion compared with other British plants. New Zealand abounds with diclinous plants and trees; and Dr Hooker calculates that out of about 756 phanerogamic plants inhabiting the islands, no less than 108 are trees, belonging to thirty-five families. Of these 108 trees, fifty-two, or very nearly half, have their sexes more or less separated. Of bushes there are 149, of which sixty-one have their sexes in the same state; whilst of the remaining 500 herbaceous plants only 121, or less than a fourth, have their sexes separated. Lastly, Professor Asa Gray informs me that in the United States there are 132 native trees (belonging to twenty-five families) of which ninety-five (belonging to seventeen families) 'have their sexes more or less separated, for the greater part decidedly separated'.

[71] With respect to the Proteaceae of Australia, Mr Bentham remarks (*Journal Linn. Soc. Bot.*, vol. xiii, 1871, pp. 58, 64) on the various contrivances by which the stigma in the several genera is screened from the action of the pollen from the same flower. For instance, in Synaphea 'the stigma is held by the eunuch (i.e., one of the stamens which is barren) safe from all pollution from her brother anthers, and is preserved intact for any pollen that may be inserted by insects and other agencies'.

[72] Kerner, *Schutzmittel des Pollens*, 1873, p. 4.

evil from impregnation occasionally failing and seeds not being produced, than in the case of short-lived plants. Hence it probably is, as Lecoq has remarked, that annual plants are rarely dioecious. /

Finally, we have seen reason to believe that the higher plants are descended from extremely low forms which are conjugated, and that the conjugating individuals differed somewhat from one another – the one representing the male and the other the female – so that plants were aboriginally dioecious. At a very early period such lowly organized dioecious plants probably gave rise by budding to monoecious plants with the two sexes borne by the same individual; and by a still closer union of the sexes to hermaphrodite plants, which are now much the commonest form.[73] As soon as plants became affixed to the ground, their pollen must have been carried by some means from flower to flower, at first almost certainly by the wind, then by pollen-devouring, and afterwards by nectar-seeking insects. During subsequent ages some few entomophilous plants have been again rendered anemophilous, and some hermaphrodite plants have had their sexes again separated; and we can vaguely see the advantages of such recurrent changes under certain conditions.

Dioecious plants, however fertilized, have a great advantage over other plants in their cross-fertilization being assured. But this advantage is gained in the case of anemophilous species at the expense of the production of an enormous superfluity of pollen, with some risk to them and to entomophilous species of their fertilization occasionally failing. Half the individuals, moreover, namely, the males, produce no / seed, and this might possibly be a disadvantage. Delpino remarks that dioecious plants cannot spread so easily as monoecious and hermaphrodite species, for a single individual which happened to reach some new site could not propagate its kind; but it may be doubted whether this is a serious evil. Monoecious anemophilous plants can hardly fail to be to a large extent dioecious in function, owing to the lightness of their pollen and to the wind blowing laterally, with the great additional advantage of occasionally or often producing some self-fertilized seeds. When they are also dichogamous, they are

[73] There is a considerable amount of evidence that all the higher animals are the descendants of hermaphrodites; and it is a curious problem whether such hermaphroditism may not have been the result of the conjugation of two slightly different individuals, which represented the two incipient sexes. On this view, the higher animals may now owe their bilateral structure, with all their organs double at an early embryonic period, to the fusion or conjugation of two primordial individuals.

necessarily dioecious in function. Lastly, hermaphrodite plants can generally produce at least some self-fertilized seeds, and they are at the same time capable, through the various means specified in this chapter, of cross-fertilization. When their structure absolutely prevents self-fertilization, they are in the same relative position to one another as monoecious or dioecious plants, with what may be an advantage, namely, that every flower is capable of yielding seeds. /

CHAPTER XI

THE HABITS OF INSECTS IN RELATION TO THE
FERTILIZATION OF FLOWERS

Insects visit the flowers of the same species as long as they can – Cause of
this habit – Means by which bees recognize the flowers of the same species
– Sudden secretion of nectar – Nectar of certain flowers unattractive to
certain insects – Industry of bees, and the number of flowers visited
within a short time – Perforation of the corolla by bees – Skill shown in the
operation – Hive-bees profit by the holes made by humble-bees – Effects
of habit – The motive for perforating flowers to save time – Flowers
growing in crowded masses chiefly perforated.

Bees and various other insects must be directed by instinct to search
flowers for nectar and pollen, as they act in this manner without
instruction as soon as they emerge from the pupa state. Their instincts,
however, are not of a specialized nature, for they visit many exotic
flowers as readily as the endemic kinds, and they often search for
nectar in flowers which do not secrete any; and they may be seen
attempting to suck it out of nectaries of such length that it cannot be
reached by them.[1] All kinds of bees and certain other insects usually
visit the flowers of the same species as long as they can, before going to
another species. This fact was observed by Aristotle with respect to
the / hive-bee more than 2,000 years ago, and was noticed by Dobbs in
a paper published in 1736 in the *Philosophical Transactions*. It may be
observed by any one, both with hive and humble-bees, in every flower-
garden; not that the habit is invariably followed. Mr Bennett watched
for several hours[2] many plants of *Lamium album*, *L. purpureum*, and
another Labiate plant, *Nepeta glechoma*, all growing mingled together
on a bank near some hives; and he found that each bee confined its

---

[1] See, on this subject, H. Müller, *Befruchtung*, etc., p. 427; and Sir J. Lubbock's
British Wild Flowers, etc., p. 20. Müller assigns (*Bienen Zeitung*, June, 1876, p. 119)
good reasons for his belief that bees and many other Hymenoptera have inherited
from some early nectar-sucking progenitor greater skill in robbing flowers than that
which is displayed by insects belonging to the other Orders.

[2] *Nature*, 4 June, 1874, p. 92.

visits to the same species. The pollen of these three plants differs in colour, so that he was able to test his observations by examining that which adhered to the bodies of the captured bees, and he found one kind on each bee.

Humble and hive-bees are good botanists, for they know that varieties may differ widely in the colour of their flowers and yet belong to the same species. I have repeatedly seen humble-bees flying straight from a plant of the ordinary red *Dictamnus fraxinella* to a white variety; from one to anther very differently coloured variety of *Delphinium consolida* and of *Primula veris*; from a dark purple to a bright yellow variety of *Viola tricolor*; and with two species of Papaver, from one variety to another which differed much in colour; but in this latter case some of the bees flew indifferently to either species, although passing by other genera, and thus acted as if the two species were merely varieties. H. Müller also has seen hive-bees flying from flower to flower of *Ranunculus bulbosus* and *arvensis*, and of *Trifolium fragiferum* and *repens*; and even from blue hyacinths to blue violets.[3]

Some species of Diptera or flies keep to the flowers / of the same species with almost as much regularity as do bees; and when captured they are found covered with pollen. I have seen *Rhingia rostrata* acting in this manner with the flowers of *Lychnis dioica*, *Ajuga reptans*, and *Vicia sepium*. *Volucella plumosa* and *Empis cheiroptera* flew straight from flower to flower of *Myosotis sylvatica*. *Dolichopus nigripennis* behaved in the same manner with *Potentilla tormentilla*; and other Diptera with *Stellaria holostea*, *Helianthemum vulgare*, *Bellis perennis*, *Veronica hederaefolia* and *chamoedrys*; but some flies visited indifferently the flowers of these two latter species. I have seen more than once a minute Thrips, with pollen adhering to its body, fly from one flower to another of the same kind; and one was observed by me crawling about within a convolvulus with four grains of pollen adhering to its head, which were deposited on the stigma.

Fabricius and Sprengel state that when flies have once entered the flowers of Aristolochia they never escape – a statement which I could not believe, as in this case the insects would not aid in the cross-fertilization of the plant; and this statement has now been shown by Hildebrand to be erroneous. As the spathes of *Arum maculatum* are furnished with filaments apparently adapted to prevent the exit of insects, they resemble in this respect the flowers of Aristolochia; and

[3] *Bienen Zeitung*, July, 1876, p. 183.

on examining several spathes, from thirty to sixty minute Diptera belonging to three species were found in some of them; and many of these insects were lying dead at the bottom, as if they had been permanently entrapped. In order to discover whether the living ones could escape and carry pollen to another plant, I tied in the spring of 1842 a fine muslin bag tightly round a spathe; and on returning in an hour's time several little flies were crawling about on the inner / surface of the bag. I then gathered a spathe and breathed hard into it; several flies soon crawled out, and all without exception were dusted with arum pollen. These flies quickly flew away, and I distinctly saw three of them fly to another plant about a yard off; they alighted on the inner or concave surface of the spathe, and suddenly flew down into the flower. I then opened this flower, and although not a single anther had burst, several grains of pollen were lying at the bottom, which must have been brought from another plant by one of these flies or by some other insect. In another flower little flies were crawling about, and I saw them leave pollen on the stigmas.

I do not know whether Lepidoptera generally keep to the flowers of the same species; but I once observed many minute moths (I believe *Lampronia* (Tinea) *calthella*) apparently eating the pollen of *Mercurialis annua*, and they had the whole front of their bodies covered with pollen. I then went to a female plant some yards off, and saw in the course of fifteen minutes three of these moths alight on the stigmas. Lepidoptera are probably often induced to frequent the flowers of the same species, whenever these are provided with a long and narrow nectary, as in this case other insects cannot suck the nectar, which will thus be preserved for those having an elongated proboscis. No doubt the Yucca moth[4] visits only the flowers whence its name is derived, for a most wonderful instinct guides this moth to place pollen on the stigma, so that the ovules may be developed on which the larvae feed. With respect to Coleoptera, I have seen Meligethes covered with pollen flying from flower to flower of the same species; and / this must often occur, as, according to M. Brisout, 'many of the species affect only one kind of plant'.[5]

It must not be supposed from these several statements that insects strictly confine their visits to the same species. They often visit other species when only a few plants of the same kind grow near together. In

[4] Described by Mr Riley in the *American Naturalist*, vol. vii, October, 1873.
[5] As quoted in *American Nat.*, May, 1873, p. 270.

a flower-garden containing some plants of Oenothera, the pollen of which can easily be recognized, I found not only single grains but masses of it within many flowers of Mimulus, Digitalis, Antirrhinum, and Linaria. Other kinds of pollen were likewise detected in these same flowers. A large number of the stigmas of a plant of Thyme, in which the anthers were completely aborted, were examined; and these stigmas, though scarcely larger than a split needle, were covered not only with pollen of Thyme brought from other plants by the bees, but with several other kinds of pollen.

That insects should visit the flowers of the same species as long as they can, is of great importance to the plant, as it favours the cross-fertilization of distinct individuals of the same species; but no one will suppose that insects act in this manner for the good of the plant. The cause probably lies in insects being thus enabled to work quicker; they have just learnt how to stand in the best position on the flower, and how far and in what direction to insert their proboscides.[6] They act on the same principle as does an artificer who has to make half-a-dozen engines, and who saves time by making consecutively each wheel and part for all of them. Insects, or at least bees, seem much influenced by habit in all their manifold operations; and / we shall presently see that this holds good in their felonious practice of biting holes through the corolla.

It is a curious question how bees recognize the flowers of the same species. That the coloured corolla is the chief guide cannot be doubted. On a fine day, when hive-bees were incessantly visiting the little blue flowers of *Lobelia erinus*, I cut off all the petals of some, and only the lower striped petals of others, and these flowers were not once again sucked by the bees, although some actually crawled over them. The removal of the two little upper petals alone made no difference in their visits. Mr J. Anderson likewise states that when he removed the corollas of the Calceolaria, bees never visited the flowers.[7] On the

[6] Since these remarks were written, I find that H. Müller has come to almost exactly the same conclusion with respect to the cause of insects frequenting as long as they can the flowers of the same species: *Bienen Zeitung*, July, 1876, p. 182.

[7] *Gardeners' Chronicle*, 1853, p. 534. Kurr cut off the nectaries from a large number of flowers of several species, and found that the greater number yielded seeds; but insects probably would not perceive the loss of the nectary until they had inserted their proboscides into the holes thus formed, and in doing so would fertilize the flowers. He also removed the whole corolla from a considerable number of flowers, and these likewise yielded seeds. Flowers which are self-fertile would naturally produce seeds under these circumstances; but I am greatly surprised that *Delphinium consolida*, as well as another species of Delphinium, and *Viola tricolor*,

other hand, in some large masses of *Geranium phaeum* which had escaped out of a garden, I observed the unusual fact of the flowers continuing to secrete an abundance of nectar after all the petals had fallen off; and the flowers in this state were still visited by humble-bees. But the bees might have learnt that these flowers with all their petals lost were still worth visiting, by finding nectar in those with only one or two lost. The colour alone of the corolla serves as an approximate guide: thus I watched for some time humble-bees which were visiting exclusively plants of / the white-flowered *Spiranthes autumnalis*, growing on short turf at a considerable distance apart; and these bees often flew within a few inches of several other plants with white flowers, and then without further examination passed onwards in search of the Spiranthes. Again, many hive-bees which confined their visits to the common ling (*Calluna vulgaris*), repeatedly flew towards *Erica tetralix*, evidently attracted by the nearly similar tint of their flowers, and then instantly passed on in search of the Calluna.

That the colour of the flower is not the sole guide, is clearly shown by the six cases above given of bees which repeatedly passed in a direct line from one variety to another of the same species, although they bore very differently coloured flowers. I observed also bees flying in a straight line from one clump of a yellow-flowered Oenothera to every other clump of the same plant in the garden, without turning an inch from their course to plants of Eschscholtzia and others with yellow flowers which lay only a foot or two on either side. In these cases the bees knew the position of each plant in the garden perfectly well, as we may infer by the directness of their flight; so that they were guided by experience and memory. But how did they discover at first that the above varieties which differently coloured flowers belonged to the same species? Improbable as it may appear, they seem, at least sometimes, to recognize plants even from a distance by their general aspect, in the same manner as we should do. On three occasions I observed humble-bees flying in a perfectly straight line from a tall larkspur (Delphinium) which was in full flower to another plant of the same species at the distance of fifteen yards which had not as yet a single flower open, and on which the buds showed only a faint tinge of

---

should have produced a fair supply of seeds when thus treated; but it does not appear that he compared the number of the seeds thus produced with those yielded by unmutilated flowers left to the free access of insects: *Bedeutung der Nektarien*, 1833, pp. 123–35.

blue. Here neither odour nor the / memory of former visits could have come into play, and the tinge of blue was so faint that it could hardly have served as a guide.[8]

The conspicuousness of the corolla does not suffice to induce repeated visits from insects, unless nectar is at the same time secreted, together perhaps with some odour emitted. I watched for a fortnight many times daily a wall covered with *Linaria cymbalaria* in full flower, and never saw a bee even looking at one. There was then a very hot day, and suddenly many bees were industriously at work on the flowers. It appears that a certain degree of heat is necessary for the secretion of nectar; for I observed with *Lobelia erinus* that if the sun ceased to shine for only half an hour, the visits of the bees slackened and soon ceased. An analogous fact with respect to the sweet excretion from the stipules of *Vicia sativa* has been already given. As in the case of the Linaria, so with *Pedicularis sylvatica, Polygala vulgaris, Viola tricolor*, some species of Trifolium, I have watched the flowers day after day without seeing a bee at work, and then suddenly all the flowers were visited by many bees. Now how did so many bees discover at once that the flowers were secreting nectar? I presume that it must have been by their odour; and that as soon as a few bees began to suck the flowers, others of the same and of different kinds observed the fact and profited by it. We shall presently see, when we treat of the perforation of the corolla, that bees are fully capable of profiting by the / labour of other species. Memory also comes into play, for, as already remarked, bees know the position of each clump of flowers in a garden. I have repeatedly seen them passing round a corner, but otherwise in as straight a line as possible, from one plant of Fraxinella and of Linaria to another and distant one of the same species; although, owing to the intervention of other plants, the two were not in sight of each other.

It would appear that either the taste or the odour of the nectar of certain flowers is unattractive to hive or to humble-bees, or to both; for there seems no other reason why certain open flowers which secrete nectar are not visited by them. The small quantity of nectar secreted by some of these flowers can hardly be the cause of their neglect, as hive-

---

[8] A fact mentioned by H. Müller (*Die Befruchtung*, etc., p. 347) shows that bees possess acute powers of vision and discrimination; for those engaged in collecting pollen from *Primula elatior* invariably passed by the flowers of the long-styled form, in which the anthers are seated low down in the tubular corolla. Yet the difference in aspect between the long-styled and short-styled forms is extremely slight.

bees search eagerly for the minute drops on the glands on the leaves of the *Prunus laurocerasus*. Even the bees from different hives sometimes visit different kinds of flowers, as is said to be the case by Mr Grant with respect to the Polyanthus and *Viola tricolor*.[9] I have known humble-bees to visit the flowers of *Lobelia fulgens* in one garden and not in another at the distance of only a few miles. The cupful of nectar in the labellum of *Epipactis latifolia* is never touched by hive- or humble-bees, although I have seen them flying close by; and yet the nectar has a pleasant taste to us, and is habitually consumed by the common wasp. As far as I have seen, wasps seek for nectar in this country only from the flowers of this Epipactis, *Scrophularia aquatica*, *Hedera helix*, *Symphoricarpus racemosa*,[10] and Tritoma; the three former plants being endemic, and the two latter exotic. As wasps are so / fond of sugar and of any sweet fluid, and as they do not disdain the minute drops on the glands of *Prunus laurocerasus*, it is a strange fact that they do not suck the nectar of many open flowers, which they could do without the aid of a proboscis. Hive-bees visit the flowers of the Symphoricarpus and Tritoma, and this makes it all the stranger that they do not visit the flowers of the Epipactis, or, as far as I have seen, those of the *Scrophulara aquatica*; although they do visit the flowers of *Scrophularia nodosa*, at least in North America.[11]

The extraordinary industry of bees and the number of flowers which they visit within a short time, so that each flower is visited repeatedly, must greatly increase the chance of each receiving pollen from a distinct plant. When the nectar is in any way hidden, bees cannot tell without inserting their proboscides whether it has lately been exhausted by other bees, and this, as remarked in a former chapter, forces them to visit many more flowers than they otherwise would. But they endeavour to lose as little time as they can; thus in flowers having several nectaries, if they find one dry they do not try the others, but as I have often observed, pass on to another flower. They work so industriously and effectually, that even in the case of social plants, of which hundreds of thousands grow together, as with the several kinds of heath, every single flower is visited, of which evidence will presently be given. They lose no time and fly very quickly from

---

[9] *Gard. Chron.*, 1844, p. 374.

[10] The same fact apparently holds good in Italy, for Delpino says that the flowers of these three plants alone are visited by wasps: *Nettarii Estranuziali, Bullettino Entomologico*, anno vi.

[11] *Silliman's American Journal of Science*, August, 1871.

plant to plant, but I do not know the rate at which hive-bees fly. Humble-bees fly at the rate of ten miles an hour, as I was able to ascertain in the case of the males from their curious habit of calling at / certain fixed points, which made it easy to measure the time taken in passing from one place to another.

With respect to the number of flowers which bees visit in a given time, I observed that in exactly one minute a humble-bee visited twenty-four of the closed flowers of the *Linaria cymbalaria*; another bee visited in the same time twenty-two flowers of the *Symphoricarpus racemosa*; and another seventeen flowers on two plants of a Delphinium. In the course of fifteen minutes a single flower on the summit of a plant of Oenothera was visited eight times by several humble-bees, and I followed the last of these bees, whilst it visited in the course of a few additional minutes every plant of the same species in a large flower-garden. In nineteen minutes every flower on a small plant of *Nemphila insignis* was visited twice. In one minute six flowers of a Campanula were entered by a pollen-collecting hive-bee; and bees when thus employed work slower than when sucking nectar. Lastly, seven flower-stalks on a plant of *Dictamnus fraxinella* were observed on 15 June, 1841 during ten minutes; they were visited by thirteen humble-bees, each of which entered many flowers. On the 22nd the same flower-stalks were visited within the same time by eleven humble-bees. This plant bore altogether 280 flowers, and from the above data, taking into consideration how late in the evening humble-bees work, each flower must have been visited at least thirty times daily, and the same flower keeps open during several days. The frequency of the visits of bees is also sometimes shown by the manner in which the petals are scratched by their hooked tarsi; I have seen large beds of Mimulus, Stachys, and Lathyrus with the beauty of their flowers thus sadly defaced.

Perforation of the corolla by bees. I have already / alluded to bees biting holes in flowers for the sake of obtaining the nectar. They often act in this manner, both with endemic and exotic species, in many parts of Europe, in the United States, and in the Himalaya; and therefore probably in all parts of the world. The plants, the fertilization of which actually depends on insects entering the flowers, will fail to produce seed when their nectar is thus stolen from the outside; and even with those species which are capable of fertilizing themselves without any aid, there can be no cross-fertilization, and this, as we know, is a serious evil in most cases. The extent to which humble-bees carry on the practice of biting holes is surprising: a remarkable case was observed by me near Bournemouth, where there were formerly extensive heaths. I took a long walk, and every now and then gathered a twig of *Erica tetralix*, and when I had got a handful examined all the flowers through a lens. This process was repeated many times; but though many hundreds were examined, I did not succeed in finding a single flower which had not been perforated. Humble-bees were at the time sucking the flowers through these perforations. On the following day a large number of flowers were examined on another heath with the same result, but here hive-bees were sucking through the holes. This case is all the more remarkable, as the innumerable holes had been made within a fortnight, for before that time I saw the bees every-where sucking in the proper manner at the mouths of the corolla. In an extensive flower-garden some large beds of *Salvia grahami*, *Stachys coccinea*, and *Pentstemon argutus* (?) had every flower perforated, and many scores were examined. I have seen whole fields of red clover (*Trifolium pratense*) in the same state. Dr Ogle found that 90 per cent of the / flowers of *Salvia glutinosa* had been bitten. In the United States Mr Bailey says it is difficult to find a blossom of the native *Gerardia pedicularia* without a hole in it; and Mr Gentry, in speaking of the introduced *Wistaria sinensis*, says 'that nearly every flower had been perforated'.[12]

As far as I have seen, it is always humble-bees which first bite the holes, and they are well fitted for the work by possessing powerful mandibles; but hive-bees afterwards profit by the holes thus made. Dr H. Müller, however, writes to me that hive-bees sometimes bite holes through the flowers of *Erica tetralix*. No insects except bees, with the single exception of wasps in the case of Tritoma, have sense enough, as far as I have observed, to profit by the holes already made. Even humble-bees do not always discover that it would be advantageous to them to perforate certain flowers. There is an abundant supply of nectar in the nectary of *Tropaeolum tricolor*, yet I have found this plant untouched in more than one garden, while the flowers of other plants had been extensively perforated; but a few years ago Sir J. Lubbock's gardener assured me that he had seen humble-bees boring through the nectary of this Tropaeolum. In the United States the common

---

[12] Dr Ogle, *Pop. Science Review*, July, 1869, p. 267. Bailey, *American Nat.*, November, 1873, p. 690. Gentry, ibid., May, 1875, p. 264.

garden Tropaeolum, as I hear from Mr Bailey, is often pierced. Müller has observed humble-bees trying to suck at the mouths of the flowers of *Primula elatior* and of an Aquilegia, and, failing in their attempts, they made holes through the corolla; but they often bite holes, although they could with very little more trouble obtain the nectar in a legitimate manner by the mouth of the corolla.

Dr W. Ogle has communicated to me a curious case. / He gathered in Switzerland 100 flower-stems of the common blue variety of the monkshood (*Aconitum napellus*), and not a single flower was perforated; he then gathered 100 stems of a white variety growing close by, and every one of the open flowers had been perforated. This surprising difference in the state of the flowers may be attributed with much probability to the blue variety being distasteful to bees, from the presence of the acrid matter which is so general in the Ranunculaceae, and to its absence in the white variety in correlation with the loss of the blue tint. According to Sprengel,[13] this plant is strongly proterandrous; it would therefore be more or less sterile unless bees carried pollen from the younger to the older flowers. Consequently the white variety, the flowers of which were always bitten instead of being properly entered by the bees, would fail to yield the full number of seeds and would be a comparatively rare plant, as Dr Ogle informs me was the case.

Bees show much skill in their manner of working, for they always make their holes from the outside close to the spot where the nectar lies hidden within the corolla. All the flowers in a large bed of *Stachys coccinea* had either one or two slits made on the upper side of the corolla, near the base. The flowers of a Mirabilis and of *Salvia coccinea* were perforated in the same manner; whilst those of Salvia grahami, in which the calyx is much elongated, had both the calyx and the corolla invariably perforated. The flowers of *Pentstemon argutus* are broader than those of the plants just named, and two holes alongside each other had here always been made just above the calyx. In these several cases the perforations were on the upper side, but in *Antirrhinum / majus* one or two holes had been made on the lower side, close to the little protuberance which represents the nectary, and therefore directly in front of and close to the spot where the nectar is secreted.

But the most remarkable case of skill and judgement known to me, is that of the perforation of the flowers of *Lathyrus sylvestris*, as described

[13] *Das Entdecke*, etc., p. 278.

by my son Francis.[14] The nectar in this plant is enclosed within a tube, formed by the united stamens, which surround the pistil so closely that a bee is forced to insert its proboscis outside the tube; but two natural rounded passages or orifices are left in the tube near the base, in order that the nectar may be reached by the bees. Now my son found in sixteen out of twenty-four flowers on this plant, and in eleven out of sixteen of those on the cultivated everlasting pea, which is either a variety of the same species or a closely allied one, that the left passage was larger than the right one. And here come the remarkable point – the humble-bees bite holes through the standard-petal, and they always operated on the left side over the passage, which is generally the larger of the two. My son remarks: 'It is difficult to say how the bees could have acquired this habit. Whether they discovered the inequality in the size of the nectar-holes in sucking the flowers in the proper way, and then utilized this knowledge in determining where to gnaw the hole; or whether they found out the best situation by biting through the standard at various points, and afterwards remembered its situation in visiting other flowers. But in either case they show a remarkable power of making use of what they have learnt by experience.' It seems probable that bees owe their skill in biting holes through flowers of all / kinds to their having long practised the instinct of moulding cells and pots of wax, or of enlarging their old cocoons with tubes of wax; for they are thus compelled to work on the inside and outside of the same object.

In the early part of the summer of 1857 I was led to observe during some weeks several rows of the scarlet kidney-bean (*Phaseolus multi-florus*), whilst attending to the fertilization of this plant, and daily saw humble- and hive-bees sucking at the mouths of the flowers. But one day I found several humble-bees employed in cutting holes in flower after flower; and on the next day every single hive-bee, without exception, instead of alighting on the left wing-petal and sucking the flower in the proper manner, flew straight without the least hesitation to the calyx, and sucked through the holes which had been made only the day before by the humble-bees; and they continued this habit for many following days.[15] Mr Belt has communicated to me (28 July, 1874) a similar case, with the sole difference that less than half of the flowers had been perforated by the humble-bees; nevertheless, all the

---

[14] *Nature*, 8 January, 1874, p. 189.
[15] *Gard. Chron.*, 1857, p. 725.

hive-bees gave up sucking at the mouths of the flowers and visited exclusively the bitten ones. Now how did the hive-bees find out so quickly that holes had been made? Instinct seems to be out of the question, as the plant is an exotic. The holes cannot be seen by bees whilst standing on the wing-petals, where they had always previously alighted. From the ease with which bees were deceived when the petals of *Lobelia erinus* were cut off, it was clear that in this case they were not guided to the nectar by its smell; and it may be doubted whether they were / attracted to the holes in the flowers of the Phaseolus by the odour emitted from them. Did they perceive the holes by the sense of touch in their proboscides, whilst sucking the flowers in the proper manner, and then reason that it would save them time to alight on the outside of the flowers and use the holes? This seems almost too abstruse an act of reason for bees; and it is more probable that they saw the humble-bees at work, and understanding what they were about, imitated them and took advantage of the shorter path to the nectar. Even with animals high in the scale, such as monkeys, we should be surprised at hearing that all the individuals of one species within the space of twenty-four hours understood an act performed by a distinct species, and profited by it.

I have repeatedly observed with various kinds of flowers that all the hive and humble-bees which were sucking through the perforations, flew to them, whether on the upper or under side of the corolla, without the least hesitation; and this shows how quickly all the individuals within the district had acquired the same knowledge. Yet habit comes into play to a certain extent, as in so many of the other operations of bees. Dr Ogle, Messrs Farrer and Belt have observed in the case of *Phaseolus multiflorus*[16] that certain individuals went exclusively to the perforations, while others of the same species visited only the mouths of the flowers. I noticed in 1861 exactly the same fact with *Trifolium pratense*. So persistent is the force of habit, that when a bee which is visiting perforated flowers comes to one which has not been bitten, it does not go to the mouth, but instantly / flies away in search of another bitten flower. Nevertheless, I once saw a humble-bee visiting the hybrid *Rhododendron azaloides*, and it entered the mouths of some flowers and cut holes into the others. Dr H. Müller informs me that in the same district he has seen some individuals of *Bombus mastrucatus*

[16] Dr Ogle, *Pop. Science Review*, April, 1870, p. 167. Mr Farrer, *Annals and Mag. of Nat. Hist.*, 4th series, vol. ii, 1868, p. 258. Mr Belt in a letter to me.

boring through the calyx and corolla of *Rhinanthus alecterolophus*, and others through the corolla alone. Different species of bees may, however, sometimes be observed acting differently at the same time on the same plant. I have seen hive-bees sucking at the mouths of the flowers of the common bean; humble-bees of one kind sucking through holes bitten in the calyx, and humble-bees of another kind sucking the little drops of fluid excreted by the stipules. Mr Beal of Michigan informs me that the flowers of the Missouri currant (*Ribes aureum*) abound with nectar, so that children often suck them; and he saw hive-bees sucking through holes made by a bird, the oriole, and at the same time humble-bees sucking in the proper manner at the mouths of the flowers.[17] This statement about the oriole calls to mind what I have before said of certain species of humming-birds boring holes through the flowers of the Brugmansia, whilst other species entered by the mouth.

The motive which impels bees to gnaw holes through the corolla seems to be the saving of time, for they lose much time in climbing into and out of large flowers, and in forcing their heads into closed ones. They were able to visit nearly twice as many flowers, as far as I could judge, of a Stachys and Pentstemon / by alighting on the upper surface of the corolla and sucking through the cut holes, as by entering in the proper way. Nevertheless each bee before it has had much practice, must lose some time in making each new perforation, especially when the perforation has to be made through both calyx and corolla. This action therefore implies foresight, of which faculty we have abundant evidence in their building operations; and may we not further believe that some trace of their social instinct that is, of working for the good of other members of the community, may here likewise play a part?

Many years ago I was struck with the fact that humble-bees as a general rule perforate flowers only when these grow in large numbers near together. In a garden where there were some very large beds of *Stachys coccinea* and of *Pentstemon argutus*, every single flower was perforated, but I found two plants of the former species growing quite separate with their petals much scratched, showing that they have been frequently visited by bees, and yet not a single flower was perforated. I found also a separate plant of the Pentstemon, and saw bees entering the mouth of the corolla, and not a single flower had been perforated.

---

[17] The flowers of the Ribes are however sometimes perforated by humble-bees, and Mr Bundy says that they were able to bite through and rob seven flowers of their honey in a minute: *American Naturalist*, 1876, p. 238.

In the following year (1842) I visited the same garden several times: on 19 July humble-bees were sucking the flowers of *Stachys coccinea* and *Sala grahami* in the proper manner, and none of the corollas were perforated. On 7 August all the flowers were perforated, even those on some few plants of the Salvia which grew at a little distance from the great bed. On 21 August only a few flowers on the summits of the spikes of both species remained fresh, and not one of these was now bored. Again, in my own garden every plant in several rows of the common bean / had many flowers perforated; but I found three plants in separate parts of the garden which had sprung up accidentally, and these had not a single flower perforated. General Strachey formerly saw many perforated flowers in a garden in the Himalaya, and he wrote to the owner to enquire whether this relation between the plants growing crowded and their perforation by the bees there held good, and was answered in the affirmative. Mr Bailey informs me that the *Gerardia pedicularia* which is so largely perforated, and *Impatiens fulva*, are both profuse flowerers. Hence it follows that the red clover (*Trifolium pratense*) and the common bean when cultivated in great masses in fields – that *Erica tetralix* growing in large numbers on heaths – rows of the scarlet kidney-bean in the kitchen-garden – and masses of any species in the flower-garden – are all eminently liable to be perforated.

The explanation of this fact is not difficult. Flowers growing in large numbers afford a rich booty to the bees, and are conspicuous from a distance. They are consequently visited by crowds of these insects, and I once counted between twenty and thirty bees flying about a bed of Pentstemen. They are thus stimulated to work quickly by rivalry, and, what is much more important, they find a large proportion of the flowers, as suggested by my son,[18] with their nectaries sucked dry. They thus waste much time in searching many empty flowers, and are led to bite the holes, so as to find out as quickly as possible whether there is any nectar present, and if so, to obtain it.

Flowers which are partially or wholly sterile unless visited by insects in the proper manner, such as / those of most species of Salvia, of *Trifolium pratense*, *Phaseolus multiflorus*, etc., will more or less completely fail to produce seeds if the bees confine their visits to the perforations. The perforated flowers of those species, which are capable of fertilizing themselves, will yield only self-fertilized seeds, and the

[18] *Nature*, 8 January, 1874, p. 189.

seedlings will in consequence be less vigorous. Therefore all plants must suffer in some degree when bees obtain their nectar in a felonious manner by biting holes through the corolla; and many species, it might be thought, would be thus exterminated. But here, as is so general throughout nature, there is a tendency towards a restored equilibrium. If a plant suffers from being perforated, fewer individuals will be reared, and if its nectar is highly important to the bees, these in their turn will suffer and decrease in number; but, what is much more effective, as soon as the plant becomes somewhat rare so as not to grow in crowded masses, the bees will no longer be stimulated to gnaw holes in the flowers, but will enter them in a legitimate manner. More seed will then be produced, and the seedlings being the product of cross-fertilization will be vigorous, so that the species will tend to increase in number, to be again checked, as soon as the plant again grows in crowded masses. /

CHAPTER XII

GENERAL RESULTS

Cross-fertilization proved to be beneficial, and self-fertilization injurious –
Allied species differ greatly in the means by which cross-fertilization is
favoured and cross-fertilization avoided – The benefits and evils of the
two processes depend on the degree of differentiation in the sexual
elements – The evil effects not due to the combination of morbid
tendencies in the parents – Nature of the conditions to which plants are
subjected when growing near together in a state of nature or under
culture, and the effects of such conditions – Theoretical considerations
with respect to the interaction of differentiated sexual elements –
Practical lessons – Genesis of the two sexes – Close correspondence
between the effects of cross-fertilization and self-fertilization, and of the
legitimate and illegitimate unions of the heterostyled plants, in comparison with hybrid unions.

The first and most important of the conclusions which may be drawn
from the observations given in this volume, is that generally cross-
fertilization is beneficial, and self-fertilization often injurious, at least
with the plants on which I experimented. Whether long-continued
self-fertilization is injurious to all plants is another and difficult
question. The truth of these conclusions is shown by the difference in
height, weight, constitutional vigour, and fertility of the offspring
from crossed and self-fertilized flowers, and in the number of seeds
produced by the parent-plants. With respect to the second of the two
propositions, namely, that self-fertilization is often injurious, we have
abundant evidence. The structure of the flowers in such plants as
Lobelia ramosa, *Digitalis purpurea*, etc., renders the aid of insects almost
indispensable for / their fertilization; and bearing in mind the prepotency of pollen from a distinct individual over that from the same
individual, such plants will almost certainly have been crossed during
many or all previous generations. So it must be, owing merely to the
prepotency of foreign pollen, with cabbages and various other plants,
the varieties of which almost invariably intercross when grown together.
The same inference may be drawn still more surely with respect to those

plants, such as of Reseda and Eschscholtzia, which are sterile with their own pollen, but fertile with that from any other individual. These several plants must therefore have been crossed during a long series of previous generations, and the artificial crosses in my experiments cannot have increased the vigour of the offspring beyond that of their progenitors. Therefore the difference between the self-fertilized and crossed plants raised by me cannot be attributed to the superiority of the crossed, but to the inferiority of the self-fertilized seedlings, due to the injurious effects of self-fertilization.

Notwithstanding the evil which many plants suffer from self-fertilization, they can be thus propagated under favourable conditions for many generations, as shown by some of my experiments, and more especially by the survival during at least half a century of the same varieties of the common pea and sweetpea. The same conclusion probably holds good with several other exotic plants, which are never or most rarely cross-fertilized in this country. But all these plants, as far as they have been tried, profit greatly by a cross with a fresh stock. Many species which bear small and inconspicuous flowers are never, or most rarely, visited by insects during the day; and Hermann Müller infers that they must be always, or almost always, / self-fertilized. But the evidence appears to me insufficient, until it can be shown that such flowers are not visited during the night by any of the innumerable kinds of small moths. From the simple fact of these small flowers expanding, and from some of them secreting nectar, it seems probable that they are at least occasionally visited and intercrossed by nocturnal insects. It is much to be desired that some one should cross and self-fertilize such plants and compare the growth, weight, and fertility of the offspring. The Rev. G. Henslow[1] remarks that the plants which have spread the most widely through the agency of man into new countries, and have there grown most vigorously, commonly bear small and inconspicuous flowers; and, as he assumes that these are always self-fertilized, he infers that this process cannot be at all injurious to plants. He believes that 'as long as a plant is self-fertilizing, it remains in the same condition, and retains its average standard, *but*

---

[1] Mr Henslow has published an elaborate review of the present work in the *Gardeners' Chronicle* from 13 January to 5 May, 1877, also in *Science and Art*, 1 May, 1877, p. 77; from which latter journal the quotation is taken. I have modified some passages in this book, and endeavoured to make others clearer, owing to Mr Henslow's criticisms, but I can by no means agree with many of his inferences. I have also profited by an able review by Hermann Müller in *Kosmos*, April, 1877, p. 57.

does not degenerate in any way. It cannot be benefited, as it cannot introduce anything new into its system, so long as it lives in the same place; hence its results are *negative.* If, however, self-fertilizing plants can migrate, and *so* obtain new peculiarities from fresh surrounding media, *then* they may acquire astonishing vigour, and even oust the native vegetation of the country they have invaded.' According to this view the male and female sexual elements must become in such cases differentiated through the action of the new / conditions; and this seems not improbable, judging from the remarkable effects of changed conditions on the reproductive system of Abutilon and Eschscholtzia.

Some few plants, owing to their structure, for instance, *Ophrys apifera,* have almost certainly been propagated in a state of nature for thousands of generations without having been once intercrossed; and whether they would profit by a cross with a fresh stock is not known. But such cases ought not to make us doubt that as a general rule crossing is beneficial and self-fertilization injurious, any more than the existence of plants which, in a state of nature are propagated asexually, that is, exclusively by rhizomes, stolons, etc.[2] (their flowers never producing seeds), should make us doubt that seminal generation must have some great advantage, as it is the common plan followed by nature. Whether any species has been reproduced asexually from a very remote period cannot, of course, be ascertained. Our sole means for forming any judgement on this head is the duration of the varieties of our fruit trees which have been long propagated by grafts or buds. Andrew Knight formerly maintained that under these circumstances they always become weakly, but this conclusion has been warmly disputed by others. A recent and competent judge, Professor Asa Gray,[3] leans to the side of Andrew Knight, which seems to me, from such evidence as I have been able to collect, the more probable view, notwithstanding many opposed facts.

With respect to the first of the two propositions at the head of this chapter, namely, that cross-fertilization is generally beneficial, we have excellent evidence. / Plants of Ipomoea were intercrossed for nine successive generations; they were then again intercrossed, and at the same time crossed with a plant of a fresh stock, that is, one brought from another garden; and the offspring of this latter cross were to the

[2] I have given several cases in *Variation under Domestication,* ch. xviii, 2nd edit., vol. ii, p. 152.
[3] *Darwiniana: Essays and Reviews pertaining to Darwinism,* 1876, p. 338.

intercrossed plants of the tenth generation in height as 100 to 78, and in fertility as 100 to 51. An analogous experiment with Eschscholtzia gave a similar result, as far as fertility was concerned. In neither of these cases were any of the plants the product of self-fertilization. Plants of Dianthus were self-fertilized for three generations, and this no doubt was injurious; but when these plants were fertilized by a fresh stock and by intercrossed plants of the same stock, there was a great difference in fertility between the two sets of seedlings, and some difference in their height. Petunia offers a nearly parallel case. With various other plants, the wonderful effects of a cross with a fresh stock may be seen in Table C. Several accounts have also been published[4] of the extraordinary growth of seedlings from a cross between two varieties of the same species, some of which are known never to fertilize themselves; so that here neither self-fertilization nor relationship even in a remote degree can have come into play. We may therefore conclude that the above two propositions are true – that cross-fertilization is generally beneficial and self-fertilization often injurious to the offspring.

That certain plants, for instance, *Viola tricolor*, *Digitalis purpurea*, *Sarothamnus scoparius*, *Cyclamen persicum*, etc., which have been naturally cross-fertilized for many or all previous generations, should suffer to an extreme degree from a single act of self-fertilization is an astonishing fact. The evil does not depend in / any corresponding degree on the pollen of the self-fertilized parents acting inefficiently on the stigmas of the same flowers; for in the case of the Ipomoea, Mimulus, Digitalis, Brassica, etc., the self-fertilized parents yielded an abundant supply of seeds; nevertheless the plants raised from these seeds were markedly inferior in many ways to their cross-fertilized brethren. Again with Reseda and Eschscholtzia the more self-sterile individuals profited in a less degree by cross-fertilization than did the more self-fertile individuals. With animals no manifest evil has been observed to follow in the first few generations from close interbreeding; but then we must remember that the closest possible interbreeding with animals, that is between brothers and sisters, cannot be considered as nearly so close a union as that between the pollen and ovules of the same flower. Whether with plants the evil from self-fertilization goes on increasing during successive generations is not as yet known; but we may infer from my experiments that the increase, if any, is far from rapid. After

[4] See *Variation under Domestication*, ch. xix, 2nd edit., vol. ii, p. 159.

plants have been propagated by self-fertilization for several generations, a single cross with a fresh stock restores their pristine vigour;
and we have a strictly analogous result with our domestic animals.[5]
The good effects of cross-fertilization are transmitted by plants to the
next generation; and judging from the varieties of the common pea, to
many succeeding generations. But this may merely be that crossed
plants of the first generation are extremely vigorous, and transmit
their vigour, like any other character, to their successors.

The means for favouring cross-fertilization and preventing self-
fertilization, or conversely for favouring / self-fertilization and preventing to a certain extent cross-fertilization, are wonderfully diversified;
and it is remarkable that these differ widely in closely allied plants,[6] –
in the species of the same genus, and sometimes in the individuals of
the same species. It is not rare to find hermaphrodite plants and others
with separated sexes within the same genus; and it is common to find
some of the species dichogamous and others maturing their sexual
elements simultaneously. The dichogamous genus Saxifraga contains
proterandrous and proterogynous species.[7] Several genera include
both heterostyled (dimorphic or trimorphic forms) and homostyled
species. Ophrys offers a remarkable instance of one species having its
structure manifestly adapted for self-fertilization, and other species as
manifestly adapted for cross-fertilization. Some con-generic species
are quite sterile and others quite fertile with their own pollen. From
these several causes we often find within the same genus species which
do not produce seeds, while others produce an abundance, when
insects are excluded. Some species bear cleistogamic flowers which
cannot be crossed, as well as perfect flowers, whilst others in the same
genus never produce cleistogamic flowers. Some species exist under
two forms, the one bearing conspicuous flowers adapted for cross-
fertilization, the other bearing inconspicuous flowers adapted for self-
fertilization, whilst other species in the same genus present only a
single form. Even with the individuals of the same species, the degree
of self-sterility varies greatly, as in Reseda. With polygamous plants,
the distribution of the sexes / differs in the individuals of the same
species. The relative period at which the sexual elements in the same
flowers are mature, differs in the varieties of Pelargonium; and

[5] *Variation under Domestication*, ch. xix, 2nd edit., vol. ii, p. 159.
[6] Hildebrand has insisted strongly to this effect in his valuable observations on the
fertilization of the Gramineae: *Monatsbericht K. Akad. Berlin*, October, 1872, p. 763.
[7] Dr Engler, *Bot. Zeitung*, 1868, p. 833.

Carrière gives several cases,[8] showing that the period varies according to the temperature to which the plants are exposed.

This extraordinary diversity in the means for favouring or preventing cross- and self-fertilization in closely allied forms, probably depends on the results of both processes being highly beneficial to the species, but in a directly opposed manner and dependent on variable conditions. Self-fertilization assures the production of a large supply of seeds; and the necessity or advantage of this will be determined by the average length of life of the plant, which largely depends on the amount of destruction suffered by the seeds and seedlings. This destruction follows from the most various and variable causes, such as the presence of animals of several kinds, and the growth of surrounding plants. The possibility of cross-fertilization depends mainly on the presence and number of certain insects, often of insects belonging to special groups, and on the degree to which they are attracted to the flowers of any particular species in preference to other flowers – all circumstances likely to change. Moreover, the advantages which follow from cross-fertilization differ much in different plants, so that it is probable that allied plants would often profit in different degrees by cross-fertilization. Under these extremely complex and fluctuating conditions, with two somewhat opposed ends to be gained, namely, the safe propagation of the species and the production of cross-fertilized, vigorous offspring, it is not surprising / that allied forms should exhibit an extreme diversity in the means which favour either end. If, as there is reason at least to suspect, self-fertilization is in some respects beneficial, although more than counterbalanced by the advantages derived from a cross with a fresh stock, the problem becomes still more complicated.

As I only twice experimented on more than a single species in a genus, I cannot say whether the crossed offspring of the several species within the same genus differ in their degree of superiority over their self-fertilized brethren; but I should expect that this would often prove to be the case from what was observed with the two species of Lobelia and with the individuals of the same species of Nicotiana. The species belonging to distinct genera in the same family certainly differ in this respect. The effects of cross- and self-fertilization may be confined either to the growth or to the fertility of the offspring, but generally extends to both qualities. There does not seem to exist any

---

[8] *Des Variétés*, 1865, p. 30.

close correspondence between the degree to which the flowers of species are adapted for cross-fertilization, and the degree to which their offspring profit by this process; but we may easily err on this head, as there are two means for favouring cross-fertilization which are not externally perceptible, namely, self-sterility and the prepotent fertilizing influence of pollen from another individual. Lastly, it has been shown in a former chapter that the effect produced by cross and self-fertilization on the fertility of the parent-plants does not always correspond with that produced on the height, vigour, and fertility of their offspring. The same remark applies to crossed and self-fertilized seedlings when these are used as the parent-plants. This want of correspondence probably depends, at least in part, on the number of seeds produced being chiefly / determined by the number of the pollen-tubes which reach the ovules, and this will be governed by the reaction between the pollen and the stigmatic secretion or tissues; whereas the growth and constitutional vigour of the offspring will be chiefly determined, not only by the number of pollen-tubes reaching the ovules, but by the nature of the reaction between the contents of the pollen-grains and ovules.

There are two other important conclusions which may be deduced from my observations: firstly, that the advantages of cross-fertilization do not follow from some mysterious virtue in the mere union of two distinct individuals, but from such individuals having been subjected during previous generations to different conditions, or to their having varied in a manner commonly called spontaneous, so that in either case their sexual elements have been in some degree differentiated. And secondly, that the injury from self-fertilization follows from the want of such differentiation in the sexual elements. These two propositions are fully established by my experiments. Thus, when plants of the Ipomoea and of the Mimulus, which had been self-fertilized for the seven previous generations and had been kept all the time under the same conditions, were intercrossed one with another, the offspring did not profit in the least by the cross. Mimulus offers another instructive case, showing that the benefit of a cross depends on the previous treatment of the progenitors: plants which had been self-fertilized for the eight previous generations were crossed with plants which had been intercrossed for the same number of generations, all having been kept under the same conditions as far as possible; seedlings from this cross were grown in competition with others /

derived from the same self-fertilized mother-plant crossed by a fresh stock; and the latter seedlings were to the former in height as 100 to 52, and in fertility as 100 to 4. An exactly parallel experiment was tried on Dianthus, with this difference, that the plants had been self-fertilized only for the three previous generations, and the result was similar though not so strongly marked. The foregoing two cases of the offspring of Ipomoea and Eschscholtzia, derived from a cross with a fresh stock, being as much superior to the intercrossed plants of the old stock, as these latter were to the self-fertilized offspring, strongly support the same conclusion. A cross with a fresh stock or with another variety seems to be always highly beneficial, whether or not the mother-plants have been intercrossed or self-fertilized for several previous generations. The fact that a cross between two flowers on the same plant does no good or very little good, is likewise a strong corroboration of our conclusion; for the sexual elements in the flowers on the same plant can rarely have been differentiated, though this is possible, as flower-buds are in one sense distinct individuals, sometimes varying and differing from one another in structure or constitution. Thus the proposition that the benefit from cross-fertilization depends on the plants which are crossed having been subjected during previous generations to some-what different conditions, or to their having varied from some unknown cause as if they had been thus subjected, is securely fortified on all sides.

Before proceeding any further, the view which has been maintained by several physiologists must be noticed, namely, that all the evils from breeding animals too closely, and no doubt, as they would say, / from the self-fertilization of plants, is the result of the increase of some morbid tendency or weakness of constitution common to the closely related parents, or to the two sexes of hermaphrodite plants. Un-doubtedly injury has often thus resulted; but it is a vain attempt to extend this view to the numerous cases given in my Tables. It should be remembered that the same mother-plant was both self-fertilized and crossed, so that if she had been unhealthy she would have transmitted half her morbid tendencies to her crossed offspring. But plants appearing perfectly healthy, some of them growing wild, or the immediate offspring of wild plants, or vigorous common garden-plants, were selected for experiment. Considering the number of species which were tried, it is nothing less than absurd to suppose that in all these cases the mother-plants, though not appearing in any way diseased, were weak or unhealthy in so peculiar a manner that their self-fertilized seedlings, many hundreds in number, were rendered inferior in height, weight, constitutional vigour, and fertility to their crossed offspring. Moreover, this belief cannot be extended to the strongly marked advantages which invariably follow, as far as my experience serves, from intercrossing the individuals of the same variety or of distinct varieties, if these have been subjected during some generations to different conditions.

It is obvious that the exposure of two sets of plants during several generations to different conditions can lead to no beneficial results, as far as crossing is concerned, unless their sexual elements are thus affected. That every organism is acted on to a certain extent by a change in its environment, will not, I presume, be disputed. It is hardly necessary to advance evidence on this head; we can perceive the difference between / individual plants of the same species which have grown in somewhat more shady or sunny, dry or damp places. Plants which have been propagated for some generations under different climates or at different seasons of the year transmit different constitutions to their seedlings. Under such circumstances, the chemical constitution of their fluids and the nature of their tissues are often modified.[9] Many other such facts could be adduced. In short, every alteration in the function of a part is probably connected with some corresponding, though often quite imperceptible change in structure or composition.

Whatever affects an organism in any way, likewise tends to act on its sexual elements. We see this in the inheritance of newly acquired modifications, such as those from the increased use or disuse of a part, and even from mutilations if followed by disease.[10] We have abundant evidence how susceptible the reproductive system is to changed conditions, in the many instances of animals rendered sterile by confinement; so that they will not unite, or if they unite do not produce offspring, though the confinement may be far from close; and of plants rendered sterile by cultivation. But hardly any cases afford more striking evidence how powerfully a change in the conditions of

[9] Numerous cases together with references are given in my *Variation under Domestication*, ch. xxiii, 2nd edit., vol. ii, p. 264. With respect to animals, Mr Brackenridge has well shown (*A Contribution to the Theory of Diathesis*, Edinburgh, 1869) that the different organs of animals are excited into different degrees of activity by differences of temperature and food, and become to a certain extent adapted to them.

[10] *Variation under Domestication*, ch. xii, 2nd edit., vol. i, p. 466.

life acts on the sexual elements, than those already given, of plants which are completely self-sterile in one country, and when brought to another, yield, even / in the first generation, a fair supply of self-fertilized seeds.

But it may be said, granting that changed conditions act on the sexual elements, how can two or more plants growing close together, either in their native country or in a garden, be differently acted on, inasmuch as they appear to be exposed to exactly the same conditions? Although this question has been already considered, it deserves further consideration from several points of view. In my experiments with *Digitalis purpurea*, some flowers on a wild plant were self-fertilized, and others were crossed with pollen from another plant growing within two or three feet's distance. The crossed and self-fertilized plants raised from the seeds thus obtained, produced flower-stems in number as 100 to 47, and in average height as 100 to 70. Therefore the cross between these two plants was highly beneficial; and how could their sexual elements have been differentiated by exposure to different conditions? If the progenitors of the two plants had lived on the same spot during the last score of generations, and had never been crossed with any plant beyond the distance of a few feet, in all probability their offspring would have been reduced to the same state as some of the plants in my experiments – such as the intercrossed plants of the ninth generation of Ipomoea – or the self-fertilized plants of the eighth generation of Mimulus – or the offspring from flowers on the same plant – and in this case a cross between the two plants of Digitalis would have done no good. But seeds are often widely dispersed by natural means, and one of the above two plants or one of their ancestors may have come from a distance, from a more shady or sunny, dry or moist place, or from a different kind of soil containing other organic or / inorganic matter. We known from the admirable researches of Messrs Lawes and Gilbert[11] that different plants require and consume very different amounts of inorganic matter. But the amount in the soil would probably not make so great a difference to the several individuals of any particular species as might at first be expected; for the surrounding species with different requirements would tend, from existing in greater or lesser numbers, to keep each species in a sort of equilibrium, with respect to what it could obtain from the soil. So it would be even with respect to moisture during dry

[11] *Journal of the Royal Agricultural Soc. of England*, vol. xxiv, part i.

seasons; and how powerful is the influence of a little more or less moisture in the soil on the presence and distribution of plants, is often well shown in old pasture fields which still retain traces of former ridges and furrows. Nevertheless, as the proportional numbers of the surrounding plants in two neighbouring places is rarely exactly the same, the individuals of the same species will be subjected to somewhat different conditions with respect to what they can absorb from the soil. It is surprising how the free growth of one set of plants affects others growing mingled with them; I allowed the plants on rather more than a square yard of turf which had been closely mown for several years, to grow up; and nine species out of twenty were thus exterminated; but whether this was altogether due to the kinds which grew up robbing the others of nutriment, I do not know.

Seeds often lie dormant for several years in the ground, and germinate when brought near the surface by any means, as by burrowing animals. They would probably be affected by the mere circumstance of having / long laid dormant; for gardeners believe that the production of double flowers and of fruit is thus influenced. Seeds, moreover, which were matured during different seasons, will have been subjected during the whole course of their development to different degrees of heat and moisture.

It was shown in the last chapter that pollen is often carried by insects to a considerable distance from plant to plant. Therefore one of the parents or ancestors of our two plants of Digitalis may have been crossed by a distant plant growing under somewhat different conditions. Plants thus crossed often produce an unusually large number of seeds; a striking instance of this fact is afforded by the Bignonia, previously mentioned, which was fertilized by Fritz Müller with pollen from some adjoining plants and set hardly any seed, but when fertilized with pollen from a distant plant, was highly fertile. Seedlings from a cross of this kind grow with great vigour, and transmit their vigour to their descendants. These, therefore, in the struggle for life, will generally beat and exterminate the seedlings from plants which have long grown near together under the same conditions, and will thus tend to spread.

When two varieties which present well-marked differences are crossed, their descendants in the later generations differ greatly from one another in external characters; and this is due to the augmentation or obliteration of some of these characters, and to the reappearance of former ones through reversion; and so it will be, as we may feel almost

379

sure, with any slight differences in the constitution of their sexual elements. Anyhow, my experiments indicate that crossing plants which have been long subjected to almost though not quite the same conditions, is the / most powerful of all the means for retaining some degree of differentiation in the sexual elements, as shown by the superiority in the later generations of the intercrossed over the self-fertilized seedlings. Nevertheless, the continued intercrossing of plants thus treated does tend to obliterate such differentiation, as may be inferred from the lessened benefit derived from intercrossing such plants, in comparison with that from a cross with a fresh stock. It seems probable, as I may add, that seeds have acquired their endless curious adaptations for wide dissemination,[12] not only that the seedlings should thus be enabled to find new and fitting homes, but that the individuals which have been long subjected to the same conditions should occasionally intercross with a fresh stock.

From the foregoing several considerations we may, I think, conclude that in the above case of the Digitalis, and even in that of plants which have grown for thousands of generations in the same district, as must often have occurred with species having a much restricted range, we are apt to over-estimate the degree to which the individuals have been subjected to absolutely the same conditions. There is at least no difficulty in believing that such plants have been subjected to sufficiently distinct conditions to differentiate their sexual elements; for we know that a plant propagated for some generations in another garden in the same district serves as a fresh stock and has high fertilizing powers. The curious cases of plants which can fertilize and be fertilized by any other individual of the same species, but are altogether sterile with their own pollen, become intelligible, if the view here propounded is correct, namely, that the individuals of the / same species growing in a state of nature near together, have not really been subjected during several previous generations to quite the same conditions.

Some naturalists assume that there is an innate tendency in all beings to vary and to advance in organization, independently of external agencies; and they would, I presume, thus explain the slight differences which distinguish all the individuals of the same species both in external characters and in constitution, as well as the greater

---

[12] See Professor Hildebrand's excellent treatise, *Verbreitungsmittel der Pflanzen*, 1873.

differences in both respects between nearly allied varieties. No two individuals can be found quite alike; thus if we sow a number of seeds from the same capsule under as nearly as possible the same conditions, they germinate at different rates and grow more or less vigorously. They resist cold and other unfavourable conditions differently. They would in all probability, as we know to be the case with animals of the same species, be somewhat differently acted on by the same poison, or by the same disease. They have different powers[13] of transmitting their characters to their offspring; and many analogous facts could be given. Now, if it were true that plants growing near together in a state of nature had been subjected during many previous generations to absolutely the same conditions, such as those just specified would be quite inexplicable; but they are to a certain extent intelligible in accordance with the views just advanced.

As most of the plants on which I experimented were grown in my garden or in pots under glass, a few words must be added on the conditions to which they were exposed, as well as on the effects of cultivation. When a species is first brought under culture, it may / or may not be subjected to a change of climate, but it is always grown in ground broken up, and more or less manured; it is also saved from competition with other plants. The paramount importance of this latter circumstance is proved by the multitude of species which flourish and multiply in a garden, but cannot exist unless they are protected from other plants. When thus saved from competition they are able to get whatever they require from the soil, probably often in excess; and they are thus subjected to a great change of conditions. It is probably in chief part owing to this cause that all plants with rare exceptions vary after being cultivated for some generations. The individuals which have already begun to vary will intercross one with another by the aid of insects; and this accounts for the extreme diversity of character which many of our long cultivated plants exhibit. But it should be observed that the result will be largely determined by the degree of their variability and by the frequency of the intercrosses; for if a plant varies very little, like most species in a state of nature, frequent intercrosses tend to give uniformity of character to it.

I have attempted to show that with plants growing naturally in the same district, except in the unusual case of each individual being surrounded by exactly the same proportional numbers of other species

[13] Vilmorin, as quoted by Verlot, *Des Variétés*, pp. 32, 38, 39.

having certain powers of absorption, each will be subjected to slightly different conditions. This does not apply to the individuals of the same species when cultivated in cleared ground in the same garden. But if their flowers are visited by insects, they will intercross; and this will give to their sexual elements during a considerable number of generations a sufficient amount of differentiation for a cross to be beneficial. Moreover, / seeds are frequently exchanged or procured from other gardens having a different kind of soil; and the individuals of the same cultivated species will thus be subjected to a change of conditions. If the flowers are not visited by our native insects, or very rarely so, as in the case of the common and sweet pea, and apparently in that of the tobacco when kept in a hothouse, any differentiation in the sexual elements caused by intercrosses will tend to disappear. This appears to have occurred with the plants just mentioned, for they were not benefited by being crossed one with another, though they were greatly benefited by a cross with a fresh stock.

I have been led to the views just advanced with respect to the causes of the differentiation of the sexual elements and of the variability of our garden plants, by the results of my various experiments, and more especially by the four cases in which extremely inconstant species, after having been self-fertilized and grown under closely similar conditions for several generations, produced flowers of a uniform and constant tint. These conditions were nearly the same as those to which plants, growing in a garden clear of weeds, are subjected, if they are propagated by self-fertilized seeds on the same spot. The plants in pots were, however, exposed to less severe fluctuations of climate than those out of doors; but their conditions, though closely uniform for all the individuals of the same generation, differed somewhat in the successive generations. Now, under these circumstances, the sexual elements of the plants which were intercrossed in each generation retained sufficient differentiation during several years for their offspring to be superior to the self-fertilized, but this superiority gradually and manifestly decreased, as was shown by the difference / in the result between a cross with one of the intercrossed plants and with a fresh stock. These intercrossed plants tended also in a few cses to become somewhat more uniform in some of their external characters than they were at first. With respect to the plants which were self-fertilized in each generation, their sexual elements apparently lost, after some years, all differentiation, for a cross between them did no more good than a cross between the flowers on the same plant. But it is a still more remarkable fact, that although the seedlings of Mimulus, Ipomoea, Dianthus, and Petunia which were first raised, varied excessively in the colour of their flowers, their offspring, after being self-fertilized and grown under uniform conditions for some generations, bore flowers almost as uniform in tint as those on a natural species. In one case also the plants themselves became remarkably uniform in height.

The conclusion that the advantages of a cross depend altogether on the differentiation of the sexual elements, harmonizes perfectly with the fact that an occasional and slight change in the conditions of life is beneficial to all plants and animals.[14] But the offspring from a cross between organisms which have been exposed to different conditions, profit in an incomparably higher degree than do young or old beings from a mere change in their conditions. In this latter case we never see anything like the effect which generally follows from a cross with another individual, especially from a cross with a fresh stock. This might, perhaps, have been expected, for the blending together of the sexual elements of two differentiated beings will affect the whole constitution at / a very early period of life, whilst the organization is highly flexible. We have, moreover, reason to believe that changed conditions generally act differently on the several parts or organs of the same individual;[15] and if we may further believe that these now slightly differentiated parts react on one another, the harmony between the beneficial effects on the individual due to changed conditions, and those due to the interaction of differentiated sexual elements, becomes still closer.

That wonderfully accurate observer, Sprengel, who first showed how important a part insects play in the fertilization of flowers, called his book *The Secret of Nature Displayed*; yet he only occasionally saw that the object for which so many curious and beautiful adaptations have been acquired, was the cross-fertilization of distinct plants; and he knew nothing of the benefits which the offspring thus receive in growth, vigour, and fertility. But the veil of secrecy is as yet far from lifted; nor will it be, until we can say why it is beneficial that the sexual elements should be differentiated to a certain extent, and why, if the differentiation be carried still further, injury follows. It is an extraordinary fact

[14] I have given sufficient evidence on this head in my *Variation under Domestication*, ch. xviii, vol. ii, 2nd edit., p. 127.
[15] See, for instance, Brackenridge, *Theory of Diathesis*, Edinburgh, 1869.

that with many species, even when growing under their natural conditions, flowers fertilized with their own pollen are either absolutely or in some degree sterile; if fertilized with pollen from another flower on the same plant, they are sometimes, though rarely, a little more fertile; if fertilized with pollen from another individual or variety of the same species, they are fully fertile; but if with pollen from a distinct species, they are sterile in all possible degrees, until utter sterility is reached. / We thus have a long series with absolute sterility at the two ends; at one end due to the sexual elements not having been sufficiently differentiated, and at the other end to their having been differentiated in too great a degree, or in some peculiar manner.

The fertilization of one of the higher plants depends, in the first place, on the mutual action of the pollen-grains and the stigmatic secretion or tissues, and afterwards on the mutual action of the contents of the pollen-grains and ovules. Both actions, judging from the increased fertility of the parent-plants and from the increased powers of growth in the offspring, are favoured by some degree of differentiation in the elements which interact and unite so as to form a new being. Here we have some analogy with chemical affinity or attraction, which comes into play only between atoms or molecules of a different nature. As Professor Miller remarks: 'Generally speaking, the greater the difference in the properties of two bodies, the more intense is their tendency to mutual chemical action. . . . But between bodies of a similar character the tendency to unite is feeble.'[16] This latter proposition accords well with the feeble effects of a plant's own pollen on the fertility of the mother-plant and on the growth of the offspring; and the former proposition accords well with the powerful influence in both ways of pollen from an individual which has been differentiated by exposure to changed conditions, or by so-called spontaneous variation. But the analogy fails when we turn to the negative or weak effects of pollen from one species on a distinct species; for although some substances which are extremely dissimilar, for instance, carbon and / chlorine, have a very feeble affinity for each other, yet it cannot be said that the weakness of the affinity depends in such cases on the extent to which the substances differ. It is not known why a certain amount of differentiation is necessary or favourable for

[16] *Elements of Chemistry*, 4th edit., 1867, part i, p. 11. Dr Frankland informs me that similar views with respect to chemical affinity are generally accepted by chemists.

the chemical affinity or union of two substances, any more than for the fertilization or union of two organisms.

Mr Herbert Spencer has discussed this whole subject at great length, and after stating that all the forces throughout nature tend towards an equilibrium, remarks, 'that the need of this union of sperm-cell and germ-cell is the need for overthrowing this equilibrium and re-establishing active molecular change in the detached germ – a result which is probably effected by mixing the slightly-different physiological units of slightly different individuals.'[17] But we must not allow this highly generalized view, or the analogy of chemical affinity, to conceal from us our ignorance. We do not know what is the nature or degree of the differentiation in the sexual elements which is favourable for union, and what is injurious for union, as in the case of distinct species. We cannot say why the individuals of certain species profit greatly, and others very little by being crossed. There are some few species which have been self-fertilized for a vast number of generations, and yet are vigorous enough to compete successfully with a host of surrounding plants. Highly self-fertile varieties sometimes arise / among plants which have been self-fertilized and grown under uniform conditions during several generations. We can form no conception why the advantage from a cross is sometimes directed exclusively to the vegetative system, and sometimes to the reproductive system, but commonly to both. It is equally inconceivable why some individuals of the same species should be sterile, whilst others are fully fertile with their own pollen; why a change of climate should either lessen or increase the sterility of self-sterile species; and why the individuals of some species should be even more fertile with pollen from a distinct species than with their own pollen. And so it is with many other facts, which are so obscure that we stand in awe before the mystery of life.

Under a practical point of view, agriculturists and horticulturists may learn something from the conclusions at which we have arrived. First, we see that the injury from the close breeding of animals and from the self-fertilization of plants, does not necessarily depend on any tendency

[17] *Principles of Biology*, vol. i, p. 274, 1864. In my *Origin of Species*, published in 1859, I spoke of the good effects from slight changes in the conditions of life and from cross-fertilization, and of the evil effects from great changes in the conditions and from crossing widely distinct forms (i.e., species), as a series of facts 'connected together by some common but unknown bond, which is essentially related to the principle of life.'

to disease or weakness of constitution common to the related parents, and only indirectly on their relationship, in so far as they are apt to resemble each other in all respects, including their sexual nature. And, secondly, that the advantages of cross-fertilization depend on the sexual elements of the parents having become in some degree differentiated by the exposure of their progenitors to different conditions, or from their having intercrossed with individuals thus exposed, or, lastly, from what we call in our ignorance spontaneous variation. He therefore who wishes to pair closely related animals ought to keep them under conditions as different as possible. Some few breeders, guided by their keen powers of / observation, have acted on this principle, and have kept stocks of the same animals at two or more distant and differently situated farms. They have then coupled the individuals from these farms with excellent results.[18] This same plan is also unconsciously followed whenever the males, reared in one place, are let out for propagation to breeders in other places. As some kinds of plants suffer much more from self-fertilization than do others, so it probably is with animals from too close interbreeding. The effects of close interbreeding on animals, judging again from plants, would be deterioration in general vigour, including fertility, with no necessary loss of excellence of form; and this seems to be the usual result.

It is a common practice with horticulturists to obtain seeds from another place having a very different soil, so as to avoid raising plants for a long succession of generations under the same conditions; but with all the species which freely intercross by the aid of insects or the wind, it would be an incomparably better plan to obtain seeds of the required variety, which had been raised for some generations under as different conditions as possible, and sow them in alternate rows with seeds matured in the old garden. The two stocks would then intercross, with a thorough blending of their whole organizations, and with no loss of purity to the variety; and this would yield far more favourable results than a mere exchange of seeds. We have seen in my experiments how wonderfully the offspring profited in height, weight, hardiness, and fertility, by crosses of this kind. For instance, plants of Ipomoea thus crossed were to the intercrossed plants of the same stock, with which they grew in competition, / as 100 to 78 in height,

---

[18] *Variation of Animals and Plants under Domestication*, ch. xvii, 2nd edit., vol. ii, pp. 98, 105.

and as 100 to 51 in fertility; and plants of Eschscholtzia similarly compared were as 100 to 45 in fertility. In comparison with self-fertilized plants the results are still more striking; thus cabbages derived from a cross with a fresh stock were to the self-fertilized as 100 to 22 in weight.

Florists may learn from the four cases which have been fully described, that they have the power of fixing each fleeting variety of colour, if they will fertilize the flowers of the desired kind with their own pollen for half-a-dozen generations, and grow the seedlings under the same conditions. But a cross with any other individual of the same variety must be carefully prevented, as each has its own peculiar constitution. After a dozen generations of self-fertilization, it is probable that the new variety would remain constant even if grown under somewhat different conditions; and there would no longer be any necessity to guard against intercrosses between the individuals of the same variety.

With respect to mankind, my son George has endeavoured to discover by a statistical investigation[19] whether the marriages of first cousins are at all injurious, although this is a degree of relationship which would not be objected to in our domestic animals; and he has come to the conclusion from his own researches and those of Dr Mitchell that the evidence as to any evil thus caused is conflicting, but on the whole points to its being very small. From the facts given in this volume we may infer that with mankind the marriages of nearly related persons, some of whose parents and ancestors had lived under very different conditions, would be much less injurious than that of persons who had always lived in the same / place and followed the same habits of life. Nor can I see reason to doubt that the widely different habits of life of men and women in civilized nations, especially among the upper classes, would tend to counterbalance any evil from marriages between healthy and somewhat closely related persons.

Under a theoretical point of view it is some gain to science to know that numberless structures in hermaphrodite plants, and probably in hermaphrodite animals, are special adaptations for securing an occasional cross between two individuals; and that the advantages from such a cross depend altogether on the beings which are united, or

[19] *Journal of Statistical Soc.*, June, 1875, p. 153; and *Fortnightly Review*, June, 1875.

their progenitors, having had their sexual elements somewhat differentiated, so that the embryo is benefited in the same manner as is a mature plant or animal by a slight change in its conditions of life, although in a much higher degree.

Another and more important result may be deduced from my observations. Eggs and seeds are highly serviceable as a means of dissemination, but we now know that fertile eggs can be produced without the aid of the male. There are also many other methods by which organisms can be propagated asexually. Why then have the two sexes been developed, and why do males exist which cannot themselves produce offspring? The answer lies, as I can hardly doubt, in the great good which is derived from the fusion of two somewhat differentiated individuals; and with the exception of the lowest organisms this is possible only by means of the sexual elements, these consisting of cells separated from the body, containing the germs of every part, and capable of being fused completely together.

It has been shown in the present volume that the / offspring from the union of two distinct individuals, especially if their progenitors have been subjected to very different conditions, have an immense advantage in height, weight, constitutional vigour and fertility over the self-fertilized offspring from one of the same parents. And this fact is amply sufficient to account for the development of the sexual elements, that is, for the genesis of the two sexes.

It is a different question why the two sexes are sometimes combined in the same individual and are sometimes separated. As with many of the lowest plants and animals the conjugation of two individuals which are either quite similar or in some degree different, is a common phenomenon, it seems probable, as remarked in the last chapter, that the sexes were primordially separate. The individual which receives the contents of the other, may be called the female; and the other, which is often smaller and more locomotive, may be called the male; though these sexual names ought hardly to be applied as long as the whole contents of the two forms are blended into one. The object gained by the two sexes becoming united in the same hermaphrodite form probably is to allow of occasional or frequent self-fertilization, so as to ensure the propagation of the species, more especially in the case of organisms affixed for life to the same spot. There does not seem to be any great difficulty in understanding how an organism, formed by the conjugation of two individuals which represented the two incipient sexes, might give rise by budding first to a monoecious and then to an hermaphrodite form; and in the case of animals even without budding to an hermaphrodite form, for the bilateral structure of animals perhaps indicates that they were aboriginally formed by the fusion of two individuals. /

It is a more difficult problem why some plants and apparently all the higher animals, after becoming hermaphrodites, have since had their sexes re-separated. This separation has been attributed by some naturalists to the advantages which follow from a division of physiological labour. The principle is intelligible when the same organ has to perform at the same time diverse functions; but it is not obvious why the male and female glands when placed in different parts of the same compound or simple individual, should not perform their functions equally well as when placed in two distinct individuals. In some instances the sexes may have been re-separated for the sake of preventing too frequent self-fertilization; but this explanation does not seem probable, as the same end might have been gained by other and simpler means, for instance dichogamy. It may be that the production of the male and female reproductive elements and the maturation of the ovules was too great a strain and expenditure of vital force for a single individual to withstand, if endowed with a highly complex organization; and that at the same time there was no need for all the individuals to produce young, and consequently that no injury, on the contrary, good resulted from half of them, or the males, failing to produce offspring.

There is another subject on which some light is thrown by the facts given in this volume, namely, hybridization. It is notorious that when distinct species of plants are crossed, they produce with the rarest exceptions fewer seeds than the normal number. This unproductiveness varies in different species up to sterility so complete that not even an empty capsule is formed; and all experimentalists have found that it is much influenced by the conditions to which the / crossed species are subjected. A plant's own pollen is strongly prepotent over that of any other species, so that if it is placed on the stigma some time after foreign pollen has been applied to it, any effect from the latter is quite obliterated. It is also notorious that not only the parent species, but the hybrids raised from them are more or less sterile; and that their pollen is often in a more or less aborted condition. The degree of sterility of various hybrids does not always strictly correspond with the degree of difficulty in uniting the parent forms. When hybrids are capable of breeding *inter se*, their descendants are more or less sterile, and they often become still more sterile in the later generations; but then close interbreeding has hitherto been practised in all such cases. The more sterile hybrids are sometimes much dwarfed in stature, and have a feeble constitution. Other facts could be given, but these will suffice for us. Naturalists formerly attributed all these results to the difference between species being fundamentally distinct from that between the varieties of the same species; and this is still the verdict of some naturalists.

The results of my experiments in self-fertilizing and cross-fertilizing the individuals or the varieties of the same species, are strikingly analogous with those just given, though in a reversed manner. With the majority of species flowers fertilized with their own pollen yield fewer sometimes much fewer seeds, than those fertilized with pollen from another individual or variety. Some self-fertilized flowers are absolutely sterile; but the degree of their sterility is largely determined by the conditions to which the parent plants have been exposed, as was well exemplified in the case of Eschscholtzia and Abutilon. The effects of pollen from the same plant are obliterated by the prepotent influence / of pollen from another individual or variety, although the latter may have been placed on the stigma some hours afterwards. The offspring from self-fertilized flowers are themselves more or less sterile, sometimes highly sterile, and their pollen is sometimes in an imperfect condition; but I have not met with any case of complete sterility in self-fertilized seedlings, as is so common with hybrids. The degree of their sterility does not correspond with that of the parent-plants when first self-fertilized. The offspring of self-fertilized plants suffer in stature, weight, and constitutional vigour more frequently and in a greater degree than do the hybrid offspring of the greater number of crossed species. Decreased height is transmitted to the next generation, but I did not ascertain whether this applies to decreased fertility.

I have elsewhere shown[20] that by uniting in various ways dimorphic or trimorphic heterostyled plants, which belong to the same un-doubted species, we get another series of results exactly parallel with those from crossing distinct species. Plants illegitimately fertilized with pollen from a distinct plant belonging to the same form, yield fewer, often much fewer seeds, than they do when legitimately fertilized with pollen from a plant belonging to a distinct form. They sometimes yield

---

[20] *The Different Forms of Flowers on Plants of the same species*, 1877, p. 240.

no seed, not even an empty capsule, like a species fertilized with pollen from a distinct genus. The degree of sterility is much affected by the conditions to which the plants have been subjected. The pollen from a distinct form is strongly prepotent over that from the same form, although the former may have been placed on the stigma many hours afterwards. / The offspring from a union between plants of the same form are more or less sterile, like hybrids, and have their pollen in a more or less aborted condition; and some of the seedlings are as barren and as dwarfed as the most barren hybrid. They also resemble hybrids in several other respects, which need not here be specified in detail – such as their sterility not corresponding in degree with that of the parent plants – the unequal sterility of the latter, when reciprocally united – and the varying sterility of the seedlings raised from the same seed-capsule.

We thus have two grand classes of cases giving results which correspond in the most striking manner with those which follow from the crossing of so-called true and distinct species. With respect to the difference between seedlings raised from cross and self-fertilized flowers, there is good evidence that this depends altogether on whether the sexual elements of the parents have been sufficiently differentiated, by exposure to different conditions or by spontaneous variation. The manner in which plants have been rendered hetero-styled is an obscure subject, but it is probable that the two or three forms first became adapted for mutual fertilization, that is for cross-fertilization, through the variation of their stamens and pistils in length, and that afterwards their pollen and ovules became co-adapted; the greater or less sterility of any one form with pollen from the same form being an incidental result.[21] Anyhow, the two or three forms of heterostyled species belong to the same species as certainly as do the two sexes of any one species. We have therefore no right to maintain that the sterility of species when first crossed and of their hybrid offspring, / is determined by some cause fundamentally dif-ferent from that which determines the sterility of the individuals both of ordinary and of heterostyled plants when united in various ways. Nevertheless, I am aware that it will take many years to remove this prejudice.

There is hardly anything more wonderful in nature than the sensitiveness of the sexual elements to external influences, and the

---

[21] This subject has been discussed in my *Different Forms of Flowers etc.*, pp. 260–8.

delicacy of their affinities. We see this in slight changes in the conditions of life being favourable to the fertility and vigour of the parents, while certain other and not great changes cause them to be quite sterile without any apparent injury to their health. We see how sensitive the sexual elements of those plants must be, which are completely sterile with their own pollen, but are fertile with that of any other individual of the same species. Such plants become either more or less self-sterile if subjected to changed conditions, although the change may be far from great. The ovules of a heterostyled trimorphic plant are affected very differently by pollen from the three sets of stamens belonging to the same species. With ordinary plants the pollen of another variety or merely of another individual of the same variety is often strongly prepotent over its own pollen, when both are placed at the same time on the same stigma. In those great families of plants containing many thousand allied species, the stigma of each distinguishes with unerring certainty its own pollen from that of every other species.

There can be no doubt that the sterility of distinct species when first crossed, and of their hybrid offspring, depends exclusively on the nature or affinities of their sexual elements. We see this in the want of any close correspondence between the degree / of sterility and the amount of external difference in the species which are crossed; and still more clearly in the wide difference in the results of crossing reciprocally the same two species; that is, when species A is crossed with pollen from B, and then B is crossed with pollen from A. Bearing in mind what has just been said on the extreme sensitiveness and delicate affinities of the reproductive system, why should we feel any surprise at the sexual elements of those forms, which we call species, having been differentiated in such a manner that they are incapable or only feebly capable of acting on one another? We know that species have generally lived under the same conditions, and have retained their own proper characters, for a much longer period than varieties. Long-continued domestication eliminates, as I have shown in my *Variation under Domestication*, the mutual sterility which distinct species lately taken from a state of nature almost always exhibit when intercrossed; and we can thus understand the fact that the most different domestic races of animals are not mutually sterile. But whether this holds good with cultivated varieties of plants is not known, though some facts indicate that it does. The elimination of sterility through long-continued domestication may probably be attributed to the varying conditions to which our domestic animals have been subjected; and no doubt it is owing to this same cause that they withstand great and sudden changes in their conditions of life with far less loss of fertility than do natural species. From these several considerations it appears probable that the difference in the affinities of the sexual elements of distinct species, on which their mutual incapacity for breeding together depends, is caused by their having been habituated for a very long period each to its own conditions, and to the sexual elements / having thus acquired firmly fixed affinities. However this may be, with the two great classes of cases before us, namely, those relating to the self-fertilization and cross-fertilization of the individuals of the same species, and those relating to the illegitimate and legitimate unions of heterostyled plants, it is quite unjustifiable to assume that the sterility of species when first crossed and of their hybrid offspring, indicates that they differ in some fundamental manner from the varieties or individuals of the same species. /

INDEX

Abutilon darwinii, self-sterile in Brazil, 333, 358; moderately self-fertile in England, 344; fertilized by birds, 371

Acacia sphaerocephala, 406

Acanthaceae, 96

Aconitum napellus, 431

Adlumia cirrhosa, 366

Adonis aestivalis, 128; measurements, 128; relative heights of crossed and self-fertilized plants, 277; self-fertile, 365

Ajuga reptans, 368

Allium cepa (blood-red var.), 369

Anagallis collina (var. *grandiflora*), 217, 367; measurements, 218; seeds, 316, 323, 325

Anderson, J., on the Calceolaria, 87; removing the corollas, 423

Anemone, 396

Anemophilous plants, 401; often diclinous, 411

Antirrhinum majus (red var.), 363; perforated corolla, 432 (white var.), 363 (peloric var.), 363

Apium petroselinum, 172; result of experiments, 277

Argemone ochroleuca, 366

Aristotle on bees frequenting flowers of the same species, 418

Aristolochia, 420

Arum maculatum, 420

Bailey, Mr, perforation of corolla, 430

Bartonia aurea, 170; measurements, 170, 171; result of experiments, 277

Bartsia odontites, 369

Beal, W. J., sterility of *Kalmia latifolia*, 359; on nectar in *Ribes aureum*, 435

Bean, the common, 435

Bees distinguish colours, 373; frequent the flowers of the same species, 418, 423; guided by coloured corolla, 423; powers of vision and discrimination, 425; memory, 426; unattracted by odour of certain flowers, 426; industry, 427; profit by the corolla perforated by humble-bees, 430; skill in working, 431; habit, 434; foresight, 436

humble, recognize varieties as of one species, 419; colour not the sole guide, 424; rate of flying, 427; number of flowers visited, 428; corolla perforated by, 429, 436; skill and judgement, 432

Belt, Mr, the hairs of *Digitalis purpurea*, 82; *Phaseolous multiflorus*, 151; not visited by bees in Nicaragua, 360; humming-birds carrying pollen, 371; secretion of nectar, 404; in *Acacia sphaerocephalus* and passion-flower, 406; perforation of corolla, 433

Bennett, A. W., on *Viola tricolor*, 123; structure of *Impatiens fulva*, 367; plants flowering in winter, 386; bees frequenting flowers of same species, 419 /

Bentham, on protection of the stigma in *Synaphea*, 415

Beta vulgaris, 228; measurements, 229, 230; crossed not exceeded by self-fertilized, 289, 367; prepotency of other pollen, 399

Bignonia, 363

Birds means of fertilization, 371

Blackley, Mr, weights of pollen of anemophilous plants, 377, 378; on anthers of rye, 378; pollen carried by wind, experiments with a kite, 408

Boraginaceae, 185

Borago officinalis, 185, 276; measurements, 186; early flowering of crossed, 293; seeds, 323; partially self-sterile, 362

Boulger, Mr, on moths frequenting Petunias, 188

Brackenridge, Mr, organism of animals affected by temperature and food, 446; different effect of changed conditions, 455

Brassica oleracea, 98; measurements, 100; weight, 101, 102; remarks on experiments, 262; superiority of crossed, 288; period of flowering, 292; seeds, 322; self-fertile, 365

napus, 395

rapa, 395

Brisout, M., insects frequenting flowers of same species, 422

Broom, 163

Brugmansia, 371; humming-birds boring the flower, 435

Bulrush, weight of pollen produced by one plant, 407, 408

Bundy, Mr, *Ribes* perforated by bees, 435

Burbidge, references on the germination of small seeds, 355

Bütschli, O., sexual relations, 412

Cabbage, 98; affected by pollen of purple bastard, 379; prepotency of other pollen, 395, 399

Cabbage, Ragged Jack, 397

Calceolaria, 87, 369

Calluna vulgaris, 424

Campanula carpathica, 174, 364

Campanulaceae, 174

Candolle, A. de, on ascending a mountain the flowers of the same species disappear abruptly, 391

Canna warscewiczi, 230; result of crossed and self-fertilized, 278; period of flowering, 294; seeds, 323, 325; highly self-fertile, 369

Cannaceae, 230

Carduus arctioides, 404

Carnation, 132

Carrière, relative period of the maturity of the sexual elements on same flower, 446

Caryophyllaceae, 130

Caspary, Professor, on *Coryalis cava*, 331; *Nymphaeaceae*, 358; *Euryale ferox*, 365; on flowers of water-lilies, 392

Cecropia, food-bodies of, 404

Centradenia floribunda,. 364

Cereals, grains of, 354

Cheeseman, Mr, on Orchids in New Zealand, 392

Chenopodiaceae, 228

Cineraria, 335

Clarkia elegans, 169; measurements, 170; early flowering of self-fertilized, 294, 296; seeds, 316

Cleistogamic flowers, 90

Coe, Mr, crossing *Phaseolous vulgaris*, 153

Colgate, R., red clover never sucked by hive-bees in New Zealand, 361

Colour, uniform, of flowers on plants self-fertilized and grown under similar conditions for several generations, 306, 307

Colours of flowers attractive to insects, 372; not the sole guide to bees, 424

Compositae, 173

Coniferae, 402

Convolvulus major, 28

tricolor, 55 /

Corolla, removal of, 423; perforation by bees, 428

Coronilla, 407

Corydalis cava, 331, 358

halleri, 331

Corydalis cava – *continued*
  intermedia, 331
  lutea, 359
  ochroleuca, 359
Corydalis solida, 358
Corylus avellana, 390
Cowslip, 219
Crinum, 396
Crossed plants, greater constitutional vigour of, 285
Cross-fertilization, 371; *see* Fertilization
Crossing flowers on the same plant, effects of, 297
Cruciferae, 98
Crüger, Dr, secretion of sweet fluid in *Mrcgraviaceae*, 407
Cuphea purpurea, 323, 362
Cycadeae, 402
Cyclamen persicum, 215; measurements, 216; early flowering of crossed, 293; seeds, 317, 323; self-sterile, 362, self-fertilization injurious, 448
  repandum, 215
Cytisus laburnum, 362

Dandelion, number of pollen-grains, 377
Darwin, C., self-fertilization in *Pisum sativum*, 161; sexual affinities, 209; on *Primula*, 219; bud variation, 298; constitutional vigour from cross parentage in common pea, 305; hybrids of *Gladiolus* and *Cistus*,06; *Phaseolus multiflorus*, 360; nectar in orchids, 407; on cross-fertilization, 440, 442, 443; inheritance of acquired modifications, 451; change in the conditionns of life beneficial to plants and animals, 459
Darwin, F., structure of *Phaseolus multiflorus*, 150; *Pteris aquilina*, 405; on nectar glands, 406; perforation of *Lathyrus sylvestris*, 432
  G., on marriages with first cousins, 465

Decaisne on *Delphinium consolida*, 129
De Candolle, nectar as an excretion, 403
Delphinium consolida, 129; measurements, 130; seeds, 322; partially sterile, 358; corolla removed, 423
Delpino, Professor, *Viola tricolor*, 123; *Phaseolus multiflorus*, 150; intercrossing of sweetpea, 156; *Lobelia ramosa*, 176; structure of the *Cannaceae*, 230; wind and water carrying pollen, 372; *Juglans regia*, 391; anemophilous plants, 401; fertilization of *Plantago*, 403; excretion of nectar, 404, 407; secretion of nectar to defend the plant, 406, 407; anemophilous and entomophilous plants, 411; dioecious plants, 417
Denny, *Pelargonium zonale*, 142
Diagram showing mean height of *Ipomoea purpurea*, 53
Dianthus caryophyllus, 132; crossed and self-fertilized, 133–6; measurements, 135–8; cross with fresh stock, 136; weight of seed, 139; colour of flowers, 139; remarks on experiments, 263, 274; early flowering of crossed, 292; uniform colour of self-fertilized, 309; seeds, 316, 319, 323, 325; few capsules, 360
Dickie, Dr, self-fertilization in *Cannaceae*, 230
Dictamnus fraxinella, 419
Digitalis purpurea, 81; measurements, 84–7; effects of intercrossing, 85, 299; superiority of crossed, 288, 452; self-sterile, 363
Dipsaceae, 172 /
Dobbs, bees frequenting flowers of same species, 419
Dodel, Dr A., sexual reproduction, 412
Dodel, Dr A., sexual reproduction, 412
Duhamel on *Raphanus sativus*, 395
Dunal, nectar as an excretion, 403
Dyer, Mr Thiselton, on *Lobelia ramosa*, 176; on *Cineraria*, 335; origin of Hermaphroditism, 413

Earley, W., self-fertilization of *Lathyrus odoratus*, 153

Eaton, Rev. A. E., on *Pringlea*, 410

Engelmann, development of sexual forms, 412

Engler, Dr, on dichogamous *Saxifraga*, 440

Entomophilous plants, 411

Epipactis latifolia, attractive only to wasps, 376, 426

Erica tetralix, 424; perforated corolla, 429, 437

Errara, M., on self-fertilization, 352

Erythrina, 360

Eschscholtzia californica, 109; measurements, 110; plants raised from Brazilian seed, 111; weight, 113; seeds, 115, 116, 315, 319, 322; experiments on, 263, 275; superiority of self-fertilized over crossed, 290; early flowering, 292, 294; artificially self-fertilized, 332; pollen from other flowers more effective, 340; self-sterile in Brazil, 343, 358; effects of changed conditions on reproductive system, 444, 449

Euphrasia officinalis, 368

Euryale amazonica, 358

  ferox, 365

Fabricius on *Aristolochia*, 420

Fabopyrum esculentum, 228; early flowering of crossed plant, 293

Faivre, Professor, self-fertilization of *Cannaceae*, 230

Farrer, T. H., papilionaceous flowers, 5; *Lupinus luteus*, 147; *Phaseolus multiflorus*, 150, 434; *Pisum sativum*, 160; cross-fertilization of *Lobelia ramosa*, 176; on *Coronilla*, 407

Fermond, M., *Phaseolus multiflorus*, 151; *P. coccineus hybridus*, 151

Fertilization, means of, 356; plants sterile, or partially so without insect-aid, 357–64; plants fertile without insect-aid, 365–9; *means of*

cross-fertilization*, 371; hummingbirds, 371; Australian flowers fertilized by honey-sucking birds, 371; in New Zealand by the *Anthornis melanura*, 371; attraction of bright colours, 372; of odours, 374; flowers adapted to certain kinds of insects, 375; large amount of pollen-grains, 377, 378; transport of pollen by insects, 379–80; structure and conspicuousness of flowers, 383; pollen from a distinct plant, 390; prepotent pollen, 394–401

Fertility, heights and weights, relative, of plants crossed by a fresh stock, self-fertilized, or intercrossed (Table C), 245–52

Fertility of plants as influenced by cross and self-fertilization (Table D), 312; *relative*, of crossed and self-fertilized parents (Table E), 314–19; *innate*, from a cross with fresh stock (Table F), 319; *relative*, of flowers crossed with pollen from a distinct plant and their own pollen (Table G), 320; of crossed and self-fertilized flowers, 324, 325

Flowering, period of, superiority of crossed over self-fertilized, 291–7

Flowers, artificial, 374

Flowers, cleistogamic, 90; white, / larger proportion smelling sweetly, 375; structure and conspicuousness of, 382; conspicuous and inconspicuous, 386; papilionaceous, 386; fertilized with pollen from a distinct plant, 390

Forsythia viridissima, 341

Foxglove, 81

Frankland, Dr, chemical affinity, 461

Fraxinus ornus, 404

Fumaria capreolata, 366

  officinalis, 366

Galium aparine, 369

Gallesio, spontaneous crossing of oranges, 396

Galto, Mr, *Limnanthes douglasii*, 146; report on the tables of measurements, 16–19, 146, 234; self-fertilized plants, 290, 291; superior vigour of crossed seedlings in *Lathyrus odoratus*, 353, 355

Gärtner, excess of pollen injurious, 24; plants fertilizing one another at a considerable distance, 152; *Lobelia fulgens*, 179, 330; sterility of *Verbascum nigrum*, 330; number of pollen-grains to fertilize *Geum urbanum*, 378; experiments with pollen, 380

Gentry, Mr, perforation of corolla, 430

Geraniaceae, 142

Geranium phaeum, 423

Gerardia pedicularia, 430, 437

Germination, period of, and relative weight of seeds from crossed and self-fertilized flowers, 352–5

Gesneria pendulina, 92; measurements, 93; seeds, 322

Gesneriaceae, 92

Geum urbanum, number of pollen-grains for fertilization, 378

Glaucium luteum, 366

Godron, intercrossing of carrot, 172; *Primula grandiflora* affected by pollen of *P. officinalis*, 380; tulips, 396

Gould, humming-birds frequenting *Impatiens*, 371

Graminaceae, 233, 445

Grant, Mr, bees of different hives visiting different kinds of flowers, 426

Gray, Asa, flowers of Drosera, 392; sexual relations of trees in United States, 414; on sexual reproduction, 442

Hallet, Major, on selection of grains of cereals, 354

Hassall, Mr, number of pollen-grains in Paeony and Dandelion, 377; weight of pollen produced by one plant of Bulrush, 407–8

Heartsease, 123

Hedychium, 364

Hedysarum onobrychis, 361

Heights, relative, of crossed and self-fertilized plants (Table A), 240–3

Heights, weights, and fertility, summary, 238–84

Henschel's experiments with pollen, 381

Henslow, Rev. G., cross-fertilization in *Sarothamnus scoparius*, 164; on self-fertilization not injurious, 441

Herbert on cross-fertilization, 7; pollen brought from distant plants, 380; spontaneous crossing of rhododendrons, 396

Hero, descendants of the plant, 47–51, 258; its self-fertilization, 349

Heterocentron mexicanum, 361

Hibiscus africanus, 140; measurements, 140; result of experiments, 277; early flowering of crossed plant, 292, 296; number of pollen-trains for fertilization, 378

Hildebrand on pollen of *Digitalis purpurea*, 82; *Thunbergia alata*, 96; experiments on *Eschscholtzia / californica*, 110; *Viola tricolor*, 123; *Lobelia ramosa*, 176; on moths frequenting Petunias, 188; *Fagopyrum esculentum*, 228; self-fertilization of *Zea mays*, 233; *Corydalis cava*, 331; *Hypecoum grandiflorum*, 331, 359; and *H. procumbens*, 331, 366; sterility of *Eschscholtzia*, 332; experiments on self-fertilization, 340; *Corydalis lutea*, 359; spontaneously self-fertilized flowers, 366; various mechanical structures to check self-fertilization, 383; early separation of the sexes, 400; on *Aristolochia*, 420; fertilization of the *Gramineae*, 445; wide dissemination of seeds, 455

Hoffmann, Professor H., self-fertilized capsules of *Papaver somniferum*, 108, 366; *Adonis aestivalis*,

129, 365; spontaneous variability of *Phaseolus multiflorus*, 151; self-fertilization of kidney-bean, 152; *Papaver alpinum*, 331; sterility of *Corydalis solida*, 358; *Linum usitatissimum*, 366; on honey-dew from a camellia, 404
Honey-dew, 404
Hooker, Dr, *Euryale ferox* and *Victoria regia*, each producing several flowers at once, 365; on sexual relation of trees in New Zealand, 414
Horse-chestnut, 401
Humble-bees, 419: *see* Bees
Humboldt, on the grains of cereals, 354
Humming-birds a means of cross-fertilization, 371
Hyacinth, 396
Hybrid plants, tendency to revert to their parent forms, 380
Hypecoum grandiflorum, 331, 359
procumbens, 331, 366

Iberis umbellata (var. *kermesiana*), 103; measurement, 104–6; cross by freshstocks, 105; remarks on experiments, 262; superiority of crossed over self-fertilized seedlings, 289; early flowering, 292; number of seeds, 315; highly self-fertile, 365; prepotency of other pollen, 394
amara, 365
Impatiens frequented by humming-birds, 371
barbigera, 366
fulva, 341, 367
noli-me-tangere, 367
pallida, 341
Inheritance, force of, in plants, 305
Insects, means of cross-fertilization, 371; attracted by bright colours, 372; by odours, 374; by conspicuous flowers, 384; dark streaks and marks as guides for, 373; flowers adapted to certain kinds, 376

Ipomoea purpurea, 28; measurements, 29–49; flowers on same plant crossed, 41–4; cross with fresh stock, 45–7; descendants of Hero, 47–51; summary of measurements, 52; diagram showing mean heights, 53; summary of observations, 53–62; of experiments, 257–9; superiority of crossed, 289; early flowering, 291, 297; effects of intercrossing, 300; uniform colour of self-fertilized, 308; seeds, 314, 322, 324; highly self-fertile, 368; prepotency of other pollen, 399
Iris, secretion of saccharine matter from calyx, 404
Isotoma, 176, 364

Juglans regia, 391

Kalmia latifolia, 359
Kerner, on protection of flowers / from crawling insects, 376; on protection of the pollen, 377; on the single daily flower of *Villarsia parnassifolia*, 392; pollen carried by wind, 408, 415
Kidney-bean, 152
Kitchener, Mr, on the action of the stigma, 64; on *Viola tricolor*, 123
Knight, A., on the sexual intercourse of plants, 7; crossing varieties of peas, 163; sexual reproduction, 442
Kohl-rabi, prepotency of pollen, 394
Kölreuter on cross-fertilization, 7; number of pollen-grains necessary for fertilization, 24; sexual affinities of *Nicotiana*, 210; *Verbascum phaeniceum*, 330; experiments with pollen of *Hibiscus vesicarius*, 378
Kuhn adopts the term cleistogamic, 90
Kurr, on excretion of nectar, 404; removal of corolla, 423

Labiatae, 93

Lactuca sativa, 173, 369; measurement, 174; prepotency of other pollen, 399

Lamium album, 391, 419

purpureum, 419

Lathyrus odoratus, 153–60; measurements, 157–60; remarks on experiments, 265; period of flowering, 295; cross-fertilization, 304; seeds, 316, 325; self-fertile, 367

grandiflorus, 155, 360

nissolia, 367

sylvestris, perforation of corolla, 432

Lawes and Gilbert, Messrs, consumption of inorganic matter by plants, 453

Laxton, Mr, crossing varieties of peas, 163, 305

Lecoq, *Cyclamen repandum*, 215; on *Fumariaceae*, 359; annual plants rarely dioecious, 415

Leersia oryzoides, 350

Leguminosae, 147; summary on the, 168

Lehmann, Professor, on seedlings from large and small seeds, 355

Leighton, Rev. W. A., on *Phaseolus multiflorus*, 151; *Acacia magnifica*, 407

Leptosiphon androsaceus, 368

Leschenaultia formosa, 364

Lettuce, 173

Lilium auratum, 341

Limnanthes douglasii, 145; measurements, 146; early flowering of crossed, 293; seeds, 316, 323; highly self-fertile, 367; prepotency of other pollen, 399

Linaria vulgaris, 9, 88; seeds, 322; self-sterile, 363

cymbalaria, 385, 426

Lindley on *Fumariaceae*, 359

Link, hypopetalous nectary in *Chironia decussata*, 404

Linum grandiflorum, 343

usitatissimum, 366

Loasaceae, 170

Lobelia erinus, 176; secretion of nectar in sunshine, 405; experiments with bees, 423

Lobelia fulgens, 179; measurements, 180–2; summary of experiments, 274; early flowering of self-fertilized, 291, 294, 295; seeds, 323; sterile unless visited by humble-bees, 364

ramosa, 176; measurements, 177, 178; early flowering of crossed, 293, 295; seeds, 325; self-sterile, 364

tenuior, 176

Loiseleur-Deslongchamp, on the grains of cereals, 354

Lotus corniculatus, 361

Lubbock, Sir J., cross-fertilization of flowers, 6; on *Viola tricolor*, 123; bees distinguishing colours, 373; instinct of bees and insects sucking nectar, 418

Lupinus luteus, 147; measurements, / 148; early flowering of self-fertilized, 294, 296; self-fertile, 367; prepotency of other pollen, 399

Lupinus pilosus, 149; self-fertile 367

Lychnis dioica, 413

Macnab, Mr, on the shorter or longer stamens of rhododendrons, 298

Mahonia aquifolium, 396

repens, 396

Malvaceae, 140

Marcgraviaceae, 407

Marck, Dr, on seedlings from large and small seeds, 355

Masters, Mr, cross-fertilization in *Pisum sativum*, 161; cabbages affected by pollen at a distance, 379

Dr Maxwell, on honey-dew, 404

Measurements, summary of, 241; Table A, 240–3; Table B, 244; Table C, 245–52

Medicago lupulina, 368

Meehan, Mr, fertilizing *Petunia violaceae* by night moth, 188

Melastomaceae, 298

Melilotus officinalis, 360

Mercurialis annua, 421

Miller, Professor, on chemical affinity, 461

Mimulus luteus, effects of crossing, 10; crossed and self-fertilized plants, 64–70; measurements, 70–8; cross with a distinct stock, 72–5; intercrossed on same plant, 75–8; summary of observations, 78–81; of experiments, 259–61; superiority of crossed plants, 286; simultaneous flowering, 294, 296; effects of intercrossing, 301; uniform colour of self-fertilized, 307; seeds, 315, 319, 322, 324; highly self-fertile, 348, 369; prepotency of other pollen, 393, 399

roseus, 63

Miner, Mr, red clover never sucked by hive-bees in the United States, 361

Mirabilis, dwarfed plants raised by using too few pollen-grains, 298; number of grains necessary for fertilization, 378

Mitchell, Dr, on first-cousins intermarrying, 465

Monochaetum ensiferum, 364

Moore, Mr, on Cinerarias, 335

Müller, Fritz, on *Posoqueria fragrans*, 5, 393; experiments on hybrid *Abutilons* and *Bignonias*, 305, 306; large number of orchidaceous genera sterile in their native home, also Bignonia and *Tabernaemontana echinata*, 331; sterility of *Eschscholtzia californica*, 332, 342; *Abutilon darwinii*, 334; experiments in self-fertilization, 340; self-sterile plants, 341; incapacity of pollen-tubes to penetrate the stigma, 342; cross-fertilization by means of birds, 371; imperfectly developed male and female Termites, 381; on ferns ants, 406; food-bodies in *Cecropia*, 406; on the glands of calyx of Malpighiaceae, 407

Müller, Hermann, fertilization of flowers by insects, 6, 7; on *Digitalis purpurea*, 82; *Calceolaria*, 87; *Linaria vulgaris*, 88; *Verbascum nigrum*, 89; the common cabbage, 98; *Papaver dubium* 107; *Viola tricolor*, 123, 124; structure of *Delphinium consolida*, 129; of *Lupinus luteus*, 147; flowers of *Pisum sativum*, 160, 161; on *Sarothamnus scoparius* not secreting nectar, 164; *Apium petroselinum*, 172; *Borago officinalis*, 185; red clover visited by hive-bees in Germany, 361; insects rarely visiting *Fumaria officinalis*, 366; comparison of lowland and alpine species, 376; structure of plants adapted to cross and self-fertilization, 381; / large conspicuous flowers more frequently visited by insects than small inconspicuous ones, 384; *Solanum* generally unattractive to insects, 389; *Lamium album*, 390, 391; on anemophilous plants, 401; fertilization of *Plantago*, 403; secretion of nectar, 407; instinct of bees sucking nectar, 418; bees frequenting flowers of the same species, 419; cause of it, 421; powers of vision and discrimination of bees, 425

Müller, Dr H., hive-bees occasionally perforate the flower of *Erica tetralix*, 430; calyx and corolla of *Rhinanthus alecterolophus* bored by *Bombus mastrucatus*, 435

Munro, Mr, some species of *Oncidium* and *Maxillaria* sterile with own pollen, 334

Myrtaceae, 414

Nägeli on odours attracting insects, 374; sexual relations, 411

Natural Selection, effect upon self-sterility and self-fertilization, 345, 346

Naudin on number of pollen-grains necessary for fertilization, 24; *Petunia violacea*, 188

Nectar regarded as an excretion, 403

Nemophila insignis, 182; measurements, 183–5; early flowering of crossed plant, 293; effects of cross and self-fertilization, 303; seeds, 316, 323

Nepeta glechoma, 419

Nicotiana glutinosa, 210

 tabacum, 203; measurements, 205–8; cross with fresh stock, 210; measurements, 212–15; summary of experiments, 266, 267, 279; superiority of crossed plants, 288–90; early flowering, 293–5; seeds, 323, 325; experiments on, 349; self-fertile, 368

Nolana prostrata, 186; measurements, 187; crossed and self-fertilized plants, 277; number of capsules and seeds, 321, 323; self-fertile, 368

Nolanaceae, 186

Nymphaea, 358, 365

Odours emitted by flowers attractive to insects, 374

Ogle, Dr, on *Digitalis purpurea*, 82; *Gesneria*, 92; *Phaseolus multiflorus*, 151, 360, 434; perforation of corolla, 429; case of the Monkshood, 431

Onagraceae, 169

Onion, prepotency of other pollen, 395

Ononis minutissima, 167; measurements, 168; seeds, 323; self-fertile, 367

Ophrys apifera, 350, 369, 408, 442

 muscifera, 385, 408

Oranges, spontaneous crossing, 395

Orchideae, 364, 369; excretion of saccharine matter, 404

Orchis, fly, 408

Origanum vulgare, 94; measurements, 95; early flowering of crossed plant, 292; effects of intercrossing, 301

Paeony, number of pollen-grains, 377

Papaveraceae, 107

Papaver alpinum, 331, 358

 argemonoides, 366

 bracteatum, 108

 dubium, 107

 rhoeas, 107

 somiferum, 108, 331, 365

 vagum, 107; measurements, / 109; number of capsules, 315; seeds, 358; prepotency of other pollen, 398

Papillae of the *Viola tricolor* attractive to insects, 124

Parsley, 172

Passiflora alata, 330, 334

 gracilis, 171; measurements, 171; crossed and self-fertilized, 276; seeds, 323; self-fertile, 365

Passifloraceae, 171, 357

Pea, common, 160, 351

Pelargonium zonale, 142; measurements, 143; effects of intercrossing, 301; almost self-sterile, 359

Pentstemon argutus, perforated corolla, 429, 431, 436

Petunia violacea, 188; measurements, 189–203; weight of seed, 196; cross with fresh stock, 196–201; relative fertility, 201–3; colour, 203; summary of experiments, 265, 274; superiority of crossed over self-fertilized, 289; early flowering, 293, 294; uniform colour of self-fertilized, 309; seeds, 316, 319, 323, 325; self-sterile, 362

Phalaris canariensis, 235; measurements, 236, 237; early flowering of crossed, 293

Phaseolus coccineus, 150

 multiflorus, 150; measurement, 152; partially sterile, 168, 360; crossed and self-fertilized, 276; early flowering of crossed, 293; seeds, 316; perforated by humble-bees, 433, 438

Phaseolus vulgaris, 153; self-fertile, 168, 367

Pisum sativum, 160; measurements,

162; seldom intercross, 169; summary of experiments, 264, 278; self-fertile, 367

Plants, crossed, greater constitutional vigour, 284

Plateau, M., on insects and artificial flowers, 374, 375

Pleroma, 364

Polemoniaceae, 182

Pollen, relative fertility of flowers crossed from a distinct plant, or with their own, 320; difference of results in *Nolana prostrata*, 321, 323; crossed and self-fertilized plants, again crossed from a distinct plant and their own pollen, 324; sterile with their own, 330–8; semi-self-sterile, 338–40; loss of, 377; number of grains in Dandelion and Paeony, 377; in *Lolium perenne, Plantage lanceolata, Scirpus lacustris*, and *Wistaria sinensis*, 378; number necessary for fertilization, 378; transported from flower to flower, 379; prepotency, 393–401; aboriginally the sole attraction to insects, 403; quantity produced by anemophilous plants, 407

Polyanthus, prepotency over cowslip, 397–8

Polygoneae, 228

Posoqueria fragrans, 5, 393

Potato, 389

Poeterium sanguisorba, 410

Potts, head of *Anthornis melanura* covered with pollen, 371

Primrose, Chinese, 225

Primula elatior, 425, 430

grandiflora, 380

mollis, 368

officinalis, 380

scotica, 362

sinensis, 225, 279; measurements, 227; early flowering of crossed, 293, 296

veris (var. *officinalis*), 219; measurements, 221; result of experiments,

267, 268; early flowering of crossed, 293; seeds, 317; self-fertility, 351; prepotency of dark red polyanthus, 397–8

Primulaceae, 215

Pringlea, 410

Proteaceae of Australia, 415

Prunus avium, 405

laurocerasus, 405

Pteris aquilina, 405 /

Radish, 395

Ranunculaceae, 128

Ranunculus acris, 365

Raphanus sativus, 365, 395

Remke, nectar-secreting glands of *Prunus avium*, 405

Reseda lutea, 117; measurements, 118, 119; result of experiments, 339; self-fertile, 365

odorata, 119; measurements, 120–3; self-fertilized scarcely exceeded by crossed, 289; seeds, 316; want of correspondence between seeds and vigour of offspring, 328; result of experiments, 336; sterile and self-fertile, 358, 365

Resedaceae, 117

Rheum rhaponticum, 403

Rhexia glandulosa, 364

Rhododendron, spontaneous crossing, 396

Rhododendron azaloides, 435

Rhubarb, 396, 403

Ribes aureum, 435

Riley, Mr, pollen carried by wind, 408; Yucca moth, 421

Rimpan, on the cross-fertilization of Rye, 341; on the self-fertility of wheat, 370

Rodgers, Mr, secretion of nectar in Vanilla, 404

Rye, experiment on pollen of, 377

Salvia coccinea, 93; measurements, 93; early flowering of crossed, 292; seeds, 315, 322; partially self-sterile, 363

Salvia coccinea – continued
  glutinosa, 430
Salvia grahami, 429, 431, 436
  tenori, 362
Sarothamnus scoparius, 163; measurements, 165–7; superiority of crossed seedlings, 284, 289; seeds, 323; self-sterile, 360
Scabiosa atro-purpurea, 172; measurements, 172, 173
Scarlet-runner, 150
Scott, J., *Papaver somiferum*, 108; sterility of Verbascum, 330; *Oncidium* and *Maxillaria*, 331; on small seeds of *Papaver*, 355; on *Primula scotica* and *Cortusa mathioli*, 362
Scrophulariaceae, 63
Seeds, size and germination of, 352
Selaginella, 413
Self-fertile varieties, appearance of, 347–51
Self-fertilization, mechanical structure to check, 383
Self-sterile plants, 329–347; wide distribution throughout the vegetable kingdom, 341; difference in plants, 342; cause of self-sterility, 343; affected by changed conditions, 344–6; necessity of differentiation in the sexual elements, 347
Senecio cruentus, 335, 364
  heritieri, 335
  maderensis, 335
  populifolius, 335
  tussilaginis, 335
Sharp, Messrs, precautions against intercrossing, 396
Snow-flake, 176
Solanaceae, 188
Solanum tuberosum, 362, 389
Specularia perfoliata, 174
  speculum, 174; measurements, 175, 176; crossed and self-fertilized, 276; early flowering of crossed, 293; seeds, 323; self-fertile, 369
Spencer, Herbert, chemical affinity, 462
Spiranthes autumnalis, 391, 424

Sprengel, C. K., fertilization of flowers by insects, 5, 6; *Viola tricolor*, 123; colours in flowers attract and guide insects, 372–4; on *Aristolochia*, 419; *Aconitum napellus*, 431; importance of insects in fertilizing flowers, 460
Stachys coccinea, 430, 431, 436 /
Stellaria media, 367
Strachey, General, perforated flowers in the Himalaya, 436
Strawberry, 396
Strelitzia fertilized by the Nectarinideae, 371
Structure of plants adapted to cross and self-fertilization, 381
Swale, Mr, garden lupine not visited by bees in New Zealand, 150
Sweetpea, 153

Tabernaemontana echinata, 331, 362
Tables of measurements of heights, weights, and fertility of plants, 240–70
Termites, imperfectly developed males and females, 382
Thunbergia alata, 96, 277, 331
Thyme, 421
Tinzmann, on *Solanum tuberosum*, 362, 389
Tobacco, 203
Transmission of the good effects of a cross to later generations, 303
Trees, separated sexes, 414
Trifolium arvense, 367, 386
  incarnatum, 361*
  minus, 368
  pratense, 361, 429, 438
  procumbens, 368
  repens, 361
Tropaeolum minus, 144; measurements, 145; early flowering of crossed, 293; seeds, 316, 323
  tricolor, 430; seeds, 323
Tulips, 396
Typha, 377, 408

Umbelliferae, 172

Urban, Ig., fertilization of *Medicago lupulina*, 368

Vandellia nummularifolia, 90, 278; seeds, 315, 322; self-fertile, 369
Vanilla, secretion of nectar, 404
Verbascum lychnitis, 89, 341, 369
    nigrum, 89, 330, 341
    phoeniceum, 330, 341, 364
    thapsus, 89; measurements, 90; self-fertile, 341, 369
Verlot on *Convolvulus tricolor*, 55; intercrossing of *Nemophila*, 183; of *Leptosiphon*, 394
Veronica agrestis, 369
    chaemoedrys, 369
    hederaefolia, 369
Vicia faba, 360, 405
    hirsuta, 367
    sativa, 367, 405
Victoria regia, 365
Villarsia parnassifolia, 392
Vilmorin on transmitting character to offspring, 456
Vinca major, 362
    rosea, 362
Viola canina, 357
    tricolor, 123; measurements, 126, 127; superiority of crossed plants, 286, 289; period of flowering, 292, 296; effects of cross-fertilization, 304; seeds, 316, 325; partially sterile, 358; corolla removed, 423

Violaceae, 123
Viscaria oculata, 130; measurement, 132; average height of crossed and self-fertilized, 276; simultaneous flowering, 295; seeds, 316, 323; self-fertile, 367

Wallace, Mr, the beaks and faces of brush-tonged lories covered with pollen, 371
Wasps attracted by *Epipactis latifolia*, 376 /
Weights, relative, of crossed and self-fertilized plants, 244, 283; and period of germination of seeds, 352–5
Wilder, Mr, fertilization of flowers with their own pollen, 341
Wilson, A. J., superior vigour of crossed seedlings in *Brassica campestris ruta baga*, 353; self-fertility of wheat, 370; on size of pollen-grains, 378
Wistaria sinensis, 378, 430

Yucca moth, 421

Zea mays, 16, 233; measurements, 16–18, 234; difference of height between crossed and self- fertilized, 288; early flowering of crossed, 293; self-fertile, 369; prepotency of other pollen, 399 /